T0091515

Quantum Anharmonic Oscillator

Quantum Anharmonic Oscillator

Alexander V Turbiner
National Autonomous University of Mexico, Mexico

Juan Carlos del Valle Rosales
National Autonomous University of Mexico, Mexico

NEW JERSEY • LONDON • SINGAPORE • BEIJING • SHANGHAI • HONG KONG • TAIPEI • CHENNAI • TOKYO

Published by

World Scientific Publishing Co. Pte. Ltd.
5 Toh Tuck Link, Singapore 596224
USA office: 27 Warren Street, Suite 401-402, Hackensack, NJ 07601
UK office: 57 Shelton Street, Covent Garden, London WC2H 9HE

Library of Congress Control Number: 2023932592

British Library Cataloguing-in-Publication Data
A catalogue record for this book is available from the British Library.

QUANTUM ANHARMONIC OSCILLATOR

ISBN 978-981-127-045-1 (hardcover)
ISBN 978-981-127-046-8 (ebook for institutions)
ISBN 978-981-127-047-5 (ebook for individuals)

For any available supplementary material, please visit
https://www.worldscientific.com/worldscibooks/10.1142/13252#t=suppl

Typeset by Stallion Press
Email: enquiries@stallionpress.com

Preface

This book is dedicated to finding an approximate solution to the quantum polynomial anharmonic oscillator, which is one of the most fundamental problems in quantum mechanics. The final result of this book is a number of highly-locally-accurate eigenfunctions for the one-dimensional case and for the d-dimensional spherically-symmetric radial case. The key components for the construction of these approximate eigenfunctions are the perturbation series expansion in the coupling constant, the semiclassical expansion in the Planck constant and the matching procedure for unifying these two expansions. The main object of study in this book is the logarithm of the wave function and the celebrated Riccati equation. The obtained results are relevant for quantum mechanics (theory of non-linear oscillations), mathematics (solutions of ODEs beyond of the higher transcendental functions) and chemistry (low-lying part of the rovibrational spectrum of diatomic molecules).

A V Turbiner
J C del Valle

July 1, 2022

Contents

Appendices of Part II **273**

Introduction

The theory of linear oscillations, where the restoring force (the Hooke's force) depends linearly on the deviation from the equilibrium position, is well developed for both one-dimensional and radial dynamics. It is intimately related to the one-dimensional and the radial harmonic oscillator potentials, which admit exact solutions in both the classical and quantum cases. The situation becomes very much different qualitatively when an anharmonicity appears, where the restoring force is no longer linear and the oscillations become non-linear[1]. This is especially clear in quantum mechanics: infinitely-many, exponentially large expectation values (observables), which were infinite for the harmonic oscillator case, become finite,

$$< e^{\alpha x^{2+\beta}} >< \infty, \quad \alpha > 0,$$

where $\alpha > 0$ and $\beta_{max} > \beta > 0$, where β_{max} is a number which characterizes the given anharmonicity. It is equal to 3, $\beta_{max} = 3$, for a quartic anharmonicity (for the quartic anharmonic oscillator). This implies, in particular, that the anharmonicity is *never* small, even though the constant of the anharmonicity — the parameter in front of the anharmonic (non-linear) term in the Hooke's force — is small (or very small, close to zero).

There are three main methods to tackle one-dimensional (radial) Schrödinger equations: (i) the perturbation theory in powers of the coupling constant or constant of anharmonicity λ, (ii) the semiclassical expansion in powers of $\hbar$ and (iii) the variational method. The first two methods (i) and (ii) have an intrinsic deficiency related to the zero radius of convergence in λ and $\hbar$, respectively. Hence, the problem of the summation of

[1]For a brief overview of the classical quartic oscillator see Appendix A.

divergent series occurs. The third one suffers from the absence of a criterion for choosing a simple trial function in order to get the variational energy as close as possible to the exact one. All three methods are mainly focused on obtaining the eigenvalues (energies). The Non-Linearization Procedure (NLP), invented in the 20th century, see for review [Turbiner (1984)], allows us to treat (i)–(iii) on the same footing: the Perturbation Theory (i) is a NLP with the harmonic oscillator as the zeroth approximation and the anharmonicity as the perturbation, the Semiclassical Expansion (ii) is the perturbation series in the NLP with respect to the derivative of momentum and the variational energy (iii) is the first two terms of the NLP perturbation theory with respect to the deviation of the original potential from the potential for which the trial function is the exact eigenfunction.

It should be emphasized that even before the establishing the NLP it had become clear that the variational trial functions which reproduce the semiclassical asymptotics at large distances [Dolgov and Turbiner (1980)] (the Stark effect on Hydrogen atom), [Turbiner (1979, 1980, 1981, 1984)] (one-dimensional anharmonic oscillators), [Turbiner (1987)] (the funnel-type potentials), [Turbiner (1988)] (two-dimensional anharmonic oscillators), [Del Valle *et al.* (2020)] (the Zeeman effect on Hydrogen atom) play an exceptionally important role, and they usually lead to highly-accurate variational energies.

There is one more widely used method: (iv) the numerical analysis of the ordinary differential equations. This method, implemented using either Fortran, C++ or MATHEMATICA codes, allows us to find the spectrum with any desired accuracy, although the domain where the eigenfunctions are exponentially small (the so-called classically forbidden domain), which is important from the physics point of view, causes serious technical difficulties. However, if the original problem has a free parameter(s), it is not completely clear how to extract the parametric dependence(s) of the obtained results. In particular, it is also unclear how to describe the coordinate dependence of the eigenfunctions.

The Planck constant $\hbar$ is a very small but dimensionful quantity. It has the meaning of the energy of a photon with unit frequency. In the quantum world, $\hbar$ is frequently used as the unit for measuring energy. However, throughout this book the energy will be measured in the unit $\left(\frac{\hbar}{(2m)^{\frac{1}{2}}}\right)$:

$$E = \frac{\hbar}{(2m)^{\frac{1}{2}}}\varepsilon,$$

where m is the mass of the oscillating particle or the reduced mass in the case of the two-body problem.

In order to describe anharmonic effects, the harmonic oscillator potential is modified by the anharmonic terms,

$$V_{aho}(x;g) = V_{ho} + \Delta V_{aho} \equiv a_2^2 x^2 + g x^3 W(gx),$$

where

$$W(u) = a_3 + a_4 u + a_5 u^2 + \cdots ,$$

is bounded from below, g can be called the coupling constant, x has the meaning of the displacement: the distance from the equilibrium position, all parameters $\{a\}$ are dimensionless. This potential includes as particular cases all known two-term anharmonic oscillators. In general, the Schrödinger equation depends on three parameters m, $\hbar$ and g. However, the special form of $V_{aho}(x;g)$ implies that these three parameters can be unified in the *effective* coupling constant,

$$\lambda = \left(\frac{\hbar^2}{m}\right)^{\frac{1}{4}} g,$$

and the problem becomes *effectively* one-parametric.

The main emphasis of this book will be on finding wave functions. Following the fundamental observation by Lazarus Fuchs (1833–1902), wavefunctions in polynomial potentials must be entire functions in the complex x-plane. This property will be exploited in order to construct the approximations by trying to avoid the appearance of singularities at finite points on the real x-line. The question of the analytic continuation of the obtained approximations to the complex x-plane is carefully avoided: our Schrödinger equation is defined on the real line, it has real potential, we are looking for real eigenfunctions and eigenvalues. We are able to realize this program: there were no instances where it was necessary to introduce the complexity.

There exists an enormous body of literature on anharmonic oscillators. This book was not intended to provide a comprehensive overview. We limited ourselves to references truly needed for the presentation following the personal taste of authors.

This book consists of two Parts with six and five Chapters, respectively, and nine Appendices: five of them assigned to Part I and four assigned to Part II.

Part I is dedicated to the one-dimensional Schrödinger equation and one-dimensional anharmonic oscillators. In Chapter 1 the Riccati equation is introduced together with the non-linear spectral problem. The Riccati equation is transformed into the Riccati-Bloch equation, which is studied using perturbation theory in powers of the (effective) coupling constant, in Chapter 2. The Generalized Riccati equation is described in Chapter 3, where the semiclassical expansion is developed. The procedure of matching the perturbation theory at small distances with the semiclassical expansion at large distances leading to the approximate expression for the eigenfunctions of the general anharmonic oscillator is presented in Chapter 4. The Quartic anharmonic oscillator and Sextic anharmonic oscillator are described in a self-contained manner in Chapters 5 and 6, respectively.

Part II is dedicated to the d-dimensional Schrödinger equation with spherically-symmetric anharmonic oscillator potentials. In Chapter 7 the radial Schrödinger equation is derived, and is used to construct the radial Riccati equation and its derivatives: the radial Riccati-Bloch and the radial Generalized Bloch equations. The two latter equations are used to construct the perturbation theory and semiclassical expansion in powers of the *effective* coupling constant λ. The procedure of matching these two expansions leads to the approximate expressions for the eigenfunctions of the general radial anharmonic oscillator and is presented in Chapter 8. The Cubic radial anharmonic oscillator, Quartic radial anharmonic oscillator and Sextic radial anharmonic oscillator are described in a self-contained manner in Chapters 9, 10 and 11, respectively. Note that Appendices E and F present the Lagrange Mesh Method (LMM) and its different modifications, which were used for obtaining benchmark results.

Personal Recollections, History, Acknowledgement (A.V. Turbiner)

When I was a student, I read the remarkable paper by Carl M. Bender and Tai Tsun Wu, *Phys. Rev.* **169** (1969) about the (quartic) anharmonic oscillator. I was deeply impressed by the simplicity of the formulation of the problem, by the richness of the hidden structures, and by the already discovered, unexpected properties. It was immediately clear that many of these structures and properties are not only the properties of some specific anharmonic oscillator, but of any arbitrary perturbed harmonic oscillator and, in general, of the scalar quantum field theory with a non-zero coupling constant. However, a systematic reliable approach to studying this problem in its full generality, in particular, in the physically important strong coupling regime was absent. My Teachers, K. A. Ter-Martirosyan and Ya. B. Zel'dovich, encouraged me to work in this direction, always warning about a small probability of success; I thank them for all that. Through the years, I benefited immensely from discussions on the subject with C. M. Bender, A. A. Migdal, A. M. Polyakov, V. L. Ginzburg, N. A. Nekrasov, M. A. Shifman, E. V. Shuryak, A. I. Vainshtein, T. T. Wu and M. Znojil, and particularly with V. I. Arnold, S.P. Novikov and M.A. Shubin, and many-many others. My deep gratitude to all of them.

Lars Brink, my lifelong friend and co-author, was, in fact, the first person whom I informed about a serious progress in the radial case and who suggested to write it all down and then to submit the first two large papers to IJMPA. After the publication of these papers, he convinced me to write the present book and Dr. K. K. Phua, World Scientific Founder and Chairman, kindly sent me a formal invitation. I was terribly depressed to learn that on October 29, 2022, Lars Brink passed away and he won't be able to see this book published. Lars was the most fine and decent person I have ever met; his love for science and dedication were exemplary.

I had the enormous luck that Dr. Juan Carlos del Valle chose me to be his MS and then PhD advisor. We had an extremely fruitful, intense collaboration throughout the years which eventually led to a solution of the quartic anharmonic oscillator and other related problems.

A Brief Recapitulation (J.C. del Valle)

After finishing my MS studies under the supervision of A.V. Turbiner, I started my PhD in 2016. On the first day of the doctoral program, he proposed tackling the anharmonic oscillators in a quantum mechanical framework. The initial idea was to explore the topic so that I would gain more research experience as well as to learn the tools and techniques developed by him. The original plan was to spend a couple of months on the subject, at most one year, and to then apply those techniques to atomic systems. Honestly, I would have never imagined that we were facing such a vibrant and vast topic. Unsurprisingly, one year was not enough to finish this line of research: it took six years, and there are still many loose ends!

Our study and many efforts resulted in the present book. Without a doubt, working on it has been an exciting and rewarding experience. I had never imagined that my PhD dissertation would contribute to a scientific book. I will be grateful for my whole life to my coauthor, A.V. Turbiner, for sharing his knowledge, encouragement, and guidance during these eight years.

Part I

The One-Dimensional Anharmonic Oscillator

Chapter 1

Generalities

1.1 The Potentials of One-Dimensional Anharmonic Oscillators

The one-dimensional quantum harmonic oscillator with potential

$$V = \frac{1}{2} m\omega^2 x^2, \tag{1.1}$$

where m is the mass of the particle and ω is its characteristic frequency, plays an exceptionally important role in the exploration of the quantum world. It describes small oscillations near an equilibrium position at $x = 0$. Sometimes, it is also called the linear oscillator. Accordingly, the Hamiltonian of the harmonic oscillator has the form

$$H = -\frac{\hbar^2}{2m}\,\partial_x^2 + \frac{1}{2} m\omega^2 x^2, \quad \partial_x \equiv \frac{d}{dx}, \tag{1.2}$$

where the first term plays the role of kinetic energy. Since the potential (1.1) grows as $|x| \to \infty$, the potential is confining, classical motion of the particle occurs in a finite domain of the configuration space and the quantum spectrum of the energies is entirely discrete. A remarkable property of the harmonic oscillator is that it can be solved *exactly*: all of its eigenstates (eigenvalues and eigenfunctions) can be found in closed analytic form.

For the sake of future convenience, we redefine the Hamiltonian (1.2) by multiplying it by factor 2 and call it the *Hamiltonian*,

$$\mathcal{H} = 2H = -\frac{\hbar^2}{m}\partial_x^2 + m\omega^2 x^2.$$

This corresponds to doubling the eigenvalues (energies). *Following this definition, the general one-dimensional quantum Hamiltonian with potential* $V(x)$ *takes the form,*

$$\mathcal{H} = -\frac{\hbar^2}{m}\partial_x^2 + V(x), \quad \partial_x \equiv \frac{d}{dx}, \tag{1.3}$$

This form of the one-dimensional Hamiltonian will be used throughout the book. In many occasions it is convenient to put $\hbar = 1$ *and* $m = 1$*. This leads to the Schrödinger equation,* $\mathcal{H}\Psi = E\Psi$*, in the form*

$$\Big(-\partial_x^2 + V(x) \Big)\, \Psi = E\, \Psi, \quad \partial_x \equiv \frac{d}{dx}. \tag{1.4}$$

This will be the standard form of the Schrödinger equation throughout the book.

The harmonic oscillator potential in (1.3) is

$$V^{(HO)} = x^2, \tag{1.5}$$

where without loss of generality we put $m\omega^2 = 1$. When the oscillations are not small, anharmonic effects begin to play role. These require us to modify the harmonic oscillator potential (1.1), (1.5) by introducing anharmonicity. There are six one-dimensional anharmonic-type potentials that have been widely explored by the scientific community:

(i) The quartic anharmonic potential

$$V = x^2 + g^2 x^4, \tag{1.6}$$

(ii) The sextic anharmonic potential

$$V = x^2 + g^4 x^6, \tag{1.7}$$

(iii) The tilted (asymmetric) quartic double well anharmonic potential

$$V = x^2(1 + agx + g^2 x^2), \tag{1.8}$$

where a is a parameter of asymmetry, $a \neq 0$ and $|a| < 2$.

(iv) The general two-term anharmonic oscillator[1]

$$V = x^2 + g^{2p-2}x^{2p}. \tag{1.9}$$

For $p = 2, 3, 4$ it is called *a quartic, sextic, octic* anharmonic oscillator, respectively, as for $p = 5$ it is called *a decatic* anharmonic oscillator. Higher values of p are usually not considered.

(v) The quartic double-well anharmonic potential

$$V = x^2(1 \pm gx)^2, \tag{1.10}$$

which corresponds to (1.8) with $a = \pm 2$.

(vi) The periodic anharmonic potential (sine-Gordon potential)

$$V = \frac{1}{g^2}(1 - \cos gx). \tag{1.11}$$

For all six potentials **(i)-(vi)** the parameter $g \geq 0$ is called the parameter of anharmonicity or the coupling constant. In the limit $g \to 0$ all six potentials are reduced to the harmonic oscillator potential

$$V_{g=0} = x^2.$$

In the limit $g \to \infty$ (the so-called (ultra)-strong coupling regime) five potentials **(i)-(v)** are reduced to a one-term oscillator potential

$$V_{g=\infty} = g^{2p-2}x^{2p}, \tag{1.12}$$

where g can be (naturally) placed equal to one, $g = 1$ without loss of generality. For $p = 2, 3, 4, 5$, this is called a *quartic, sextic, octic, decatic* oscillator, respectively. These are particular cases of the so-called power-like potentials,

$$V = |x|^{2p}, \tag{1.13}$$

where $p > 0$ is a real number. The potential with $p = 1/2$ is called the *symmetric linear triangular potential*: it has numerous applications. In particular, it serves as the basis for the so-called connection formulas in one-dimensional WKB.

---

[1]Sometimes, for $p > 2$ it is called the generalized anharmonic oscillator.

All six potentials **(i)-(vi)** are particular cases of the general anharmonic oscillator potential

$$V(x) = \frac{1}{g^2}\hat{V}(g\,x) = \frac{1}{g^2}\sum_{k=2} a_k g^k x^k =$$
$$a_2 x^2 + a_3 g x^3 + a_4 g^2 x^4 + \cdots + a_{2p} g^{2p-2} x^{2p} + \cdots \,, \qquad (1.14)$$

cf. [Brezin *et al.* (1977b)], eq. (8) therein, for $x \in (-\infty, \infty)$, where $g \geq 0$ is the coupling constant and a_k, $k = 2, 3, ..., 2p, \ldots$ are dimensionless parameters; $a_2 = m\,\omega^2$, hence, it is proportional to the square of frequency ω, however, for most cases for the sake of convenience we set $a_2 = 1$. This potential is invariant with respect to a simultaneous change $x \to -x$ and $g \to -g$. As a consequence of this invariance the energy E in the Schrödinger equation (1.4) depends on g^2. It is assumed that like in the polynomial potentials **(i)-(v)** the parameter a_{2p} in front of the leading term x^{2p} is positive, thus, the potential is confining. For **(vi)**, $p = \infty$: it is a non-polynomial potential.

The general anharmonic potential (1.14) is assumed always *to be bounded from below.* Furthermore, we assume that the reference point for x (the origin) is chosen in such a way that the minimum of the potential at $x = 0$ is global (or one among globals). By choosing appropriately the reference point for the potential to be $V(0) = 0$, one can always make the potential non-negative: $V(x) \geq 0$ for any x. It is worth noting once again that the minimum at the origin $V(0) = 0$ can be global like is the case for the potentials **(i)-(iv)**, however, there can exist degenerate minima like is the case for the potentials **(v)-(vi)**. It must be also emphasized that *any* two-term anharmonic oscillator potential,

$$V(x) = a_2 x^2 + \lambda x^{2p}, a_2 > 0,$$

can be written in the form (1.14) first by putting $a_k = 0$ for $k = 3, 4, \ldots,$ $(2p-1), (2p+1), \ldots$ and then defining $\lambda \equiv g^{2p-2}$, see (1.9). By also setting $a_2 = 0$ we can reproduce any one-term oscillator (power-like) potential, λx^{2p}, see (1.12).

The goal of Part I is to study the energies (eigenvalues) and wave functions (eigenfunctions) of the one-dimensional anharmonic potential (1.14).

1.2 From the Schrödinger Equation to the Riccati Equation

The Schrödinger operator (or, equivalently, the Hamiltonian) (1.3) with potential (1.14) is defined as

$$\mathcal{H}_x = -\frac{\hbar^2}{m}\partial_x^2 + \frac{1}{g^2}\hat{V}(gx), \tag{1.15}$$

where $\hbar$ is the Planck constant, m is the mass of the system, and the coupling constant g has dimension $[x]^{-1}$. For any polynomial potential (1.14) that grows as $|x| \to \infty$, it has an infinite discrete spectrum. Furthermore, if V has a single global minimum, it does not contain any non-analytic terms in g. Needless to say that for $p = 3, 4, \ldots$ and $g \neq 0$, the spectral problem — the Schrödinger equation — the central object of Part I,

$$\mathcal{H}_x \Psi = E\Psi, \quad \int |\Psi|^2 dx < \infty, \quad x \in (-\infty, \infty), \tag{1.16}$$

is non-solvable in closed analytic form. Hence, the energies and wave functions can only be found approximately. In general, $|\Psi|$ decays as $|x| \to \infty$, it can have several extrema and, according to the oscillation theorem, see e.g. [Landau and Lifshitz (1977)], it can have several simple zeroes (nodes).

The general formalism that we will use to study the Schrödinger equation (1.16) is based on writing the wave function as an exponential function (the exponential representation)

$$\Psi = e^{-\frac{1}{\hbar}\Phi}, \tag{1.17}$$

where the function Φ is called the *phase* of the wavefunction or, saying differently, the logarithm of the wavefunction. To guarantee the normalizability of a solution of the Schrödinger equation (1.16) the phase Φ should grow being positive at large distances, $|x| \to \infty$. It may contain a number of logarithmic singularities at finite real x as the reflection of the existence of simple zeroes (nodes) of the eigenfunction. Lazarus Fuchs (1833–1902) proved that for any polynomial potential all eigenfunctions are entire functions, see [Eremenko and Gabrielov (2009)].

Making substitution of (1.17) into the Schrödinger equation (1.16) we arrive at the celebrated Riccati equation, see e.g. [Landau and Lifshitz (1977)],

$$\boxed{\frac{1}{m}(\hbar y' - y^2) = E - \frac{1}{g^2}\hat{V}(gx), \quad y = \Phi',} \tag{1.18}$$

where y is, in fact, the logarithmic derivative of the wavefunction. We are looking for solutions of (1.18), which at large distances, $x \to +\infty$ should be positive and behave like $y \to +\infty$ at $x \to +\infty$, while as $x \to -\infty$, they are negative and $y \to -\infty$. This implies the square-integrability of the wavefunction. These conditions play the role of the boundary conditions for (1.18). Naturally, there must exist x_0 such that the function y changes sign, $y(x_0) = 0$, where the phase Φ has a minimum (maximum) and the eigenfunction Ψ has a maximum (minimum). It can happen several times. If the function y changes the sign once at $x = x_0$, in this point x_0 the phase Φ has a minimum and the eigenfunction Ψ has a maximum. In general, at origin $y(0) = \text{const}$. In the case of symmetric potentials $V(x) = V(-x)$ the function y is antisymmetric, therefore, $y(0) = 0$. This holds for any wavefunction which does not vanish at $x = 0$, hence, the function y is non-singular at $x = 0$.

If the eigenfunction vanishes at $x = 0$ and has a simple zero at $x = 0$, the function y has a simple pole at $x = 0$ with residue (-1). The position of the zero and pole is unchanged under perturbation similar to what happens for negative parity states $\Psi(x) = -\Psi(-x)$ of the general even potential $V(x) = V(-x)$. In order to consider this concrete case it makes sense to introduce a new function

$$y = -\frac{1}{x} + \tilde{y},$$

and substitute it to (1.18). This leads to a modification of the Riccati equation (1.18),

$$\frac{1}{m}\left(\hbar\tilde{y}' + \frac{2}{x}\tilde{y} - \tilde{y}^2\right) = E - \frac{1}{g^2}\hat{V}(gx), \quad \tilde{y}(0) = 0. \tag{1.19}$$

This form of the Riccati equation will be exploited later.

In general, by transforming the Schrödinger equation, which is a linear spectral problem, into the Riccati equation, which represents the *non-linear* spectral problem, we arrive at much less studied subject. The requirements that $y \to +\infty$ as $x \to +\infty$ and $y \to -\infty$ as $x \to -\infty$ lead to the quantization of energy E in the right hand side of the equations (1.18) or (1.19). An analogue of the oscillation theorem, see e.g. [Landau and Lifshitz (1977)], which states that for the Schrödinger equation the eigenfunction Ψ_N of the Nth excited state has N simple zeroes, now states that for the Riccati equation the function y_N of Nth excited state has N simple poles at real x with residues equal to (-1).

The equations (1.18), (1.19) contain the Planck constant $\hbar$ in front of the leading (first) derivative term and g^{2p-2} in front of the dominant anharmonic term x^{2p} in (1.14): this leads to a bifurcation at $\hbar = 0$ and to the divergence of the perturbation theory in powers of g^2. If we get rid of the explicit $\hbar$ dependence in the equations (1.18), (1.19), we will be no longer able to take *explicitly* the limit $\hbar \to 0$. There are two ways to do so, which will be covered in the two coming Chapters.

Chapter 2

Riccati-Bloch Equation, Weak/Strong Coupling Regime

2.1 Riccati-Bloch Equation

Let us substitute into the Riccati equation (1.18) the following new definitions for the variable $x \to v$, for the function $y \to \mathcal{Y}$ and for the energy $E \to \varepsilon$:

$$x = \left(\frac{\hbar^2}{m}\right)^{1/4} v, \quad y = (m\hbar^2)^{1/4}\mathcal{Y}(v), \quad E = \frac{\hbar}{m^{1/2}}\varepsilon, \tag{2.1}$$

respectively, and also the effective coupling constant

$$\lambda = \left(\frac{\hbar^2}{m}\right)^{1/4} g. \tag{2.2}$$

Note that

$$\lambda v = gx.$$

Finally, we arrive at the so-called *Riccati-Bloch* (RB) equation

$$\boxed{\partial_v \mathcal{Y} - \mathcal{Y}^2 = \varepsilon(\lambda) - \frac{1}{\lambda^2}\hat{V}(\lambda v), \quad \partial_v \equiv \frac{d}{dv}, v \in (-\infty, \infty),} \tag{2.3}$$

where the boundary conditions are

$$\mathcal{Y}(\pm\infty) = \pm\infty, \tag{2.4}$$

and the potential

$$\frac{1}{\lambda^2}\hat{V}(\lambda v) = a_2 v^2 + a_3 \lambda v^3 + a_4 \lambda^2 v^4 + \cdots + a_{2p}\lambda^p v^{2p} + \cdots,$$

cf. (1.14). The equation (2.3) is one of the key equations that we are going to study. Note that if $\mathcal{Y}(v)$ is non-singular at $v = 0$,

$$\partial_v \mathcal{Y}(v)\Big|_{v=0} - \mathcal{Y}^2(0) = \varepsilon.$$

In general,

$$\partial_v^{2p} \left(\partial_v \mathcal{Y} - \mathcal{Y}^2\right)\Big|_{v=0} = a_{2p}(2p)!\lambda^p. \tag{2.5}$$

For future convenience, we denote

$$\mathcal{Y}(0) = \alpha.$$

· By making the substitution (2.1) into the modified Riccati equation (1.19), we get the *modified Riccati-Bloch* (mRB) equation

$$\partial_v \tilde{\mathcal{Y}} + \frac{2}{v}\tilde{\mathcal{Y}} - \tilde{\mathcal{Y}}^2 = \varepsilon(\lambda) - \frac{1}{\lambda^2}\hat{V}(\lambda v), \quad \partial_v \equiv \frac{d}{dv}, v \in (-\infty, \infty), \tag{2.6}$$

where the boundary conditions are

$$\tilde{\mathcal{Y}}(\pm\infty) = \pm\infty. \tag{2.7}$$

Interestingly, the same equation occurs for radial potentials in three-dimensional space $D = 3$ over the half-line $x \in [0, \infty)$. Since $\tilde{\mathcal{Y}}(v)$ can not be singular at $v = 0$, $\tilde{\mathcal{Y}}(0) = 0$ and

$$\partial_v \tilde{\mathcal{Y}}(v)\Big|_{v=0} = \varepsilon.$$

The RB and mRB equations (2.3), (2.6) have no explicit $\hbar$-dependence: *they describe dynamics in a "quantum", $\hbar$-dependent coordinate v (2.1) instead of x, which is governed by the $\hbar$-dependent, effective coupling constant λ (2.2).* Also the RB and mRB equations (2.3), (2.6) have no explicit dependence on mass m. This implies that the original Schrödinger equation depends effectively on a *single* parameter λ (2.2): the three parameters $\hbar, m, g$ have been combined into λ.

2.2 Weak Coupling Expansion

Let us develop the Perturbation Theory (PT) for the ground state in powers of λ for the RB equation (2.3) with potential (1.14), setting for simplicity $a_2 = 1$. Thus, we assume that $\mathcal{Y}$ has no simple poles for real v. In general, using PT requires λ to be small hoping for a finite radius of convergence. In order to avoid a discussion about the convergence of this PT, we will consider λ to be a formal parameter and we will postpone the question of convergence for later. We have to note that the domain of small λ is called the *weak coupling domain* and the PT in powers of λ is called the *weak coupling* expansion.

There are several ways to develop the PT for the Schrödinger equation. We will use one of them called the *Non-Linearization procedure* [Turbiner (1984)], see the next Section for the general description. Due to the special form of the potential (1.14) the calculations for the RB equation can be carried out algebraically, the results appear as follows for the energy,

$$\varepsilon = \sum_{n=0}^{\infty} \lambda^n \varepsilon_n, \quad \varepsilon_0 = 1, \quad \varepsilon_1 = 0,$$

$$\varepsilon_2 = \frac{3}{4}a_4 - \frac{11}{16}a_3^2, \varepsilon_3 = 0, \ldots \tag{2.8}$$

and for the logarithmic derivative

$$\mathcal{Y} = \sum_0^{\infty} \lambda^n \mathcal{Y}_n(v), \quad \mathcal{Y}_0 = v, \quad \mathcal{Y}_1 = \frac{a_3}{2}(v^2+1),$$

$$\mathcal{Y}_2 = \frac{v}{2}\left[\left(a_4 - \frac{a_3^2}{4}\right)v^2 + \left(\frac{3}{2}a_4 - \frac{7}{8}a_3^2\right)\right],$$

$$\mathcal{Y}_3 = \left(\frac{1}{16}a_3^3 - \frac{1}{4}a_3a_4 + \frac{1}{2}a_5\right)v^4 + \left(\tfrac{13}{32}a_3^3 - \tfrac{9}{8}a_3a_4 + a_5\right)v^2$$

$$+ \left(\frac{5}{8}a_3^3 - \frac{3}{2}a_3a_4 + a_5\right), \ldots \tag{2.9}$$

where, in general, $\mathcal{Y}_n = P_{n+1}(v)$ where P_{n+1} is an $(n+1)$th degree polynomial with the property that $\mathcal{Y}_n(-v) = (-1)^{n+1}\mathcal{Y}_n(v)$. All odd corrections $\varepsilon_{2k+1} = 0$. It is clear that the expansion (2.8) is simultaneously the perturbation series in powers of g and the semiclassical expansion in powers of $\hbar^{\frac{1}{2}}$, since the coefficients $\varepsilon_n(\{a\})$ are parameters — they do not depend on g and $\hbar$. Contrary to that, the expansion (2.9) is a PT expansion in powers of g only, since the corrections $\mathcal{Y}_n(v)$ are $\hbar$-dependent. Hence, it is not a semiclassical expansion in powers of $\hbar^{\frac{1}{2}}$. The expansion (2.9) mimics the asymptotic expansion at small v of Eq. (2.3),

$$\mathcal{Y} = \alpha + (\varepsilon + \alpha^2)v + \alpha(\varepsilon + \alpha^2)v^2 + \frac{\varepsilon^2 + \alpha^2(4\varepsilon + 3\alpha^2) - a_2}{3}v^3 + \cdots, \tag{2.10}$$

where the parameters α, ε can be found approximately only. For even potentials the ground state function is even, its logarithmic derivative $\mathcal{Y}$ is odd, hence, $\alpha = 0$. The asymptotic expansion (2.10) gets the simplified form

$$\mathcal{Y} = \varepsilon v + \frac{\varepsilon^2 - a_2}{3} v^3 + O(v^5), \tag{2.11}$$

The expansion (2.11) mimics also the perturbation theory in powers of λ for $\mathcal{Y}$ (2.9): the coefficients in (2.11) can be found in the form of the expansion in powers of λ. It must be emphasized that the expansion (2.11) has the form of a Taylor expansion even for the one-term potentials

$$V = \lambda^{2p-2} v^{2p}, \quad p = 2, 3, \ldots,$$

where the perturbation theory in powers of λ can *not* be developed.

In general, for the case of even potentials: $\hat{V}(\lambda v) = \hat{V}(-\lambda v)$ in (2.3), where all odd terms are absent: $a_3 = a_5 = \ldots = 0$, the potential has the explicit form

$$\hat{V}(u) = u^2 + a_4 u^4 + a_6 u^6 + \cdots a_{2p} u^{2p} + \cdots, \quad u \equiv \lambda v, \tag{2.12}$$

cf.(1.14), see potentials (i), (ii), (iv), (vi); the situation gets simplified drastically. It is evident that in this case the function $\mathcal{Y}$ is antisymmetric, $\mathcal{Y}(-v) = -\mathcal{Y}(v)$. To exploit this property, one introduces a new function of the form

$$\mathcal{Y}(v) = v\hat{\mathcal{Y}}(v^2). \tag{2.13}$$

The RB equation for the function $\hat{\mathcal{Y}}$ becomes

$$2v_2 \partial_{v_2} \hat{\mathcal{Y}}(v_2) + \hat{\mathcal{Y}}(v_2) - v_2 \hat{\mathcal{Y}}^2(v_2) = \varepsilon(\lambda) - \frac{1}{\lambda^2} \hat{V}(\lambda^2 v_2), \tag{2.14}$$

where

$$\partial_{v_2} \equiv \frac{d}{dv_2}, \quad v_2 \equiv v^2.$$

The PT in powers of λ for the energy ε remains the same as in (2.9), while

$$\hat{\mathcal{Y}} = \sum_{n=0}^{\infty} \lambda^{2n} \hat{\mathcal{Y}}_{2n}, \; \varepsilon_0 = 1, \; \hat{\mathcal{Y}}_0 = 1,$$

$$\varepsilon_2 = \frac{3}{4} a_4, \; \hat{\mathcal{Y}}_2 = \frac{a_4}{4} (2\, v_2 + 3), \tag{2.15}$$

$$\varepsilon_4 = \frac{3}{16} \left(10 a_6 - 7 a_4^2\right),$$

$$\hat{\mathcal{Y}}_4 = \frac{1}{16}\left[\left(8a_6 - 2a_4^2\right)v_2^2 + \left(20a_6 - 11a_4^2\right)v_2 + \left(30a_6 - 21a_4^2\right)\right],\ldots$$

where, in general, $\hat{\mathcal{Y}}_{2n} = P_n(v_2)$, where P_n is an nth degree polynomial.

In the same way one can develop the PT for the mRB equation (2.6) with even potential (1.14) assuming that $\tilde{\mathcal{Y}}$ has no simple poles, see e.g. [Turbiner (1984)]. In order to do this let us transform equation (2.6) by changing the unknown function, see (2.13),

$$2v_2\partial_{v_2}\hat{\mathcal{Y}}(v_2) + 3\,\hat{\mathcal{Y}}(v_2) - v_2\hat{\mathcal{Y}}(v_2)^2 = \varepsilon(\lambda) - \frac{1}{\lambda^2}\hat{V}(\lambda^2 v_2),$$
$$\partial_{v_2} \equiv \frac{d}{dv_2},\quad v_2 \equiv v^2, \tag{2.16}$$

cf.(2.14). As a result, we get

$$\varepsilon = \sum_{n=0}^{\infty}\lambda^{2n}\varepsilon_{2n},\qquad \varepsilon_0 = 3,\qquad \varepsilon_2 = \frac{15}{4}a_4,$$
$$\varepsilon_4 = \frac{15}{8}\left(14a_6 - 11a_4^2\right),\ldots \tag{2.17}$$

$$\hat{\mathcal{Y}} = \sum_{0}^{\infty}\lambda^{2n}\hat{\mathcal{Y}}_{2n}(v_2),\quad \hat{\mathcal{Y}}_{n=0} = 1,\ \hat{\mathcal{Y}}_2 = \frac{a_4}{2}\left(v_2 + \frac{5}{2}\right),$$
$$\hat{\mathcal{Y}}_4 = \frac{1}{2}\left[\frac{1}{4}\left(4a_6 - a_4^2\right)v_2^2 + \frac{1}{8}\left(28a_6 - 17a_4^2\right)v_2 + \frac{5}{8}\left(14a_6 - 11a_4^2\right)\right],\ldots \tag{2.18}$$

where, in general, $\hat{\mathcal{Y}}_{2n} = P_n(v_2)$ where P_n is an nth degree polynomial.

It is clear that in general the PT expansion (2.17) for the energy ε is simultaneously the perturbation series in powers of g^2 and the semiclassical expansion in powers of $\hbar$, the coefficients $\varepsilon_n(\{a\})$ are parameters which do not depend on g and $\hbar$. Contrary to that, the expansion (2.18) (as well as (2.15)) is the PT expansion in powers of g^2 only, since the corrections $\hat{\mathcal{Y}}_{2n}(v_2)$ depend on $\hbar$: $\hat{\mathcal{Y}}_{2n}(v_2) = \hat{\mathcal{Y}}_{2n}(v_2;\hbar)$. Hence, it is not a semiclassical expansion in powers of $\hbar$.

This concludes the study of the weak coupling regime for the ground state.

2.3 Non-Linearization Procedure and Perturbation Theory

In this Section we will develop a general formalism for the perturbative expansion of a Riccati-Bloch equation. Sometimes, this is called a multiplicative perturbation theory e.g. [Turbiner (1984)] as opposed to the celebrated Rayleigh-Schrödinger perturbation theory see e.g. [Landau and Lifshitz (1977)], which is an example of an additive perturbation theory.

2.3.1 *Ground State (No Nodes)*

Let us take the RB equation (2.3)

$$\partial_v \mathcal{Y}(v) - \mathcal{Y}^2(v) = \varepsilon\,(\lambda_f) - V(v;\lambda_f), \tag{2.19}$$

with potential $V(v;\lambda_f)$ which admits a formal Taylor expansion

$$V(v;\lambda_f) = \sum_{n=0}^{\infty} V_n(v)\lambda_f^n. \tag{2.20}$$

Here λ_f is formal parameter and the coefficient functions $V_n(v)$, $n = 0, 1, \ldots$, are real functions in v. Let us now assume that the unperturbed equation, which occurs formally at $\lambda_f = 0$,

$$\partial_v \mathcal{Y}_0(v) - \mathcal{Y}_0^2(v) = \varepsilon_0 - V_0(v), \tag{2.21}$$

can be solved explicitly. This can always be achieved via the *inverse problem*: we take a normalizable function Ψ_0, then calculate its logarithmic derivative $\partial_x \log \Psi_0$, which defines the function $\mathcal{Y}_0(v)$, and then we calculate the r.h.s. in (2.21): the potential $V_0(v)$ and ε_0. Evidently, if for some reasons we know $\mathcal{Y}_0(v)$, the wave function Ψ_0 can be recovered,

$$\Psi_0(v) = \exp\left(-\int^v \mathcal{Y}_0(s)\,ds\right). \tag{2.22}$$

This leads to a constraint on the choice of $\mathcal{Y}_0(s)$: the function $\Psi_0(v)$ should be normalizable. Now, we can develop the PT in powers of λ_f,

$$\varepsilon(\lambda_f) = \sum_{n=0}^{\infty} \varepsilon_n \lambda_f^n \tag{2.23}$$

and

$$\mathcal{Y}(v) = \sum_{n=0}^{\infty} \mathcal{Y}_n(v)\lambda_f^n. \tag{2.24}$$

By substituting (2.23) and (2.24) into the RB equation (2.19) it is easy to see that by collecting terms of order λ_f^n, the nth correction $\mathcal{Y}_n(v)$ and ε_n satisfy the first order linear differential equation

$$\partial_v \left(\Psi_0^2\, \mathcal{Y}_n(v)\right) = \left(\varepsilon_n - Q_n(v)\right)\Psi_0^2, \tag{2.25}$$

where

$$Q_1(v) = V_1(v), \quad Q_n(v) = V_n(v) - \sum_{k=1}^{n-1} \mathcal{Y}_k(v)\, \mathcal{Y}_{n-k}(v), \quad n = 2, 3, \ldots. \tag{2.26}$$

Note that the second term in $Q_n, n > 1$ is made from lower order corrections $\mathcal{Y}_i(v), i = 1, 2, \ldots (n-1)$. Remarkably, the equation (2.25) describes one-dimensional electrostatics with variable permittivity Ψ_0^2: $\mathcal{Y}_n(v)$ plays the role of the electric field and the r.h.s. is a charge distribution.

Evidently, the general solution $\mathcal{Y}_n(v)$ of the equation (2.25) is given by

$$\mathcal{Y}_n(v) = \frac{1}{\Psi_0^2}\left(\int_{-\infty}^{v} (\varepsilon_n - Q_n)\Psi_0^2\, ds\right), \tag{2.27}$$

where ε_n is a constant to define. Its value can be fixed by imposing the boundary conditions. Since we are interested in finding bound states we have to impose the condition of the absence of current of particles at large distances: for both limits $v \to -\infty$ and $v \to \infty$,

$$\mathcal{Y}_n(v)\, \Psi_0^2\Big|_{v\to\{-\infty,\infty\}} \to 0, \tag{2.28}$$

as a boundary condition [Turbiner (1984)]. One can easily verify for the solution (2.27) that as v tends to $-\infty$ the condition (2.28) is satisfied automatically, while for the other limit $v = +\infty$ the constant ε_n should be chosen as follows

$$\varepsilon_n = \frac{\int_{-\infty}^{\infty} Q_n\, \Psi_0^2 dv}{\int_{-\infty}^{\infty} \Psi_0^2 dv}, \tag{2.29}$$

in order to satisfy (2.28). This expression defines the nth energy correction. Finally, the nth correction $\mathcal{Y}_n(v)$ is given by

$$\mathcal{Y}_n(v) = \frac{1}{\Psi_0^2}\int_{-\infty}^{v} (\varepsilon_n - Q_n)\Psi_0^2 ds, \tag{2.30}$$

with ε_n from (2.29).

Such a perturbative approach is called *the perturbation theory* in the *Non-Linearization Procedure* [Turbiner (1984)]. We will continue to simply call it the *Non-Linearization Procedure.* In contrast with the Rayleigh-Schrödinger PT, the knowledge of the entire spectrum of the unperturbed problem is not required to explicitly construct the perturbative corrections in (2.23) and (2.24). It is sufficient to know only the ground state wave function of the unperturbed problem that we are studying. In general, this approach leads to a closed analytic expressions for both the corrections ε_n and $\mathcal{Y}_n(v)$ in the form of nested integrals. However, for the particular case of the anharmonic oscillator with potential (1.14) all corrections $\mathcal{Y}_n(v)$ are polynomials in v, thus, all integrals in (2.30) can be evaluated analytically, and they can be found using linear algebra means, see (2.8)–(2.9) and (2.17)–(2.18).

Therefore, the Non-linearization Procedure is an efficient method for calculating any finite number of PT corrections (2.23), (2.24), see [Turbiner (1984)] for a review, and [Del Valle and Turbiner (2019, 2020)] for the generalization to radial d-dimensional potentials.

Interestingly, in this framework the convergence of the perturbation series (2.23) and (2.24) is guaranteed as long as the first correction $\mathcal{Y}_1(v)$ is bounded,

$$|\mathcal{Y}_1(v)| \leq Const. \tag{2.31}$$

We have presented a brief overview of the Non-Linearization Procedure applied to the ground state. In principle, this approach can be modified to study the excited states by admitting a number of simple poles with residues equal to (-1) in $\mathcal{Y}_n(v)$. In this case the pole positions are found in a PT as the expansion in powers of λ_f. However, in practice it is more convenient to proceed to the excited states differently. It will be realized in the next Section.

2.3.2 *Excited States (The Case of K Nodes)*

In order to study the excited states in the Non-Linearization Procedure let us take the wave function in a representation, which we call *generalized exponential representation,*

$$\Psi^{(K)} = P(x)e^{-\frac{1}{\hbar}\Phi(x)}, \ P(x) = \prod_{i=1}^{K}(x - x_i), \tag{2.32}$$

where we assume that the pre-exponential factor $P(x)$ is a polynomial with real coefficients of degree K having K real roots; we will continue to call the function Φ the *phase*. It is worth noting that for the ground state $P = 1$ or, equivalently, $K = 0$, the representation (2.32) is reduced to (1.17). Integer $K = 0, 1, 2, \ldots$ is the quantum number, it numerates eigenstates.

Making substitution the (2.32) into the Schrödinger equation (1.16) we arrive at a generalized Riccati equation, see e.g. [Turbiner (1984)],

$$\frac{1}{m}\left(\hbar\, y' - y^2\right) - \frac{1}{m}\frac{\hbar^2\, P'' - 2\hbar y P'}{P} = E - \frac{1}{g^2}\hat{V}(gx), \quad y = \Phi', \tag{2.33}$$

cf.(1.18), with boundary conditions $y \to \pm\infty$ as $x \to \pm\infty$, respectively. The coefficients of the polynomial $P(x)$ are found by imposing the condition of the *exact* cancellation of the numerator and denominator in the ratio in the l.h.s. of (2.33). This implies that the zeroes of denominator coincide with corresponding zeroes of numerator. As a result, the potential $\hat{V}(gx)$ has no simple poles corresponding to nodes of eigenfunctions.

By making the substitution (2.1) into (2.33), in particular, replacing $x \to v$, we arrive at the *general Riccati-Bloch* (gRB) equation,

$$\mathcal{Y}_v' - \mathcal{Y}^2 - \frac{P_v'' - 2\mathcal{Y} P_v'}{P} = \varepsilon(\lambda_f) - V(v;\lambda_f), \tag{2.34}$$

cf. (2.3), see [Turbiner (1984)] with a potential $V(v;\lambda_f)$ instead of $\frac{1}{g^2}\,\hat{V}(gx)$, see (2.20), and with the same boundary conditions (2.4) as for the case of the ground state, where $P = 1$ in (2.32); all derivatives are taken with respect to v, since now on we drop subindex v to simplify notations. The pre-factor P is found by imposing the condition that the simple zeroes of P (its simple roots) must coincide with the roots of the numerator $(P'' - 2\mathcal{Y}P')$. For example, for the harmonic oscillator $V = v^2$, where the function $\mathcal{Y} = v$, the ratio is equal to $(2K)$ and $P_K(v)$ is the Hermite polynomial of degree K.

Let us assume that $V(v;\lambda_f)$ in (2.34) admits a formal Taylor expansion (2.20):

$$V(v;\lambda_f) = \sum_{n=0}^{\infty} V_n(v)\,\lambda_f^n,$$

and develop the PT for the Kth excited state as a Taylor expansion in powers of λ_f,

$$\varepsilon^{(K)}(\lambda_f) = \sum_{n=0}^{\infty} \varepsilon_n^{(K)}\lambda_f^n, \tag{2.35}$$

and

$$\mathcal{Y}^{(K)}(v) = \sum_{n=0}^{\infty} \mathcal{Y}_n^{(K)}(v)\lambda_f^n, \tag{2.36}$$

and

$$P^{(K)}(v) = \sum_{n=0}^{\infty} P_n^{(K)}(v)\lambda_f^n. \tag{2.37}$$

The unperturbed problem corresponds to the potential V_0,

$$\mathcal{Y}_0' - \mathcal{Y}_0^2 - \frac{P_0'' - 2\mathcal{Y}_0 P_0'}{P_0} = \varepsilon_0 - V_0. \tag{2.38}$$

This comes from (2.34) at $\lambda_f = 0$. To simplify notations from now on we drop the index (K) in this Chapter whenever this does not lead to confusion. Note that the potential V_0 can be built by choosing Ψ_0 and then $\mathcal{Y}_0$ and P_0, appropriately. Note that (2.38) can be rewritten as

$$\mathcal{Y}_0' - \mathcal{Y}_0^2 - (\varepsilon_0 - V_0) = \frac{P_0'' - 2\mathcal{Y}_0 P_0'}{P_0}. \tag{2.39}$$

It can be considered as the indication to exact cancellation of the ratio in the r.h.s.: l.h.s. is not singular at nodes of the original eigenfunction. This representation will be used in future considerations to simplify the equations which define corrections.

In order to calculate higher corrections it is convenient to multiply both sides of the gRB equation (2.34) by P,

$$\left(\mathcal{Y}' - \mathcal{Y}^2\right) P - P'' + 2\mathcal{Y}P' = \left(\varepsilon(\lambda_f) - V(v;\lambda_f)\right) P, \tag{2.40}$$

and then to substitute (2.20), (2.35), (2.36), (2.37) into it, collect all terms of order λ_f^n and set their sum equal to zero. At $n = 0$ we arrive at the unperturbed equation (2.38), (2.39). As for the first correction at $n = 1$ we obtain

$$\mathcal{Y}_1' - 2\mathcal{Y}_0\mathcal{Y}_1 - \frac{P_1'' - 2\mathcal{Y}_0 P_1' - [\mathcal{Y}_0' - \mathcal{Y}_0^2 - (\varepsilon_0 - V_0)]\, P_1 - 2\, P_0'\, \mathcal{Y}_1}{P_0}$$
$$= \left(\varepsilon_1 - V_1\right), \tag{2.41}$$

while for $n = 2$

$$\mathcal{Y}_2' - 2\mathcal{Y}_0\mathcal{Y}_2 - \frac{P_2'' - 2\mathcal{Y}_0 P_2' - [\mathcal{Y}_0' - \mathcal{Y}_0^2 - (\varepsilon_0 - V_0)]P_2 - 2P_0'\mathcal{Y}_2}{P_0}$$
$$= \left(\varepsilon_2 - \tilde{Q}_2\right), \tag{2.42}$$

where

$$\tilde{Q}_2 = V_2 - \mathcal{Y}_1^2 + \frac{2\mathcal{Y}_1 P_1'}{P_0}$$
$$+ \frac{P_1}{P_0^2}\left[P_1'' - 2\mathcal{Y}_0 P_1' - P_1\left(\mathcal{Y}_0' - \mathcal{Y}_0^2 - \varepsilon_0 + V_0\right) - 2P_0'\mathcal{Y}_1\right]. \quad (2.43)$$

Note that for both equations (2.41), (2.42) there is a certain self-similarity: the left-hand-sides are the same, while the right-hand-sides are different. This property remains for all equations which define the higher order corrections.

For arbitrary n the equation to find the corrections of order n takes the self-similar form

$$\mathcal{Y}_n' - 2\mathcal{Y}_0\mathcal{Y}_n - \frac{P_n'' - 2\mathcal{Y}_0 P_n' - 2P_0'\mathcal{Y}_n}{P_0}$$
$$+ \frac{P_0'' - 2\mathcal{Y}_0 P_0'}{P_0^2} P_n = \left(\varepsilon_n - \tilde{Q}_n\right), \quad (2.44)$$

where $\tilde{Q}_n = \tilde{Q}_n^{()}$ plays the role of a perturbative potential of the nth order. It is made from the lower order corrections $\mathcal{Y}_i(v), P_i^{(K)}$, $i = 1, 2, \ldots (n-1)$. If for some reasons the pre-factor $P^{(K)}(v)$ is unchanged under perturbation, meaning $P_i^{(K)} = 0$ for $i > 0$, as happens for the ground state $p = 0$ the potential $\tilde{Q}_n$ coincides with the one given by (2.26),

$$\tilde{Q}_n^{(K)} = Q_n.$$

If the potential $V(v; \lambda_f)$ in (2.34) contains only two terms in its formal Taylor expansion (2.20)

$$V(v; \lambda_f) = V_0(v) + V_1(v)\lambda_f, \quad (2.45)$$

the perturbation potential $\tilde{Q}_n^{(K)}$ can be written explicitly in a compact form,

$$\tilde{Q}_n^{(K)}(v) = -\sum_{k=1}^{n-1} \mathcal{Y}_k(v)\, \mathcal{Y}_{n-k}(v)$$
$$-\frac{1}{P_0}\left[\sum_{k=1}^{n-1} P_k \left(\sum_{i=0}^{n-k} \mathcal{Y}_i(v)\mathcal{Y}_{n-k-i}(v) - \partial_v \mathcal{Y}_k(v) + \varepsilon_{n-k}\right) - 2\mathcal{Y}_k(v)\partial_v P_{n-k}\right],$$
$$(2.46)$$

for $n > 1$. For a general potential with Taylor expansion (2.20),

$$\tilde{Q}_n^{(N)}(v) = V_n - \sum_{k=1}^{n-1} \mathcal{Y}_k(v)\,\mathcal{Y}_{n-k}(v)$$

$$-\frac{1}{P_0}\left[\sum_{k=1}^{n-1} P_k\left(\sum_{i=0}^{n-k} \mathcal{Y}_i(v)\mathcal{Y}_{n-k-i}(v) - \partial_v \mathcal{Y}_k(v) + \varepsilon_{n-k} - V_{n-k}\right)\right.$$

$$\left. -2\mathcal{Y}_k(v)\partial_v P_{n-k}\right]. \tag{2.47}$$

By multiplying both sides of (2.44) by Ψ_0^2, we obtain the equivalent but much simpler form of this equation,

$$\partial_v\left\{\Psi_0^2\left[\mathcal{Y}_n - \partial_v\left(\frac{P_n}{P_0}\right)\right]\right\} = \left(\varepsilon_n - \tilde{Q}_n\right)\Psi_0^2. \tag{2.48}$$

Its solution can be found explicitly

$$\Psi_0^2\left[\mathcal{Y}_n - \partial_v\left(\frac{P_n}{P_0}\right)\right] = \int_{-\infty}^{v}\left(\varepsilon_n - \tilde{Q}_n\right)\Psi_0^2 dv. \tag{2.49}$$

By imposing the boundary conditions (2.4) one can get the nth energy correction in closed analytic form

$$\varepsilon_n = \frac{\int_{-\infty}^{\infty} \tilde{Q}_n \Psi_0^2\, dv}{\int_{-\infty}^{\infty} \Psi_0^2\, dv}. \tag{2.50}$$

In order to find the nth correction $P_n^{(K)}(v)$ to the pre-factor $P(v)$ in (2.32) let us denote, as the first step, the roots of the unperturbed pre-factor $P_0^{(N)}(v)$ as $v_k^{(0)}, k = 1, 2, \ldots K$, hence,

$$P_0^{(K)}(v_k^{(0)}) = 0.$$

For the sake of convenience, we normalize $P_0^{(K)}(v)$ in such a way that the coefficient in front of the leading term of degree v^K is equal to one,

$$P_0^{(K)}(v) = v^K + c_{K-1}\,v^{K-1} + \cdots + c_0, \tag{2.51}$$

here the coefficients $c_0, c_1, \ldots c_{K-1}$ are the elementary symmetric polynomials of its roots. It is natural to impose the requirement that the number of roots (nodes) must remain unchanged under perturbation, hence, the corrections $P_n^{(K)}(v), n = 1, 2, \ldots$ should be in the form of a polynomial of degree $(K-1)$,

$$P_n^{(K)}(v) \equiv R_{K-1}(v) = c_{K-1}^{(n)} v^{K-1} + \cdots + c_0^{(n)}. \tag{2.52}$$

Thus, we will study how the elementary symmetric polynomials of the roots (nodes) $c_0, \ldots, c_{K-1}$ in (2.51) will change under perturbation instead of the roots themselves. From a practical point of view it is much easier to calculate their changes than those of the roots themselves. By analysing the equation (2.49) one can find the values of the correction $P_n^{(K)}(v)$ to the pre-factor $P_0^{(K)}(v)$ at the roots (nodes) $v = v_k^{(0)}$ of the unperturbed problem:

$$P_n^{(K)}(v_k^{(0)}) = \frac{1}{\left(\partial_v P_0^{(K)}\right)\Big|_{v=v_k^{(0)}} e^{-2\Phi_0^{(K)}(v_k^{(0)})}} \times \int_{-\infty}^{v_k^{(0)}} \left(\varepsilon_n - \tilde{Q}_n^{(K)}\right) \Psi_0^2 \, dv \equiv A_k^{(n)}. \tag{2.53}$$

In turn, in order to find the coefficients $c_i^{(n)}, i = 0, 1, \ldots (K-1)$ in $R_{K-1}(v)$ (2.52) it is necessary to solve an inhomogeneous system of K linear equations,

$$c_{K-1}^{(n)} (v_k^{(0)})^{K-1} + c_{K-2}^{(n)} (v_k^{(0)})^{K-2} + \cdots + c_1^{(n)} (v_k^{(0)}) + c_0^{(n)} = A_k^{(n)}, \tag{2.54}$$

where $k = 1, 2, \ldots, K$. The matrix of coefficients of this systems coincides with the Vandermonde matrix, whose entries are the roots $v_k^{(0)}, k = 1, 2, \ldots, K$ of the polynomial $P_0^{(K)}(v)$. This is a typical linear algebra problem, that is covered in any textbook on linear algebra. Let us note that for $p = 1$ the pre-factor $P_0^{(1)}(v)$ is a monomial, the system of equations is reduced to a single equation and all corrections $P_n^{(1)}$ are constants.

2.3.3 *Anharmonic Oscillator (Excited States)*

formalism developed in Section 2.3.2 to the general anhar-
otential (1.14),

$$V(v; \lambda) = \frac{1}{\lambda^2} \hat{V}(\lambda v),$$

where the Taylor coefficients in the expansion (2.20) of the general potential become monomials,

$$V_n = a_n \lambda^{n-2} v^n, \; a_2 = 1, \tag{2.55}$$

for $n = 2, 3, \ldots$.

By analysing the equations (2.41),(2.42),(2.44) one will conclude that the correction $\mathcal{Y}_n$ for any integer n is a polynomial in v of order $(n+1)$. This implies that the procedure for constructing the corrections $\varepsilon_n, \mathcal{Y}_n, P_n$ can be carried out using linear algebra means. Furthermore, if all a's in the potential (1.14) are chosen to be equal to one, all coefficients in $\varepsilon_n, \mathcal{Y}_n, P_n$ are rational numbers!

Let us take the gRB equation (2.34) with the potential (2.55) and develop the PT (2.35),(2.36),(2.37),

$$\varepsilon^{(K)}(\lambda) = \sum_{n=0}^{\infty} \varepsilon_n^{(K)} \lambda^n, \quad \mathcal{Y}^{(K)}(v) = \sum_{n=0}^{\infty} \mathcal{Y}_n^{(K)}(v) \lambda^n,$$

$$P^{(K)}(v) = \sum_{n=0}^{\infty} P_n^{(K)}(v)\, \lambda^n,$$

for the Kth excited state, $K = 0, 1, 2, \ldots$. The unperturbed problem (at $\lambda = 0$) corresponds to the harmonic oscillator with potential $V_2 = v^2$,

$$\mathcal{Y}_0' - \mathcal{Y}_0^2 - \frac{P_0'' - 2\mathcal{Y}_0 P_0'}{P_0} = \varepsilon_0 - v^2. \tag{2.56}$$

One can immediately check by making a direct calculation that

$$\varepsilon_0^{(K)} = 1 + 2K, \; \mathcal{Y}_0^{(K)} = v, \; P_0^{(K)} = H_K(v), \tag{2.57}$$

where $H_K(v)$ is the Hermite polynomial of order K. The exact cancellation of the ratio in l.h.s. of (2.56) is derived from the Hermite equation,

$$\frac{P_0'' - 2v P_0'}{P_0} = -2K.$$

From equations (2.41), (2.42) one can find the 1st and 2nd corrections,

$$\varepsilon_1^{(K)} = 0, \qquad \mathcal{Y}_1^{(K)} = \frac{a_3}{2}(v^2 + K + 1),$$
$$P_1^{(K)} = a_3\, K \left[\frac{(K-1)(K-2)}{3} H_{K-3}(v) + \frac{(3K+1)}{2} H_{K-1}(v)\right], \tag{2.59}$$

$$\varepsilon_2^{(K)} = \frac{1}{16}\left[12(2K^2+2K+1)a_4 - (30K^2+30K+11)a_3^2\right],$$
$$\mathcal{Y}_2^{(K)} = \frac{v}{2}\left[\left(a_4 - \frac{a_3^2}{4}\right)v^2 + \frac{1}{8}\left((8K+12)a_4 - (6K+7)a_3^2\right)\right],$$
$$P_2^{(K)} = \frac{K!}{18(K-6)!}\, a_3^2\, H_{K-6}(v)$$
$$+ \frac{K!}{48(K-4)!}(12a_4 + (24K-7)a_3^2) H_{K-4}(v)$$
$$+ \frac{K!}{8(K-2)!}\left(8a_4 K + (9K^2 - 5K - 1)a_3^2\right) H_{K-2}(v). \tag{2.60}$$

Several higher order corrections more can be calculated explicitly as well. By analysing the equation (2.44) with the potential (2.47) one can see that all odd energy corrections vanish, $\varepsilon_{2k+1} = 0$. The nth correction to $\mathcal{Y}$ is a polynomial of degree n in v: $\mathcal{Y}_n^{(K)} = \mathrm{Pol}_n(v)$; the coefficients in front of the terms of leading degrees do not depend on the quantum number K. Surprisingly, the sum of all these terms of leading degrees is a certain compact analytic expression, which has a meaning of classical momentum at zero energy, see Chapter 3, 5, 6 for detailed discussion. The nth correction to the pre-factor $P^{(K)}$ is a linear superposition of $(n+1)$ Hermite polynomials of decreasing degrees starting from $(K-2)$ for even n and $(K-1)$ for odd n.

2.4 Strong Coupling Expansion

Let us take the RB equation (2.3) for a polynomial potential of $(2p)$th degree, where for simplicity all coefficients a's are set equal to one,

$$\partial_v \mathcal{Y} - \mathcal{Y}^2 = \varepsilon(\lambda) - v^2 - \lambda v^3 - \cdots - \lambda^{2p-3} v^{2p-1} - \lambda^{2p-2} v^{2p}. \tag{2.61}$$

Making scaling transformation (it is the so-called Symanzik scaling transformation),

$$v = \lambda^{-\frac{p-1}{p+1}}\,\tilde{v},\ y = \lambda^{\frac{p-1}{p+1}}\,\tilde{\mathcal{Y}},$$

we will arrive at the *rescaled RB* equation (rRB) of the form,

$$\boxed{\begin{aligned}&\partial_{\tilde{v}}\tilde{\mathcal{Y}} - \tilde{\mathcal{Y}}^2\\ &= \tilde{\varepsilon}(\lambda) - \frac{1}{\lambda^{4\,\frac{p-1}{p+1}}}\,\tilde{v}^2 - \frac{1}{\lambda^{5\,\frac{p-1}{p+1}-1}}\,\tilde{v}^3 - \cdots - \frac{1}{\lambda^{\frac{2}{p+1}}}\tilde{v}^{2p-1} - \tilde{v}^{2p},\end{aligned}} \tag{2.62}$$

where

$$\tilde{\varepsilon}(\lambda) = \lambda^{-2\frac{p-1}{p+1}}\,\varepsilon(\lambda),$$

is the new, rescaled energy. The equation (2.62) corresponds to the RB equation with the polynomial potential of the degree p,

$$V(\tilde{v}) = \frac{1}{\tilde{\lambda}^{2p}}\hat{V}(\tilde{\lambda}\,\tilde{v}), \quad \tilde{\lambda} = \lambda^{\frac{2}{p+1}},$$

and with energy $\tilde{\varepsilon}$, which depends evidently on $\tilde{\lambda}$. Its final form is

$$\partial_{\tilde{v}}\tilde{\mathcal{Y}} - \tilde{\mathcal{Y}}^2 = \tilde{\varepsilon}(\tilde{\lambda}) - \frac{1}{\tilde{\lambda}^{2p}}\hat{V}(\tilde{\lambda}\,\tilde{v}), \tag{2.63}$$

cf. (2.19).

Now one can develop the PT in the rRB equation (2.63) with respect to the parameter $1/\tilde{\lambda}$ in a similar way as was done in Section 2.3.1 for the ground state of the RB equation (2.19) with potential (2.20),

$$\varepsilon(\lambda) = \sum_{n=0}^{\infty} \varepsilon_n \frac{1}{\tilde{\lambda}^n}, \tag{2.64}$$

and

$$\mathcal{Y}(v) = \sum_{n=0}^{\infty} \mathcal{Y}_n(v) \frac{1}{\tilde{\lambda}^n}, \tag{2.65}$$

cf. (2.23), (2.24). The formulas (2.29), (2.30) continue to hold.

As for the excited states, the rRB equation (2.63) can be rewritten in a form similar to the gRB equation (2.34) which involves the pre-factor P, defined in (2.32). Then the PT with respect to the parameter $1/\tilde{\lambda}$ can be developed as a Taylor expansion in powers of λ,

$$\varepsilon^{(K)}(\lambda) = \sum_{n=0}^{\infty} \varepsilon_n^{(K)} \frac{1}{\tilde{\lambda}^n}, \tag{2.66}$$

and

$$\mathcal{Y}^{(K)}(v) = \sum_{n=0}^{\infty} \mathcal{Y}_n^{(K)}(v) \frac{1}{\tilde{\lambda}^n}, \tag{2.67}$$

and

$$P^{(K)}(v) = \sum_{n=0}^{\infty} P_n^{(K)}(v) \frac{1}{\tilde{\lambda}^n}. \tag{2.68}$$

The formulas (2.49), (2.50) as well as (2.53), (2.54) hold. This PT is called the *strong coupling expansion.*

In the (ultra)-strong coupling domain, the unperturbed problem corresponds to the potential $V = v^{2p}$ with integer $p = 2, 3, \ldots$. It appears as the term of highest degree in the anharmonic oscillator potential. This implies that the remaining terms in the anharmonic oscillator potential are subordinate to it, and the PT with respect to the parameter $1/\tilde{\lambda}$ must be convergent.

It is well known that the Schrödinger equation with power-like potential

$$V = v^{2p}, \quad p \geq 1, \tag{2.69}$$

is characterized by a purely discrete spectrum and for $p > 1$ it can not be solved *exactly*. Hence, the PT (2.64), (2.65) and (2.66), (2.67), (2.67) can not be developed constructively as was done for the weak-coupling domain, where the parameter of perturbation was λ and unperturbed problem was the harmonic oscillator with $p = 1$. However, if the eigenfunctions for the potential v^{2p} were to be known with high local accuracy at any point in the coordinate space the PT in $1/\tilde{\lambda}$ could be developed approximately but also with high accuracy. This will be discussed in Chapters 5, 6.

Chapter 3

Generalized Bloch Equation, Semiclassical Expansion

In this Chapter it will be derived the so-called Generalized Bloch equation, which has no dependence on the Planck constant $\hbar$ similarly to the Riccati-Bloch equation, derived in Chapter 2. The generalized Bloch equation will allow us to construct the asymptotic expansion at large distances and the semiclassical expansion of the solution. By matching the expansion at small distances (or perturbation theory) with the expansion at large distances (or semiclassical expansion) leads to the remarkable simple, approximate expression for the eigenfunction, which holds for any anharmonic oscillator.

3.1 Generalized Bloch Equation

There exists an alternative way to transform the original Riccati equation (1.18),

$$\frac{1}{m}(\hbar y' - y^2) = E - \frac{1}{g^2}\hat{V}(gx), \quad y = \Phi',$$

into a certain non-linear equation without an explicit dependence on $\hbar$ and mass m other than (2.3). This is achieved by a change of variable into the *classical variable*[1]

$$u = g\,x = \lambda v, \tag{3.1}$$

and by introducing a new unknown function

$$\mathcal{Z}\,(u) = \frac{g}{(m)^{1/2}}\,y, \tag{3.2}$$

[1]In particular, this variable was introduced in [Brezin *et al.* (1977b)].

while keeping the energy and the effective coupling constant

$$E = \frac{\hbar}{m^{1/2}}\,\varepsilon, \quad \lambda = \left(\frac{\hbar^2}{m}\right)^{1/4} g,$$

the same as in (2.1) and (2.2). The resulting $\mathcal{Z}(u)$ satisfies the non-linear differential equation

$$\boxed{\lambda^2\,\partial_u \mathcal{Z}(u) - \mathcal{Z}^2(u) = \lambda^2\,\varepsilon(\lambda) - \hat{V}(u), \quad \partial_u \equiv \frac{d}{du},} \tag{3.3}$$

c.f. (2.3), with boundary conditions

$$\partial_u \mathcal{Z}(u)\Big|_{u=0} = \varepsilon, \quad \mathcal{Z}(\pm\infty) = \pm\infty, \tag{3.4}$$

if we set $\mathcal{Z}(0) = 0$. The first condition in (3.4) certainly holds for even potentials, $\hat{V}(u) = \hat{V}(-u)$. In general, this condition at $u = 0$ is of the form

$$\lambda^2\,\partial_u \mathcal{Z}(u)\Big|_{u=0} - \mathcal{Z}^2(0) = \lambda^2\,\varepsilon(\lambda),$$

if we exclude, for the sake of simplicity, the case $\mathcal{Z}(0) = \infty$. The second condition in (3.4) guarantees the square integrability of the wave function. Note that ε and λ play the role of energy and effective coupling constant in both the v-space (2.1) and u-space (3.1) dynamics. Also note that equation (3.3) requires a regularization if the coupling constant g tends to zero, $g \to 0$, and the potential becomes the harmonic oscillator potential $V = x^2$. In this limit the effective coupling $\lambda \to 0$ and, formally, the potential $\hat{V}(u)$ "jumps" to u^2. This regularization eventually leads to the original RB equation (2.3). To avoid this complication we will *not* take this limit and always assuming that $\lambda \neq 0$, thus, $\hat{V}(u) \neq u^2$. Hence, in (3.3) the parameter λ is taken as a formal parameter needed to construct the PT.

Equation (3.3) is called the *(one-dimensional) Generalized Bloch (GB) Equation* in [Escobar-Ruiz *et al.* (2016, 2017); Shuryak and Turbiner (2018)]. Its general solution contains a certain number of simple poles in the u-variable and has the form

$$\mathcal{Z}_N(u) = -\lambda^2 \sum_{i=1}^{N} \frac{1}{u - u_i} + \tilde{\mathcal{Z}}_N(u). \tag{3.5}$$

The number of poles N can also be identified with the quantum number, which measures the excitation number and takes the values $N = 0, 1, 2, \ldots$. As for the ground state, where $N = 0$, there are no simple poles and the sum in the r.h.s. in (3.5) is absent. The positions of the poles $u_i, i = 1, 2, \ldots, N$ coincide with the zeroes (nodes) of the eigenfunction of the Nth excited state.

Ground state. Let us consider the case of the ground state $N = 0$. Naturally, from equation (3.3) one can construct the asymptotic expansion for small u,

$$\mathcal{Z}(u) = \varepsilon\, u + \frac{(\varepsilon^2 - 1)}{3\lambda^2}\, u^3 - \frac{a_3}{4\lambda^2}\, u^4 - \frac{a_4\, 3\lambda^2 - 2\,\varepsilon\,(\varepsilon^2 - 1)}{15\lambda^4} u^5 + \cdots\,, \tag{3.6}$$

which is similar to (2.10). Looking at this expansion it becomes clear that we should not put $\lambda = 0$ as was previously mentioned. In the particular case of the polynomial confining potential of degree $(2p)$,

$$\hat{V}(u) = \sum_{k=2}^{2p} a_k u^k, \quad p \text{ integer}, \quad a_{2p} > 0, \tag{3.7}$$

cf. (1.14), (2.61), the asymptotic expansion at large u can also be constructed from equation (3.3). It has the form

$$\mathcal{Z}(u) = a_{2p}^{1/2}\, u^p + \frac{a_{2p-1}}{2a_{2p}^{1/2}}\, u^{p-1} + \frac{(4a_{2p}a_{2p-2} - a_{2p-1}^2)}{8a_{2p}^{3/2}} u^{p-2} + \cdots\,. \tag{3.8}$$

Note that in the first $(2p)$ terms in the expansion the coefficients do not depend on the energy ε, while the first p terms in (3.8) grow as $u \to \pm\infty$. This implies that these terms remain the same for *any* excited state. Furthermore, the coefficients in front of the first $[\frac{p}{2}]+1$ growing with u terms in (3.8) do *not* depend on the effective coupling constant λ. Surprisingly all those growing with u terms are reproduced exactly, if the expansion of the classical momentum at large $|u|$ is taken, see below (3.11). Later on this property will be demonstrated in detail for quartic and sextic anharmonic oscillators.

Power-like potentials. It is interesting to check what would happen for one-term, power-like even potentials, where all $a_2 = a_3 = \ldots = a_{2p-1} = 0$ except for $a_{2p} \neq 0$. In this case the potential is dramatically simplified becoming, cf.(3.7),

$$V = a^2 u^{2p}, a^2 \equiv a_{2p}. \tag{3.9}$$

One can easily verify that in the expansion (3.8) all terms of positive degrees vanish, except for the one of leading degree p, as well as all terms of negative degrees $-2, -3, \ldots - (p-1), -(p+1)$,

$$\mathcal{Z} = \pm a u^p + \frac{\lambda^2 p}{2}\frac{1}{u} - \frac{\lambda^2 \varepsilon}{\pm 2a}\frac{1}{u^p} - \frac{\lambda^2 p(p+2)}{\pm 8a}\frac{1}{u^{p+2}} + \cdots, \tag{3.10}$$

where for even p the plus sign is chosen for positive $u > 0$ and the minus sign for negative $u < 0$ to guarantee the square integrability of the ground state function. For odd p the signs in (3.10) should all be positive.

The property that the first $(2p)$ coefficients in the expansion (3.8) (and (3.10)) do not depend on energy and can be found exactly in terms of the parameters of the potential, plays a crucially important role in the construction of the approximation for the ground state wavefunction, see Section 3.3.

3.2 Perturbation Theory for the Generalized Bloch Equation (Ground State)

Now we develop the PT in powers of λ for equation (3.3) assuming that the solution $\mathcal{Z}$ has no simple poles at real u. This is equivalent to saying that we consider the ground state. Evidently the expansion of the energy ε in powers of λ (2.8) remains the same as in the Riccati-Bloch equation (2.3),

$$\varepsilon = \sum_n \lambda^n \varepsilon_n,$$

unlike the expansion for $\mathcal{Z}$,

$$\mathcal{Z} = \sum_n \lambda^n \mathcal{Z}_n(u).$$

Here the first corrections can be found explicitly by making simple algebraic manipulations

$$\mathcal{Z}_0 = \sqrt{\hat{V}(u)}, \mathcal{Z}_1 = 0, \mathcal{Z}_2 = \frac{1}{2}\left(\log\sqrt{\hat{V}(u)}\right)'_u - \frac{\varepsilon_0}{2\sqrt{\hat{V}(u)}}, \ldots \tag{3.11}$$

where $\varepsilon_0 = 1$, cf.(2.9). It must be emphasized that the corrections $\mathcal{Z}_n(u), n = 0, 1, 2, \ldots$ do not depend on the Planck constant $\hbar$. Note that in the standard WKB approach, see e.g. [Landau and Lifshitz (1977)], ε_0 is replaced by ε, which depends on g and $\hbar$, and, in general, it can *not* be found exactly. It is worth presenting three particular examples:

(i) the quartic anharmonic oscillator $\hat{V} = u^2 + u^4$, where

$$\mathcal{Z}_0 = u\sqrt{1+u^2}, \quad \mathcal{Z}_2 = \frac{1}{4}\left(\log u^2(1+u^2)\right)'_u - \frac{1}{2u\sqrt{1+u^2}},$$

(ii) the sine-Gordon potential $\hat{V} = \sin^2 u$, where

$$\mathcal{Z}_0 = \sin u, \quad \mathcal{Z}_2 = \frac{1}{2}\cot u - \frac{1}{2\sin u},$$

and,

(iii) the quartic oscillator $\hat{V} = u^4$, where

$$\mathcal{Z}_0 = u|u|, \quad \mathcal{Z}_2 = \frac{1}{u} - \frac{\varepsilon}{2u|u|},$$

here $\varepsilon \approx 1.0604$ has the meaning of the ground state energy of the quartic oscillator.

One can immediately recognize that $\mathcal{Z}_0$ in (3.11) is, in fact, the classical momentum at zero energy and at $\lambda = 1$. In turn, $\int \mathcal{Z}_0\, du$ is the classical action at zero energy and at $\lambda = 1$. Furthermore, by changing the argument of $\mathcal{Z}_0$ from u to v by using the relation $u = \lambda v$ (3.1), one can see that $\mathcal{Z}_0(\lambda v)$ is the generating function for the leading terms (of highest degrees in v) of the $\mathcal{Y}_n(v)$ corrections of the expansion (2.9), while $\mathcal{Z}_2(\lambda v)$ is the generating function for the next, subleading terms (of one-before-the-highest degrees in v) of the $\mathcal{Y}_n(v)$ corrections etc. From another side,

$$\int \mathcal{Z}_2\, du = \frac{1}{4}\log \hat{V}(u) - \frac{\sqrt{\hat{V}(u)}}{\hat{V}'(u)}, \tag{3.12}$$

is related to the logarithm of the determinant in the path integral formalism, for discussion see the next Section. In general, the expansion (3.11) is the *true* semiclassical expansion in powers of $\hbar^{1/2}$,

$$\mathcal{Z} = \sum_n \hbar^{\frac{n}{2}} \left(g^n \, \mathcal{Z}_n(gx) \right),$$

while

$$\varepsilon(\hbar^{1/2}\, g) = \sum_n (\hbar^{1/2}\, g)^n \varepsilon_n.$$

In closing it must be stated that if the original potential (1.14) has two or more degenerate global minima, e.g. the potential **(v)** (1.10) $V_{dw} = x^2(1 \pm gx)^2$ and $\hat{V}_{dw}(u) = u^2(1 \pm u)^2$, in addition to the Taylor expansion in powers of λ for energy, which is summed up to the *perturbative* energy: $E_{PT}(\lambda) = \sum E_n \lambda^n$, exponentially-small terms occur. These are summed up into the so-called *non-perturbative* energy E_{NPT}. Hence, the so-called *trans-series* expansion in λ occurs, see e.g. [Shifman (2015)], and also [Shuryak and Turbiner (2018)] and references therein for the case of the double-well potential V_{dw}. Eventually, the total energy $E = E_{PT} + E_{NPT}$. In the RB equation (2.3), where the energy depends on the single parameter $\lambda = \hbar^{1/2}\, g$, it has the form

$$\varepsilon(\hbar^{1/2}\, g) = \varepsilon_{PT}(\hbar^{1/2}\, g) + \varepsilon_{NPT}(\hbar^{1/2}\, g).$$

The semiclassical expansion in powers of $\hbar^{1/2}$ (2.8) (the Taylor expansion) becomes more general semiclassical expansion in the form of the trans-series in $\hbar^{1/2}$, hence, having the exponentially small terms in addition to the Taylor expansion. A similar situation occurs for the eigenfunctions: they can be written as the product of perturbative and non-perturbative functions [Shuryak and Turbiner (2018)].

3.3 Semiclassical Approximation of Path Integrals, the "Flucton" Paths

In this Section the semiclassical expansion will be re-derived in the path integral formalism. Since this topic lies aside of the mainstream presentation, not much details will be given.

Following Feynman, see [Feynman and Hibbs (1965)], the probability amplitude for a quantum system to evolve from point x_i to point x_f in time t can be expressed as a functional integral over all paths, starting and ending at these points. Feynman also famously pointed out that by moving this expression to Euclidean (imaginary) time $\tau = it$ defined on a circle with "Matsubara time" circumference related to temperature by

$$\beta = \frac{\hbar}{T}, \tag{3.13}$$

one can use path integrals in statistical mechanics [Feynman (1972)]. Specifically, the partition function is given by the integral over the periodic paths with time period β.

In this Section, based on the presentation in the unpublished paper [Turbiner and Shuryak (2021)], we will use this formalism in the zero temperature limit only, in which $\beta \to \infty$ and where the path integral naturally describes the ground state, to construct the semiclassical expansion — the expansion in powers of $\hbar$. Its density matrix — the probability density for finding the quantum particle at a certain position $P(x_0)$ — is given by an integral over periodic paths, where the initial and final points coincide, $x_i = x_f = x_0$. In the $\beta \to \infty$ limit this corresponds to the square of the ground state wave function $P(x_0) = |\psi_0(x_0)|^2$. Thus,

$$\psi_0^2(x_0) = \int \mathcal{D}x e^{-\frac{S_E}{\hbar}}, \tag{3.14}$$

where S_E is the Euclidian action, if $\psi_0(x)$ is a real, positive function. Needless to say that the square-root of the rhs should correspond to the exact solution of the Schrödinger equation.

Development of the semiclassical theory of the ground state, based on special classical paths called *fluctons*, was started in the paper [Shuryak (1988)]. Significant development of this theory has then been made in two papers [Escobar-Ruiz *et al.* (2016, 2017)], focused on the ground state of a number of quantum-mechanical problems, harmonic and anharmonic oscillators, as well as the double-well and sine-Gordon potentials.[2]

A flucton is a path possessing the least action among all paths passing via the *observation point* x_0. Therefore it satisfies the classical (Newtonian) equation of motion

$$m\ddot{x}(\tau) = \frac{\partial V}{\partial x}, \tag{3.15}$$

[2]For first application of this semiclassical approximation to finite temperatures for some of these examples, see [Shuryak and Torres-Rincon (2020)].

where dots indicate derivatives over *Euclidean* time τ. Note that the usual minus sign in the r.h.s. is absent: one can view this as motion in the *inverted* potential $(-V(x))$.

As usual in one-dimensional classical mechanical problems in order to find trajectory in (3.15) in easy way, one can employ conservation of energy, which in this case takes the form,

$$\frac{m}{2}\dot{x}(\tau)^2 - V(x) = E. \tag{3.16}$$

The maximum of the inverted potential $(-V(x))$ is conveniently set to zero, so a particle with $E = 0$ may stay at the maximum for infinite time. This simple idea defines the shape of the *flucton*.

Before giving its explicit form, let us do some redefinitions. For finite β, the time variable τ is defined on a circle. The only condition for the paths is that they must pass through the observation point x_0, but it does not matter at what moment in time this happens. Therefore, one can choose it to be zero

$$x(\tau = 0) = x_0, \tag{3.17}$$

with two symmetric arms, for positive and negative τ, describing the *path relaxation*. In the zero temperature limit, or stated differently at $\beta \to \infty$, which is the limit we are interested, the domain is $\tau \in (-\infty, \infty)$ and the asymptotic coordinate values correspond to the position of the potential minimum, defined as

$$x(\tau \to \pm\infty) = 0.$$

The explicit functional shape follows readily from conservation of energy (with $m = 1, E = 0$)

$$\tau = \int_{x_0}^{x(\tau)} \frac{dx}{\sqrt{2V(x)}}. \tag{3.18}$$

Remark. *At this point let us break presentation and, following the wisdom of the new variables introduced in Section 3.2, let us describe the Euclidean action. With the classical coordinate* $u = g\,x$ *and for the potential defined* *as $V(x) = \hat{V}(u)/g^2$ it takes the form*

$$\frac{S}{\hbar} = \frac{1}{\hbar g^2} \int d\tau \left(\frac{1}{2} \dot{u}(\tau)^2 + \hat{V}(u) \right), \tag{3.19}$$

see (3.14), in which the coupling constant is united with the Planck constant (leading to effective coupling $\lambda^2 = \hbar g^2$), but both of them are absent in the Equation of Motion (EoM). The path integral now takes the form

$$P(u_0) = \int \mathcal{D}u \; e^{-\frac{1}{\hbar g^2} \int d\tau \left(\frac{1}{2} \dot{u}(\tau)^2 + \hat{V}(u) \right)}, \tag{3.20}$$

where, we remind, the dependence on the observation point $u_0 = gx_0$ comes from the condition (3.17) which all paths must obey.

The change of variable in (3.19) to the *classical variable* u has important consequences. In particular, the integral in the r.h.s. of (3.18) becomes

$$\tau = \int_{u_0}^{u(\tau)} \frac{du}{\sqrt{2\hat{V}(u)}}, \tag{3.21}$$

independent on g and $\hbar$, where $u_0 = gx_0$ is the starting point in u-space. Therefore, in such notations the flucton shape is universal: it does not depend on the coupling constant.

For an example, for the potential $\hat{V} = u^2/2 + u^4$, the integral (3.21) can be evaluated analytically, its inverse can be found also in closed analytic form, giving

$$u_{fl}(\tau) = \frac{1}{\sqrt{2} \sinh \left(arccsch \left(\sqrt{2}\, u_0 \right) + \tau \right)}. \tag{3.22}$$

Three examples of flucton paths for different $u(0)$ are plotted in Fig. 3.1. Inserting (3.22) into the (Euclidean) action (3.19), one gets

$$S_{fl}(u_0) = \frac{1}{6\hbar g^2} \left(-1 + (1 + 2u_0^2)^{3/2} \right), \tag{3.23}$$

and therefore the explicit form of density matrix $P(u_0) \sim exp(-S_{fl}(u_0))$. This result, of course, reproduces the standard WKB expression for the ground state wave function at zero energy $E = 0$, cf. (3.11).

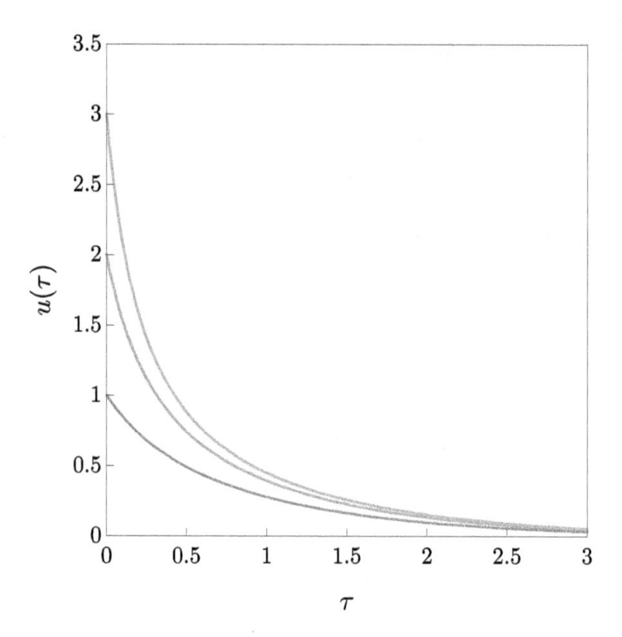

Fig. 3.1. Examples of flucton paths $u_{fl}(\tau)$ (3.22), when $u(0) = 1, 2, 3$ versus Euclidean time τ (blue, yellow, green curves, respectively). Only half of the path, for positive τ is shown: the path at negative τ is its mirror image, $u(-\tau) = u(\tau)$.

As another example, for instance, for the quartic oscillator, $\hat{V} = u^4$, the integral in the r.h.s. of (3.21) can also be found in closed analytic form, leading to

$$u_{fl}(\tau) = \frac{u_0}{1 + \sqrt{2} u_0 \tau}, \tag{3.24}$$

which is not much different than the flucton trajectory for the anharmonic oscillator, cf. (3.22). Three examples of flucton paths for the quartic oscillator for different $u(0)$ are plotted in Fig. 3.2. Putting (3.24) into the (Euclidean) action (3.19), one gets

$$S_{fl} = \frac{\sqrt{2}}{3\,\hbar\, g^2}\, u_0^3, \tag{3.25}$$

cf. (3.23), and therefore the explicit expression for density matrix $P(u_0) \sim exp(- S_{fl})$.

However, unlike the standard WKB, the flucton version of the semiclassical theory allows us to derive systematically the expansion in the semiclassical corrections as a series in powers of $\hbar^{1/2}$, see [Escobar-Ruiz *et al.*

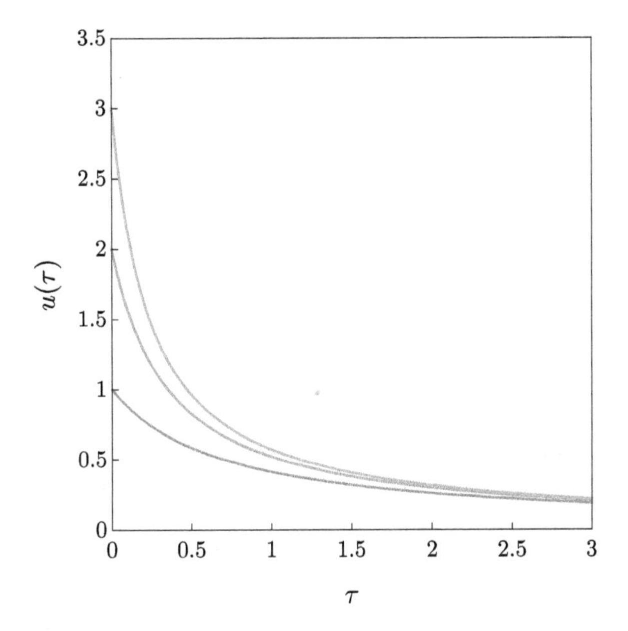

Fig. 3.2. Examples of flucton paths $u_{fl}(\tau)$ (3.24), when $u(0) = 1, 2, 3$ versus Euclidean time τ (blue, yellow, green curves, respectively). Only half of the path, for positive τ is shown: the path at negative values is its mirror image.

(2016, 2017)]. This works as follows. An arbitrary path can be viewed as a classical flucton path modified by a *quantum fluctuation*,

$$u(\tau) = u_{fl}(\tau) + \hbar^{1/2} q(\tau). \tag{3.26}$$

By plugging this into the action one can develop a systematic expansion in powers of q

$$\frac{S - S_{fl}}{\hbar} = \frac{1}{\hbar g^2} \int d\tau \left(\frac{\dot{q}^2}{2} + \frac{1}{2!} \frac{\partial^2 \hat{V}(u = u_{fl})}{\partial u^2} q^2(u) \right. \\ \left. + \frac{\hbar^{1/2}}{3!} \frac{\partial^3 \hat{V}(u = u_{fl})}{\partial u^3} q^3(u) + \cdots \right). \tag{3.27}$$

Note that in the first two terms (quadratic in q) both the coupling constant g and the Planck constant $\hbar$ are absent, while the subsequent nonlinear terms contain growing powers of $\hbar^{1/2}$. If only the quadratic terms in q are kept, the resulting EoM is linear, defining the so called *fluctuation operator* $\hat{O}$. In this approximation the functional integral over fluctuations is Gaussian, yielding the *determinant* of the operator $\hat{O}$. In the notations we use, it becomes obvious that the operator, its eigenvalue spectrum, and their product — the determinant — are all universal, independent on g and $\hbar$.

It is important to note, that unlike instanton and many other classical trajectories (e.g. solitons), the flucton background does not have *any* zero modes of $\hat{O}$, and therefore its inversion is straightforward. Indeed, there is *no* symmetry corresponding to time translation, because by definition the paths should satisfy condition (3.17), which implies that $q(0) = 0$.

The eigenvalue equation

$$\hat{O}u_\lambda(\tau) = \lambda\, u_\lambda(\tau),$$

is second order in the derivatives and thus similar to Schrödinger equation with an effective potential given by the second derivative at the background. For the example at hand, $\hat{V} = u^2/2 + u^4$ (the quartic anharmonic oscillator),

$$\hat{V}''(u = u_{fl}) = 1 + 12u_{fl}^2,$$

where u_{fl} is presented in (3.22), hence, the effective potential is equal to 1 at $\tau \to \pm\infty$ and larger than 1 at the origin, $\tau = 0$. Clearly, there are no bound states and all states are scattering states. With standard quantization in a box, all those can be found. This direct diagonalization approach has been used in the works [Escobar-Ruiz *et al.* (2016, 2017)]. However, this is no longer necessary, since in Section 3.2 it was derived analytic expression for the determinant for arbitrary potential(!), see (3.12).

Higher order terms $O(q^3, q^4 ...)$ in the expansion (3.27) can be viewed as vertices of Feynman diagrams: the growing powers of $\hbar^{1/2}$ in front of them show that it is a true semiclassical expansion. The propagators which occur in such diagrams are nothing but the Green function, which is the inverse of the fluctuation operator $\hat{O}$. In fact, several of them were calculated in Section II, see (3.11). These terms correspond to 1-,2-,3-... loop results. It is worth noting that the one-loop contribution, see Fig. 3.3, is given (formally) by a single diagram, cf. [Escobar-Ruiz *et al.* (2016)], Fig. 3 therein and [Escobar-Ruiz *et al.* (2017)], Fig. 2 therein.

In turn, the two-loop contribution is the sum of three diagrams, see Fig. 3.4, cf. [Escobar-Ruiz *et al.* (2016)], Fig. 4 therein and [Escobar-Ruiz *et al.* (2017)], Fig. 3 therein, etc.

A non-trivial fact observed in calculations of concrete systems [Escobar-Ruiz *et al.* (2016, 2017)] is that each individual diagram can be quite complicated, it can contain highly transcendental expressions, not always can be calculated analytically, while in the sum of all individual diagrams of a given

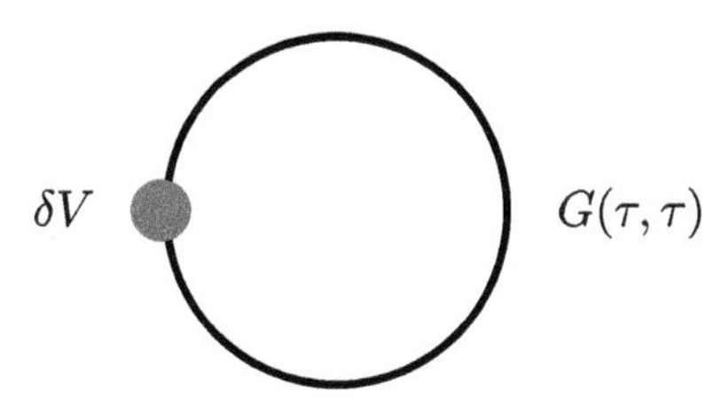

Fig. 3.3. Symbolic Feynman one-loop diagram, including variation of the fluctuation potential $\delta V \propto V''$ as a vertex and the simplified *single-loop* Green function $G(\tau,\tau)$, see [Escobar-Ruiz *et al.* (2016, 2017)].

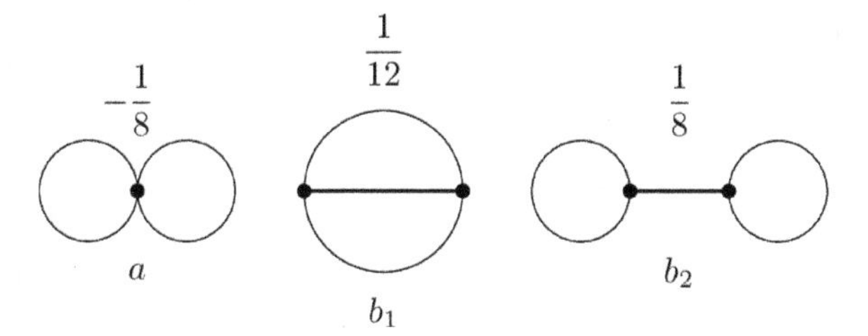

Fig. 3.4. Feynman diagrams a, b_1, b_2 contributing to the two-loop correction $B_1 = a + b_1 + b_2$. The signs of contributions and symmetry factors are indicated, see [Escobar-Ruiz *et al.* (2016, 2017)].

order with correct symmetry factors taken into account some *mysterious* cancellations occur. It may lead to a sufficiently simple final expressions, which sometimes can be even written in closed analytic form. A certain advantage of the formalism developed in Section 3.2 is that the expansion (3.11) deals with the sum of Feynman diagrams and we do *not* see all these complications due to individual diagrams.

Chapter 4

Matching Perturbation Theory and Semiclassical Expansion

In this short Chapter we will elaborate the main result of Part I: the uniform approximation of the eigenfunctions of the one-dimensional AHO with a polynomial potential of even degree $(2p)$. First, we consider the ground state function and construct its approximation in the form of a prefactor multiplied by an exponential both with several free parameters, which are fixed by using the standard variational procedure. Second, we take the ground state function approximation with arbitrary parameters, then (i) we modify their prefactor accordingly, and then (ii) we multiply it by a polynomial of degree K with real coefficients, whose all roots are real and simple (no roots coincide), which plays the role of extra prefactor. The coefficients of the polynomial are fixed by imposing the orthogonality to the approximations of lower energy eigenstates with lower degree polynomials, which are assumed to be known already. In turn, the free parameters of the modified ground state *type* approximation are fixed by making the minimization of the variational energy.

Let us consider the polynomial potential of degree $(2p)$, see (1.14) or (3.7). We will look for the eigenfunction of the Kth excited state in the generalized exponential representation (2.32),

$$\Psi^{(K)} = P_K(x)\, e^{-\frac{1}{\hbar}\Phi_K(x)}, P_K(x) = \prod_{i=1}^{K}(x - x_i),$$

where all roots x_i are assumed real. Our main focus is on how to find the approximate expression for $\Phi_K(x)$.

By taking the perturbation theory for the logarithmic derivative of the ground state wavefunction $\mathcal{Y}$ (2.9) at small distances (or, said differently, the asymptotic expansions (2.10), (2.11)) and by *matching* it with the semi-classical expansion of $\mathcal{Z}$ (3.11) at large distances (or, said differently, the asymptotic expansion (3.8)), with an emphasis on the exact reproduction of all growing terms as $|x| \to \infty$, we can arrive practically unambiguously(!) at an approximate expression for the ground state eigenfunction. By multiplying it by a Kth degree polynomial in r we will end up with an approximation for the Kth excited state. Finally, it has the following explicit form

$$\Psi^{(K)}_{approximation}(x) = P_K(x) \times [prefactor(x, g; \{\tilde{b}\})]$$
$$\times \exp\left(- \frac{A + \tilde{a}_1 x + \hat{V}(x, g; \{\tilde{a}\})}{\sqrt{\frac{1}{x^2}\hat{V}(x, g;\ \{\tilde{b}\})}} + \frac{A}{\tilde{b}_0^2} \right). \quad (4.1)$$

Here $\hat{V}(x, g; \{\tilde{a}\})$, $\hat{V}(x, g; \{\tilde{b}\})$ are two polynomial potentials of degree $(2p)$, both of them are of the form of the original one, which was defined in (1.14) with coefficients $a_2, a_3, \ldots, a_{2p}$ (denoted here as $\{a\}$), but with a different sets of coefficients $\{\tilde{a}\}$ or $\{\tilde{b}\}$, respectively, instead. In general, $\{\tilde{a}\}, \{\tilde{b}\}$ and $A, \tilde{a}_1$ are $(4p - 1)$ free (variational) parameters subject to (p) constraints: for the phase Φ (1.17) and its (reduced) logarithmic derivative $\mathcal{Z}$ (3.2) all (p) growing terms at large distances $x(u)$ in (3.8) do not depend on energy and should be reproduced exactly. This guarantees the absence of exponentially-large deviation of (4.1) from the exact eigenfunction at large distances $|x| \to \infty$. This requirement leads to constraints on coefficients $\{\tilde{a}\}, \{\tilde{b}\}$. The prefactor $[prefactor(x, g; \{\tilde{b}\})]$ in (4.1) depends on the concrete potential under consideration, it is usually defined by the second correction $\mathcal{Z}_2$ in (3.11) (hence, by the determinant), see also (3.12), where the parameters $\{a\}$ are replaced by $\{\tilde{b}\}$. Here $P_K(x)$ is a polynomial of degree K, whose coefficients are found unambiguously by imposing the orthogonality condition to the functions $\Psi^{(L)}(x)$, see (4.1), of previous states with $L = 0, 1, 2, \ldots, (K - 1)$. By taking $\Psi^{(K)}(x)$ (4.1) to be the zeroth order approximation for the non-linearization procedure, see Chapter 2, Section 2.3,

$$\Psi^{(K)}(x) \equiv P_K^{(0)}(x) e^{-\phi_K^{(0)}},$$

see SubSections 2.3.1 and 2.3.3, one can develop a perturbation theory explicitly with respect to the deviation of the function $\Psi^{(K)}(x)$ (4.1) from the exact eigenfunction. Mainly due to the correct asymptotic behavior of the wavefunction (4.1), this perturbation theory is rapidly convergent. In particular,

- the second correction to the energy

$$|E^{(2)}| \ll E^{(0)} + E^{(1)},$$

- the first correction to the (nodal) prefactor

$$|P_K^{(1)}(x)| \ll |P_K^{(0)}(x)|,$$

  which implies that the coefficients in $P_K^{(1)}(x)$ are much smaller than the corresponding ones in $P_K^{(0)}(x)$, meaning that the positions of the nodes are almost unchanged,
- and the first correction to the phase

$$|\phi^{(1)}| \ll |\phi^{(0)}| \text{ for } \forall x \in \mathbf{R},$$

  uniformly.

In all concrete calculations

$$|E^{(2)}| \sim 10^{-9} - 10^{-10}$$

(or better) and the ratio

$$|\phi^{(1)}/\phi^{(0)}| \lesssim 10^{-6},$$

see e.g. Chapters 5-6 of Part I.

Formula (4.1) is the **central** formula of Part I of this book. A similar formula will be presented in Part II for an arbitrary radial anharmonic oscillator. As an illustration let us present two examples. It is easy to check that for the harmonic oscillator (HO),

$$V_{ho} = a_0^2 x^2,$$

cf. (1.1), formula (4.1) becomes exact, it contains no free parameters and it is of the form

$$\Psi_{ho} = P_K \, e^{-\frac{a_0}{2} x^2},$$

hence, the prefactor in (4.1) is equal to one, and P_K becomes the Kth Hermite polynomial.

Another example, which is worth mentioning, is the quartic symmetric anharmonic oscillator s(5.1), see Chapter 5,

$$V_{aho}^{(4)} = a_0^2 x^2 + a_2^2 g^2 x^4,$$

where one can set $a_0 = a_2 = 1$ without loss of generality. By matching the two asymptotic expansions at small and large distances (equivalently, the perturbation theory in x and the new semiclassical expansion) as described above, we arrive at the following extremely compact function for the ($K = 2N+p$)th-excited state with quantum numbers (N, p), $N = 0, 1, 2, \ldots, p = 0, 1$[1] [Turbiner and del Valle (2021a)],

$$\Psi^{(N,p)}_{(approximation)} = \frac{x^p P_{N,p}(x^2; g^2)}{(B^2 + g^2\, x^2)^{\frac{1}{4}} \left(B + \sqrt{B^2 + g^2\, x^2}\right)^{2N+p+\frac{1}{2}}}$$
$$\exp\left(-\frac{A + (B^2+3)x^2/6 + g^2\, x^4/3}{\sqrt{B^2 + g^2\, x^2}} + \frac{A}{B}\right), \quad (4.2)$$

as a reduction of (4.1) where two constraints (emerging from (3.8)) are imposed, see [Turbiner and del Valle (2021a)]. Here $P_{N,p}$ is some polynomial of degree N in x^2 with positive roots. The parameters $A = A_{N,p}(g^2)$, $B = B_{N,p}(g^2)$ are two variational parameters. It was shown in [Turbiner and del Valle (2021a)] that for the six lowest states $N = 0, 1, 2$ and $p = 0, 1$ the variational energies for coupling constants $g \in [0, \infty)$ are obtained with an *unprecedented* accuracy of 10-11 significant digits with the variational trial function with two free parameters. Variational (optimal) parameters $A = A_{N,p}(g^2)$, $B = B_{N,p}(g^2)$ are easily fitted [Turbiner and del Valle (2021a)], see Chapter 5 for details. This fit leads to a slight reduction in accuracy for the energy (which is the expectation value for the Hamiltonian) to 9-10 significant digits for any positive coupling constant g. Formula (4.2) remains valid for the quartic oscillator, $a_0 = 0$,

$$V^{(4)} = a_2^2\, g^2 x^4, \quad (4.3)$$

[1]In this case K is the *principal* quantum number

cf. (2.69),(3.9), with new parameters $A = A_{N,p}(g^2)$, $B = B_{N,p}(g^2)$. In the case of the double-well potential,

$$V_{dw}^{(4)} = -a_0^2 x^2 + a_2^2 g^2 x^4,$$

formula (4.2) needs to be modified in order to take into account tunneling between the two symmetric harmonic wells. The derivation of formula (4.2) and its further modification for the double-well potential will be the subject of the next Chapter 5.

Chapter 5

Quartic Anharmonic Oscillator

In this Chapter the knowledge presented in Chapters 2 and 3 for the general polynomial anharmonic oscillator will be applied to the one-dimensional quantum single-well symmetric anharmonic oscillator with potential $V(x) = x^2 + g^2x^4$. In this case the Perturbation Theory (PT) in powers of g^2 (weak coupling regime) and the semiclassical expansion in powers of $\hbar$ for the energies (eigenvalues) coincide. This is related to the fact that the dynamics in x-space and in (gx)-space corresponds to the same energy spectrum with effective coupling constant $\hbar g^2$. The two equations which govern the dynamics in those two spaces, the Riccati-Bloch (RB) and the Generalized Bloch (GB) equations, respectively, are derived. The PT in g^2 for the logarithmic derivative of the wave function leads to the PT (with polynomial in x coefficients) of the RB equation and to the true semi-classical expansion in powers of $\hbar$ of the GB equation, which corresponds to a loop expansion of the density matrix in the path integral formalism. A 2-parametric interpolation of these two expansions leads to a uniform approximation of the wavefunction in x-space with an unprecedented accuracy of $\sim 10^{-6}$ locally and an unprecedented accuracy $\sim 10^{-9} - 10^{-10}$ for the energy associated with it for any $g^2 > 0$. This Chapter is written in maximally self-contained way.

5.1 Quartic Anharmonic Oscillator: Introduction and Generalities

Overview. It is common knowledge that the quartic symmetric anharmonic oscillator (pendulum), both classical and quantum, with potential

$$V(x) = x^2 + g^2x^4 = x^2(1+g^2x^2) = \frac{1}{g^2}\left((gx)^2 + (gx)^4\right) \equiv \frac{1}{g^2}\,\hat{V}_4(gx), \tag{5.1}$$

cf. (1.14), where

$$\hat{V}_4(u) = u^2 + u^4, \quad u = gx, \tag{5.2}$$

cf. (3.7), is among the most fundamental problems in physics. In particular, it was one of the first problems that the newly-the-born quantum mechanics tried to tackle. It appears in all branches of physics, it is present in practically all textbooks on quantum mechanics. It reveals a great richness of non-trivial properties reflecting a complexity of Nature. Many of these properties are universal, they remain to be present in more complicated problems, they are of the general nature.

Enormous efforts were dedicated to the exploration of this problem: thousands of articles were published on this subject. In general, these efforts will not be reflected in this Chapter: definitely, this problem alone deserves to be the subject of a separate book. It will be mentioned a very few papers and their respective results, which are the most relevant to a formalism the present authors are going to explore. All that reflects the personal vision of the authors. It is worth noting that a particular direction of study performed is the field-theoretical treatment of (5.1) as $(0+1)$ quantum field theory, it was considered as a precursor to understanding of the realistic four-dimensional quantum field theory $\lambda\phi^4$, see e.g.[Lipatov (1976)], this will be mentioned briefly. This Chapter is intended to be maximally self-contained.

In a seminal paper by Bender and Wu [Bender and Wu (1969)] the theory of the weak coupling regime, where the coupling constant g^2 is small, was created. In this remarkable paper it was shown that in the weak coupling domain

- the problem is, in fact, *algebraic*: practically everything can be calculated by linear algebra means,
- the perturbation theory

$$\varepsilon = \sum_0^\infty \varepsilon_k\, g^{2k}, \tag{5.3}$$

for any eigenvalue has zero radius of convergence, the coefficients ε_k are always rational numbers with alternating signs and their asymptotics at

$k \to \infty$ is given by

$$\varepsilon_k = (-1)^{k+1}\, 4\, \pi^{-3/2} \left(\frac{3}{2}\right)^{k+1/2} \Gamma\left(k + \frac{1}{2}\right) \left[1 - \frac{95}{72}\frac{1}{k} + O\left(\frac{1}{k^2}\right)\right], \quad (5.4)$$

for the ground state,[1] this series is Borel summable, see for discussion [Graffi *et al.* (1970)],[2]

- there exist infinitely-many, symmetric with respect to real line (complex-conjugated), square-root branch points in the complex $m^2 = g^{-4/3}$ plane of a given eigenvalue, accumulating to m^2 tends to infinity (the so-called *horn structure* of the singularities, see Fig. 5.1; recently, their existence was proved rigorously in [Eremenko and Gabrielov (2009)]) (as the algebraic ramification points) as a reflection of the Landau-Zener theory of level crossings, see [Landau and Lifshitz (1977)].[3]

Each singularity occurs at a point of level crossing of the states of the same parity.[4] It indicates a highly-complicated analytic structure of the eigenvalues in the coupling constant m^2 (or g^2). The problem how to sum up the asymptotic series (5.3) to get reliable results for $g^2 \neq 0$ remains, in general, unsolved.

As a certain way to overtake the asymptotic nature of the PT (5.3) it was proposed to study the strong coupling expansion in inverse (rational) powers of the coupling constant:

$$E = g^{\frac{2}{3}} \sum_{k=0}^{\infty} b_k g^{-\frac{4k}{3}}, \quad (5.5)$$

which is evidently convergent with a finite radius of convergence. However, there is a technical problem related to a question how to find the coefficients b_k constructively. Many years after the publication of the paper by Bender and Wu [Bender and Wu (1969)] in [Turbiner and Ushveridze (1988)] a sort

[1]This formula is called the *Bender-Wu formula.* It can be generalized to any excited state of the quartic AHO and to any two-term anharmonic oscillator. It was re-derived and verified using quantum field theory methods in [Brezin *et al.* (1977a,b)]
[2]It implies that (5.3) does not contains the (hidden) exponentially-small terms, the Borel image coincides with the exact energy.
[3]The exact location of singularities is still unknown.
[4]The quartic symmetric anharmonic oscillator with potential (5.1) is Z_2-reflectionally symmetric: $(x \to -x)$. Hence, each state is characterized by its parity, positive or negative. Singularity at level crossing occurs for the eigenstates with definite parity only, positive or negative.

of duality between the weak

$$E = \sum_{k=0}^{\infty} \varepsilon_k g^{2k}, \cdot$$

and the strong coupling expansions

$$E = g^{\frac{2}{3}} \sum_{k=0}^{\infty} b_k g^{-\frac{4k}{3}},$$

was discovered: the coefficient b_k in the strong coupling expansion can be represented as an infinite, slowly convergent sum of the $\varepsilon_k, k = 0, 1, 2, \ldots$ coefficients of the weak coupling expansion and vice versa. Based on this property the first nine coefficients $b_n, n = 0, 1, \ldots 8$ for the ground state were calculated reliably [Turbiner and Ushveridze (1988)]. They hint that the radius of convergence of (5.5) is of the order of one. This radius is defined by the distance in the $m^2 \equiv g^{-4/3}$ complex plane between the origin and the nearest square-root branch point, see Fig. 5.1.

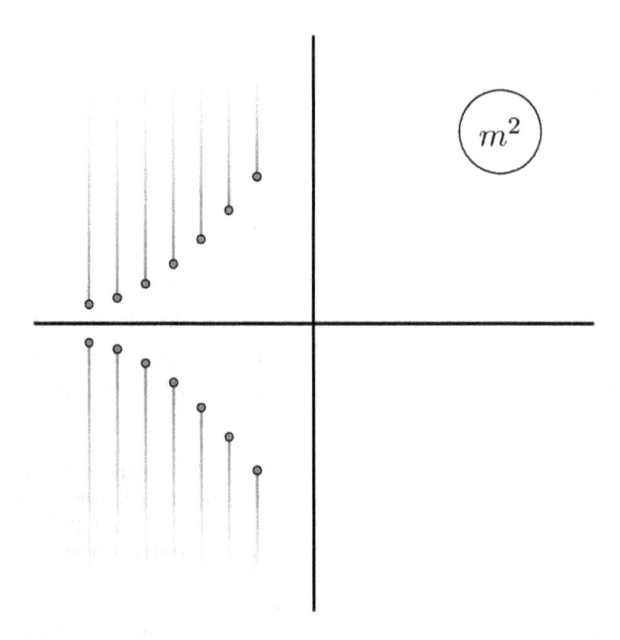

Fig. 5.1. Quartic anharmonic oscillator: Analytic structure of the complex plane $m^2 = g^{-4/3}$ of energy. Following the Symanzik scaling relations (5.9) the potential in the form $V = m^2 x^2 + x^4$ is chosen for convenience.

The way to go around all above-mentioned complications is to study the eigenfunctions $\Psi(x; g^2)$. An outstanding property of eigenfunctions of *any* polynomial anharmonic oscillator, discovered by Lazarus Fuchs (1833–1902) as early as around 1870 and proved rigorously in [Eremenko and Gabrielov (2009)] for quartic anharmonic oscillator (5.1), is that all the eigenfunctions are the *entire* functions in x when the coupling constant g^2 is kept fixed. This implies that an eigenfunction has no singularities at finite x having at most zeroes. All these zeroes are simple ones: finite number of them are situated on real axis accordingly to the oscillation theorem, see e.g. [Landau and Lifshitz (1977)], while the infinitely many simple zeroes are situated on imaginary axis symmetrically with respect to the origin forming complex-conjugated pairs [Eremenko *et al.* (2008)], see Fig. 5.2.

Generalities. The problem of bound states in non-relativistic quantum mechanics is governed by the time-independent Schrödinger equation

$$\mathcal{H} = -\frac{\hbar^2}{2m}\partial_x^2 + V(x), \quad \mathcal{H}\Psi = E\Psi, \quad \int_{-\infty}^{\infty} |\Psi|^2 dx < \infty, \quad \partial_x \equiv \frac{d}{dx},$$

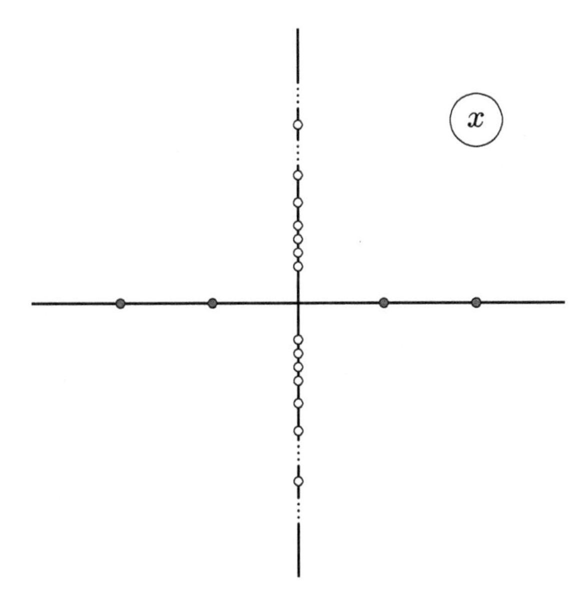

Fig. 5.2. Quartic oscillator $V = x^4$: the real and complex zeroes of the eigenfunction of the 4th excited state taken as an example. Two simple zeroes on the imaginary axis closest to the origin are situated at $x_0 = \pm 2.6721\, i$ (A. Eremenko, *private communication*). A similar picture occurs for any eigenfunction of the quartic oscillator as well as for any quartic anharmonic oscillator (5.1). As $g^2 \to 0$ all complex zeroes move along the imaginary axis to infinity symmetrically and at $g^2 = 0$ all of them disappear.

cf. (1.4), where the potential V is of the form (5.1), $x \in \mathbf{R}$. We consider the real Hamiltonian

$$\mathcal{H} = -\frac{\hbar^2}{2m}\partial_x^2 + a_2\, x^2 + g^2\, x^4, \tag{5.6}$$

hence, $\hbar, m, g^2$ and a_2 are real parameters, and we are looking for real eigenvalues and square-integrable, non-singular at real x eigenfunctions. The Hamiltonian is evidently a Hermitian operator with infinite discrete spectra. The questions related to the analytic continuation to the complex x and g^2 planes will be postponed for later. The Schrödinger equation in its final form reads,

$$-\frac{\hbar^2}{2m}\partial_x^2\,\Psi + (a_2 x^2 + g^2 x^4)\Psi = E\Psi, \tag{5.7}$$

where the eigenstates are denoted as follows: the wavefunction $\Psi = \Psi(x; a_2, g^2)$ and the energy $E = E(a_2, g^2)$. Making the rescaling of the variable: $x \to g^{-1/3}x$, we arrive at the Schrödinger equation for a similar potential with different parameters

$$-\frac{\hbar^2}{2m}\partial_x^2\,\Psi + (g^{-4/3}a_2 x^2 + x^4)\Psi = g^{-2/3}E\Psi. \tag{5.8}$$

This allows us to derive relations between eigenstates in these potentials, eventually leading to

$$\begin{aligned} E(a_2, g^2) &= g^{2/3}\, E(g^{-4/3}\, a_2, 1), \\ \Psi(x; a_2, g^2) &= \Psi(g^{1/3}\, x; g^{-4/3}\, a_2, 1). \end{aligned} \tag{5.9}$$

These relations are called the Symanzik scaling relations.

By taking an exponential representation for the wavefunction

$$\Psi = e^{-\frac{1}{\hbar}\phi},$$

cf. (1.17), substituting it into (5.7) and putting since now on $a_2 = 1$, we arrive at the well-known Riccati equation

$$\hbar\, y' - y^2 = E - x^2 - g^2 x^4, \quad y = \phi', \tag{5.10}$$

cf. (1.18), where for the sake of convenience $m = 1/2$. This equation contains $\hbar$ in front of the leading derivative term and g^2 in front of the anharmonic term: this leads to a bifurcation in $\hbar$ and to the divergence of the PT in powers of g^2,[5] respectively. Now let us get rid of the explicit $\hbar$ dependence in this equation. There are two ways to do it: via the Riccati-Bloch equation and via the Generalized Bloch equation.

5.2 Riccati-Bloch Equation for the Quartic Anharmonic Oscillator

5.2.1 *Ground State*

Let us insert in (5.10) a new variable v, a new function $\mathcal{Y}$ and a new energy,

$$x = \hbar^{1/2}\, v, y = \hbar^{1/2} \mathcal{Y}\,(v)\,, E = \hbar\varepsilon, \tag{5.11}$$

cf. (2.1), respectively, and also the effective coupling

$$\lambda = \hbar^{1/2} g, \tag{5.12}$$

cf. (2.2). We arrive at the so-called *Riccati-Bloch* (RB) equation

$$\partial_v \mathcal{Y} \quad \mathcal{Y}^2 = \varepsilon(\lambda^2) - v^2 - \lambda^2 v^4, \quad \partial_v \equiv \frac{d}{dv}, v \in (-\infty, \infty), \tag{5.13}$$

cf. (2.3). This equation has no $\hbar$-dependence: *it describes dynamics in a "quantum", $\hbar$-dependent coordinate v (5.11) instead of x, it is governed by the $\hbar$-dependent, effective coupling constant λ (5.12).*

Following the prescriptions of Chapter 2 we develop the Perturbation Theory (PT) in (5.13), taking the ground state for simplicity, see e.g. [Turbiner (1984)],

$$\varepsilon = \sum_0^\infty \lambda^{2n} \varepsilon_n, \quad \varepsilon_0 = 1, \varepsilon_1 = \frac{3}{4}, \ldots \tag{5.14}$$

$$\mathcal{Y} = \sum_0^\infty \lambda^{2n} \mathcal{Y}_n(v), \quad \mathcal{Y}_0 = v, \mathcal{Y}_1 = \frac{v}{2}\left(v^2 + \frac{3}{2}\right), \ldots, \tag{5.15}$$

cf. (2.8), (2.9), where $\mathcal{Y}_n = vP_n(v^2)$ with P_n as an nth degree polynomial. It becomes clear that (5.14) is simultaneously the perturbation series in powers of g^2 and the semiclassical expansion in powers of $\hbar$, since the coefficients ε_n are numbers. Contrary to that, the expansion (5.15) is the PT expansion in powers of g^2 only, since the corrections $\mathcal{Y}_n(v)$ are $\hbar$-dependent.

[5] Following the Dyson instability argument [Dyson (1952)]

Hence, (5.15) is not a semiclassical expansion in powers of $\hbar$. Expansion (5.15) mimics the asymptotic expansion at small v in Eq.(5.13),

$$\mathcal{Y} = \varepsilon\, v + \frac{\varepsilon^2 - 1}{3} v^3 + \cdots\,, \tag{5.16}$$

where ε is given by the expansion (5.14).

The potential of the quartic anharmonic oscillator (5.1) is symmetric, $V(x) = V(-x)$, or $\hat{V}(u) = \hat{V}(-u)$ in (5.13),

$$\hat{V}(u) = u^2 + u^4.$$

Therefore, the Hamiltonian (5.6) is Z_2-invariant, $x \to -x$. It is evident that in this case the function $\mathcal{Y}$ in (5.13) should be antisymmetric, $\mathcal{Y}(-v) = -\mathcal{Y}(v)$. Hence, the function $\mathcal{Y}$ admits a natural representation

$$\mathcal{Y}(v) = v\hat{\mathcal{Y}}(v^2). \tag{5.17}$$

The RB equation (5.13), written for the function $\hat{\mathcal{Y}}$, is converted into the form

$$2\, v_2 \partial_{v_2} \hat{\mathcal{Y}}(v_2) + \hat{\mathcal{Y}}(v_2) - v_2\, \hat{\mathcal{Y}}^2(v_2) = \varepsilon(\lambda) - v_2 - \lambda^2 v_2^2, \tag{5.18}$$

where

$$\partial_{v_2} \equiv \frac{d}{dv_2}, v_2 \equiv v^2,$$

cf. (2.14). Now the PT in powers of λ^2 can be developed for (5.18). If the expansion for the energy ε remains the same as in (5.15), the expansion for $\hat{\mathcal{Y}}$ gets simplified in comparison with the expansion for $\mathcal{Y}$,

$$\begin{aligned}
\hat{\mathcal{Y}} &= \sum_{n=0}^{\infty} \lambda^{2n} \hat{\mathcal{Y}}_n, \varepsilon_0 = 1, \hat{\mathcal{Y}}_0 = 1,\\
\varepsilon_1 &= \frac{3}{4}, \hat{\mathcal{Y}}_1 = \frac{1}{4}\left(2\, v_2 + 3\right),\\
\varepsilon_2 &= -\frac{21}{16}, \quad \hat{\mathcal{Y}}_2 = -\frac{1}{16}\left(2v_2^2 + 11v_2 + 21\right), \ldots
\end{aligned} \tag{5.19}$$

where, in general, the n-th correction $\hat{\mathcal{Y}}_n$ is of a polynomial form being a polynomial of degree n,

$$\hat{\mathcal{Y}}_n = Pol_n(v_2) = a_0^{(n)} + a_1^{(n)} v_2 + a_2^{(n)} v_2^2 + \cdots + a_k^{(n)}\, v_2^k + \cdots + a_n^{(n)}\, v_2^n. \tag{5.20}$$

By substituting the Taylor expansions for ε (5.14) and $\hat{\mathcal{Y}}$ (5.19) into the equation (5.18) and collecting terms of order λ^{2n} we arrive at the following equation to determine the nth correction,

$$2v_2\partial_{v_2}\hat{\mathcal{Y}}_n + \hat{\mathcal{Y}}_n - 2\,v_2\,\hat{\mathcal{Y}}_0\hat{\mathcal{Y}}_n = \varepsilon_n - Q_n, \tag{5.21}$$

where

$$\varepsilon_0 = 1, \hat{\mathcal{Y}}_0 = 1,$$

corresponds to the ground state of the harmonic oscillator, which is the solution to the unperturbed equation (5.18) at $\lambda^2 = 0$; here

$$Q_1 = v_2^2, \quad Q_n = -v_2\sum_{i=1}^{n-1} \hat{\mathcal{Y}}_i\hat{\mathcal{Y}}_{n-i},\ n = 2, 3, \ldots. \tag{5.22}$$

play the role of the *(effective) perturbation potentials.* It can be immediately verified this assumption about the polynomial nature for nth correction (5.20). Furthermore, the effective perturbation potential Q_n has the form of a polynomial of degree $(n+1)$. As a result the relation

$$\varepsilon_n = a_0^{(n)},$$

which provides a connection between the energy correction and $\hat{\mathcal{Y}}_n(0)$. By inserting the polynomials $Pol_{n-k}(v_2)$, $k = 0, 1, \ldots, n$, see (5.20), into equation (5.21) with Q_n given by (5.22) we obtain the following system of recurrence relations:

$$\begin{aligned}
-2a_n^{(n)} &= \sum a_k^{(k)} a_{n-k}^{(n-k)}, \\
-2a_{n-1}^{(n)} &= -(2n+1)\,a_n^{(n)} + 2\sum a_{k-1}^{(k)}\,a_{n-k}^{(n-k)}, \\
&\ \ \vdots
\end{aligned}$$

This system can be solved iteratively, from nth coefficient $a_n^{(n)}$ down to the zeroth coefficient $a_0^{(n)}$. The procedure of solving the recurrence relations can be easily programmed in MATHEMATICA, see Appendix B.1, where an explicitly written 8-line code is presented. In reasonable CPU time one can get practically any finite number of corrections.[6] The only

[6]It was tested to obtain $n = 400$ corrections, an interested reader can repeat this calculation by using the MATHEMATICA code presented in App.B.1.

limitation is related to the memory needed to store the rational coefficients of $\hat{\mathcal{Y}}_k, k = 1, 2, \dots n$ and the large RAM needed to carry out the large arithmetic operations. As an example, the first fifteen corrections ε_n are presented in Table 5.1, the first seven corrections $\hat{\mathcal{Y}}_n$ are presented in Table 5.2.

Interestingly, several coefficients standing in front of the highest degrees in the polynomial (5.20) can be found explicitly by using the method of

Table 5.1. Quartic AHO, the Ground State $(0, 0)$: First 15 corrections in λ^2 for ε, see (5.14).

Correction	Value	Correction	Value
ε_0	1	$-\varepsilon_8$	$\frac{1030495099053}{4194304}$
ε_1	$\frac{3}{4}$	ε_9	$\frac{54626982511455}{16777216}$
$-\varepsilon_2$	$\frac{21}{16}$	$-\varepsilon_{10}$	$\frac{6417007431590595}{134217728}$
ε_3	$\frac{333}{64}$	ε_{11}	$\frac{413837985580636167}{536870912}$
$-\varepsilon_4$	$\frac{30885}{1024}$	$-\varepsilon_{12}$	$\frac{116344863173284543665}{8589934592}$
ε_5	$\frac{916731}{4096}$	ε_{13}	$\frac{8855406003085477228503}{34359738368}$
$-\varepsilon_6$	$\frac{65518401}{32768}$	$-\varepsilon_{14}$	$\frac{1451836748576538293163705}{274877906944}$
ε_7	$\frac{2723294673}{131072}$	ε_{15}	$\frac{1275616828027135000067360049}{1099511627776}$

Table 5.2. Quartic AHO, the Ground State $(0, 0)$: First 7 corrections for $\hat{Y}(v_2)$, see (5.19).

Correction	Value
$\hat{Y}_0$	1
$\hat{Y}_1$	$\frac{v_2}{2} + \varepsilon_1$
$-\hat{Y}_2$	$\frac{v_2^2}{8} + \frac{11v_2}{16} - \varepsilon_2$
$\hat{Y}_3$	$\frac{v_2^3}{16} + \frac{21v_2^2}{32} + \frac{45v_2}{16} + \varepsilon_3$
$-\hat{Y}_4$	$\frac{5v_2^4}{128} + \frac{163v_2^3}{256} + \frac{1159v_2^2}{256} + \frac{8669v_2}{512} - \varepsilon_4$
$\hat{Y}_5$	$\frac{7v_2^5}{256} + \frac{319v_2^4}{512} + \frac{823v_2^3}{128} + \frac{19359v_2^2}{512} + \frac{33171v_2}{256} + \varepsilon_5$
$-\hat{Y}_6$	$\frac{21v_2^6}{1024} + \frac{1255v_2^5}{2048} + \frac{17405v_2^4}{2048} + \frac{143783v_2^3}{2048} + \frac{752825v_2^2}{2048} + \frac{19425763v_2}{16384} - \varepsilon_6$
$\hat{Y}_7$	$\frac{33v_2^7}{2048} + \frac{2477v_2^6}{4096} + \frac{21951v_2^5}{2048} + \frac{953675v_2^4}{8192} + \frac{3440609v_2^3}{4096} + \frac{133239549v_2^2}{32768} + \frac{411277893v_2}{32768} + \varepsilon_7$

generating functions by starting again from the nth coefficient $a_n^{(n)}$ down to the zeroth coefficient $a_0^{(n)}$,

$$Z_0 = \sum_{n=0}^{\infty} \lambda^{2n}\, v_2^n\, a_n^{(n)} = \sqrt{1+\lambda^2 v_2}, \tag{5.23}$$

$$Z_1 = \sum_{n=1}^{\infty} \lambda^{2n}\, v_2^{n-1}\, a_{n-1}^{(n)} = \frac{1+2\lambda^2 v_2 - \sqrt{1+\lambda^2 v_2}}{2v_2(1+\lambda^2 v_2)}, \tag{5.24}$$

$$\vdots$$

$$Z_k \doteq \sum_{n=k}^{\infty} \lambda^{2n}\, v_2^{n-k}\, a_{n-k}^{(n)}, \tag{5.25}$$

$$\vdots$$

$$Z_{\infty-k} \doteq \sum_{i=k}^{\infty} \lambda^{2i}\, v_2^k\, a_k^{(i)}, \tag{5.26}$$

$$\vdots$$

$$Z_{\infty-1} \equiv \sum_{i=1}^{\infty} \lambda^{2i}\, v_2\, a_1^{(i)} = v_2\, \frac{\varepsilon^2 - 1}{3}, \tag{5.27}$$

which is the sum of coefficients in front of v_2 in $\hat{\mathcal{Y}}_k, k = 1, 2, \ldots, \infty$

$$Z_{\infty} \equiv \sum_{i=0}^{\infty} \lambda^{2i}\, a_0^{(i)} = \varepsilon, \tag{5.28}$$

which is the sum of the coefficients in front of $(v_2)^0$ in $\hat{\mathcal{Y}}_k, k = 0, 1, 2, \ldots, \infty$, cf. Z_∞ and $Z_{\infty-1}$ with the asymptotic expansion (5.16). Asymptotically, at large n the coefficients behave like

$$a_{n-k}^{(n)} \sim \frac{n!}{(n-k)!}.$$

Hence, for finite fixed k, $k = 0, 1, 2, \ldots$ the sum of coefficients is convergent, while for finite $(n-k) = 0, 1, 2, \ldots$ all these sums are divergent. This is similar to the situation with the double-logarithmic approximation in Quantum Field Theory, see for discussion [Turbiner (1984)] and references therein.

5.2.2 *Padé-Borel Summation*

The divergent series (5.28) is of particular interest since it formally leads to the ground state energy ε. When a sufficiently large number of coefficients $\varepsilon_n = a_0^{(n)}$ is known exactly, different techniques of summation can be used to obtain an accurate value of the sum (5.28) for concrete values of λ^2. As was previously mentioned, the series (5.28) is Borel summable [Graffi *et al.* (1970)]. Thus, a certain summation technique related to use of the Padé approximations, see the book [Baker and Graves-Morris (2010)], can be naturally used. This is called the *Páde-Borel approximation*, it is a two-step procedure.

(1) Construct the Borel transform of (5.28) given by

$$\mathrm{BS}(t\,\lambda^2) = \sum_{n=0}^{\infty} \frac{\varepsilon_n}{n!} (t\lambda^2)^n, \tag{5.29}$$

where t is an auxiliary variable. Since the $n!$ in denominator cancels the factorial growth of the coefficients ε_n, see the asymptotic formula (5.4), the radius of convergence of the Borel transformed series (5.29) will be finite. The function BS is usually called *the Borel image.*

(2) Calculate the *inverse* Borel transform of $\mathrm{BS}(t, \lambda^2)$, defined by the following integral

$$\varepsilon(\lambda^2) = \int_0^{\infty} \mathrm{BS}(t\,\lambda^2)\, e^{-t}\, dt. \tag{5.30}$$

One can easily check that by inserting (5.29) into (5.30) and then changing the order of integration and summation, we arrive formally to (6.27).

In order to have the equality (5.30) be hold the Borel transform (5.29) should be known with infinitely many PT coefficients (5.14). In reality, the finite number of the PT coefficients can be found constructively. For the quartic AHO the first 15 PT coefficients in (5.28) are presented in Table 5.1 although a computational procedure presented in App.A allows to compute several hundred PT coefficients easily. In general, a closed analytic expression for the Borel image (5.29) is unknown, but one can approximate it by using, for instance, the diagonal Padé approximants:

$$\mathrm{BS}(t\,\lambda^2) \approx P[J/J](t\,\lambda^2) \equiv \frac{\sum_{j=0}^{J} A_j\,(t\,\lambda^2)^j}{\sum_{j=0}^{J} B_j (t\,\lambda^2)^j}. \tag{5.31}$$

Table 5.3. Quartic AHO, the Ground State Energy: Padé-Borel Summation results for different λ^2 *versus* the order of the approximant $P[J/J] \equiv P_J^J$.E_{exact} are found in LMM method, see App.F, all printed digits are correct.

	λ^2		
$P[J,J]$	0.1	1	10
$P[5,5]$	1.065 285 509 194 086 058	1.392 624 658 693 619 616	2.387 681 684 619 605 696
$P[10,10]$	1.065 285 509 543 717 787	1.392 635 616 629 641 616	2.447 612 656 611 677 655
$P[15,15]$	1.065 285 509 543 717 688	1.392 635 616 640 634 601	2.448 672 659 692 693 674
$P[20,20]$	1.065 285 509 543 717 688	1.392 635 616 641 651 629	2.449 610 605 683 696 673
$P[25,25]$	1.065 285 509 543 717 688	1.392 635 616 641 653 600	2.449 616 620 685 659 665
$P[50,50]$	1.065 285 509 543 717 688	1.392 635 616 641 653 602	2.449 174 069 566 047 903
$P[100,100]$	1.065 285 509 543 717 688	1.392 635 616 641 653 602	2.449 174 072 118 382 361
$P[150,150]$	1.065 285 509 543 717 688	1.392 635 616 641 653 602	2.449 174 072 118 386 918
Exact	1.065 285 509 543 717 688	1.392 635 616 641 653 602	2.449 174 072 118 386 918

Here A_j and B_j are coefficients that are determined by imposing the constraint that the first $(2J+1)$ Taylor coefficients in the expansion of the Padé approximant $P[J/J]$ coincide with the first $(2J+1)$ PT coefficients in (5.14) divided by $n!$; schematically, at $(t\,\lambda^2) \to 0$,

$$\mathrm{BS}(t\,\lambda^2) - P[J/J](t\,\lambda^2) = O\left((t\,\lambda^2)^{2J+1}\right), \tag{5.32}$$

for a given J.[7] As a consequence, we can obtain the approximate value of the energy by making the inverse Borel transform of the Padé approximant $P[J/J]$. As a result we arrive at the expression called the *Padé-Borel approximant.* In general, this procedure carries the name the *Padé-Borel summation*, see e.g. [Brezin *et al.* (1977a)].

Calculations of the ground state energy for the coupling constants $\lambda^2 = 0.1, 1, 10$ for different values of J with the Padé approximant $P[J/J]$ are presented in Table 5.3. In general, the accuracy improves with increasing J. However, the order of the approximant, needed to get given accuracy, increases dramatically with growth of λ^2. For example, for $\lambda^2 = 0.1$ in order to reach an accuracy of 19 significant figures, $J = 15$ is required (equivalently, 30 coefficients of the PT (5.14) should be known exactly), while for $\lambda^2 = 1$ the order $J = 50$ (or 100 PT coefficients) is required, as

[7]It is known [Baker and Graves-Morris (2010)] that for convergent series the diagonal Páde approximants lead usually to the fast convergence.

for $\lambda^2 = 10$ a value of $J = 150$ is needed, hence, 300 coefficients of PT (5.14) should be known constructively.[8] It does not seem feasible to get the similar accuracy for $\lambda^2 = 100$ in the Padé-Borel summation.

The mathematical details concerning the convergence of the Padé-Borel procedure can be found in [Graffi *et al.* (1970)].

5.2.3 *Excited States*

Taking the generalized exponential representation (2.32),

$$\Psi = f(x)\, e^{-\frac{1}{\hbar}\,\Phi(x)},$$

and plugging it into the Schrödinger equation (1.16) we arrive at a generalized Riccati equation (2.33),

$$\frac{1}{m}\,(\hbar\, y' - y^2) - \frac{1}{m}\,\frac{\hbar^2\, f'' - 2\,\hbar\, y f'}{f} = E - x^2 - g^2\, x^4, \quad y = \Phi',$$

for the quartic anharmonic oscillator. After introducing a new variable v, a new energy ε and a new function $\mathcal{Y}$, see (2.1), and also the effective coupling λ in the equation (2.33), we arrive at the gRB equation for the quartic anharmonic oscillator,

$$\mathcal{Y}_v' - \mathcal{Y}^2 - \frac{f_v'' - 2\mathcal{Y} f_v'}{f} = \varepsilon - v^2 - \lambda^2 v^4, \tag{5.33}$$

where the subindex v indicates that the derivative is taken with respect to v, which will be dropped from now on. As a consequence of the Z_2 invariance $(v \to -v)$ of the Hamiltonian $\mathcal{H}(v)$, see (5.6) with $(x \to v)$, any eigenstate is characterized by definite parity $(-)^p, p = 0, 1$. Hence, for any eigenfunction

$$\mathcal{Y}(-v) = -\mathcal{Y}(v), f(-v) = (-)^p\, f(v),$$

where $p = 0$ corresponds to the positive parity states and $p = 1$ is for the negative parity states. This implies the existence of the representation

$$\mathcal{Y}(v) = v\hat{\mathcal{Y}}(v^2), f(v) = v^p\, \hat{f}(v^2). \tag{5.34}$$

[8] In order to avoid the error accumulation the PT coefficients should be known with absolute accuracy, hence, in the form of rational numbers.

After plugging (5.34) into the gRB equation (5.33) and introducing a new variable $v_2 = v^2$, we arrive at

$$2v_2\partial_{v_2}\hat{\mathcal{Y}} + (1+2p)\hat{\mathcal{Y}} - v_2\hat{\mathcal{Y}}^2 - \\ 2\frac{2\,v_2\,\partial^2_{v_2}\hat{f} + (1+2p)\,\partial_{v_2}\hat{f} - 2\,v_2\,\hat{\mathcal{Y}}\,\partial_{v_2}\hat{f}}{\hat{f}} = \varepsilon - v_2 - \lambda^2 v_2^2, \quad (5.35)$$

where

$$\partial_{v_2} \equiv \frac{d}{dv_2},$$

cf. (5.18) and (5.33). Namely this form of the gRB equation will be used to study the PT for the excited states of the quartic anharmonic oscillator. It must be emphasized that in this formalism the principal quantum number K of the excited state is equal to the number of symmetric nodes ($2N$), where $N = 0, 1, 2, \ldots$, plus the parity $p = 0, 1$ of the state,

$$K = 2N + p. \quad (5.36)$$

This implies that if the prefactor $\hat{f}(v_2)$ is a polynomial with positive roots, it should be of degree N. For $N = 0$, the prefactor $\hat{f}(v_2) = 1$.

Now we are ready to develop the PT for the $(2N + p)$-th excited state as a Taylor expansion in powers of λ^2,

$$\varepsilon^{(K)}(\lambda) = \sum_{n=0}^{\infty} \varepsilon_n^{(K)}\,\lambda^{2n}, \quad (5.37)$$

and

$$\hat{\mathcal{Y}}^{(K)}(v_2) = \sum_{n=0}^{\infty} \hat{\mathcal{Y}}_n^{(K)}(v_2)\lambda^{2n}, \quad (5.38)$$

and

$$\hat{f}^{(K)}(v_2) = \sum_{n=0}^{\infty} \hat{f}_n^{(K)}(v_2)\,\lambda^{2n}. \quad (5.39)$$

cf. (2.35), (2.36), (2.37), respectively. We have to substitute the Taylor expansions (5.37), (5.38), (5.39) into equation (5.35), we collect terms of the order of λ^{2n} and then we set the final coefficient in front of this term be equal to zero. It must be emphasized that for the ground state of the positive ($K = 0$) and negative ($K = 1$) parities, the prefactor is constant,

$\hat{f}^{(0)} = \hat{f}^{(1)} = 1$; it is not deformed under the perturbation: all corrections can be set equal to zero, $\hat{f}_i^{(0)} = \hat{f}_i^{(1)} = 0$ for $i > 0$.

The unperturbed problem corresponds to the coefficient in front of λ^0. Zeroing this coefficient leads to the equation,

$$2v_2\partial_{v_2}\hat{\mathcal{Y}}_0^{(K)} + (1+2p)\hat{\mathcal{Y}}_0^{(K)} - v_2\big(\hat{\mathcal{Y}}_0^{(K)}\big)^2 \\ -2\frac{2\,v_2\partial^2_{v_2}\hat{f}_0^{(K)} + (1+2p)\partial_{v_2}\hat{f}_0^{(K)} - 2\,v_2\,\hat{\mathcal{Y}}_0^{(K)}\partial_{v_2}\hat{f}_0^{(K)}}{\hat{f}_0^{(K)}} = \varepsilon_0 - v_2. \tag{5.40}$$

Since this equation describes the celebrated harmonic oscillator, its exact solution can be written explicitly in the form:

$$\begin{aligned} \varepsilon_0^{(K)} &= (2K+1) = (4N+2p+1), \\ \hat{\mathcal{Y}}_0^{(K)} &= 1, \\ \hat{f}_0^{(K)} &= L_N^{(-1/2+p)}(v_2),\ N = [K/2], \end{aligned} \tag{5.41}$$

where the symbol [] means the integer part, $L_N^{(\alpha)}(x)$ is the Nth associated Laguerre polynomial with index $\alpha = (-1/2+p)$ with the standard normalization $L_0^{(\alpha)}(x) = 1$ [Bateman (1953)]. It manifests the explicit solution of non-perturbed problem. The ratio in the l.h.s. of (5.40) admits the exact cancellation in the case when $\hat{\mathcal{Y}}_0^{(N)} = 1$, it is equal to $(4N)$, thus,

$$-\frac{2v_2\partial^2_{v_2}\hat{f}_0^{(N)} + (1+2p)\partial_{v_2}\hat{f}_0^{(N)} - 2v_2\partial_{v_2}\hat{f}_0^{(N)}}{\hat{f}_0^{(N)}} = 2N, \tag{5.42}$$

which is the eigenvalue of the Laguerre operator.

In order to find the equation for the nth corrections it is convenient to multiply the gRB equation (5.35) by $\hat{f}$

$$\left(2\,v_2\,\partial_{v_2}\hat{\mathcal{Y}} + (1+2p)\,\hat{\mathcal{Y}} - v_2\,\hat{\mathcal{Y}}^2\right)\,\hat{f} \\ -4\,v_2\,\partial^2_{v_2}\hat{f} - 2(1+2p)\,\partial_{v_2}\hat{f} + 4v_2\hat{\mathcal{Y}}\partial_{v_2}\hat{f} = (\varepsilon - v_2 - \lambda^2 v_2^2)\,\hat{f}, \tag{5.43}$$

cf. (2.40), then substitute into it the Taylor expansions (5.37), (5.38), (5.39) and after that collect terms of order λ^{2n}. At $n = 0$ we arrive at the unperturbed equation (5.40), cf. (2.38), (2.39). As for the first correction at $n = 1$

we obtain

$$2\,v_2\,\partial_{v_2}\hat{\mathcal{Y}}_1 + (1+2p)\,\hat{\mathcal{Y}}_1 - 2\,v_2\hat{\mathcal{Y}}_1$$
$$-2\frac{2\,v_2\partial^2_{v_2}\hat{f}_1 + (1+2p-2v_2)\partial_{v_2}\hat{f}_1 + 2N\,\hat{f}_1 - 2(v_2\,\partial_{v_2}\hat{f}_0)\,\hat{\mathcal{Y}}_1}{\hat{f}_0}$$
$$= \left(\varepsilon_1 - v_2^2\right), \tag{5.44}$$

while for $n=2$

$$2v_2\,\partial_{v_2}\hat{\mathcal{Y}}_n + (1+2p)\,\hat{\mathcal{Y}}_n - 2\,v_2\,\hat{\mathcal{Y}}_n$$
$$-2\frac{2v_2\,\partial^2_{v_2}\hat{f}_2 + (1+2p-2v_2)\partial_{v_2}\hat{f}_2 + 2\,N\,\hat{f}_2 - 2(v_2\partial_{v_2}\hat{f}_0)\hat{\mathcal{Y}}_2}{\hat{f}_0}$$
$$= \left(\varepsilon_2 - \tilde{Q}_2\right), \tag{5.45}$$

where

$$\tilde{Q}_2 = -v_2\hat{\mathcal{Y}}_1^2 + \frac{\begin{array}{c}(1+2p)\hat{\mathcal{Y}}_1\hat{f}_1 - (\varepsilon_1 - v_2^2)\hat{f}_1 \\ +v_2\left(2\hat{f}_1\partial_{v_2}\hat{\mathcal{Y}}_1 + 4\hat{\mathcal{Y}}_1\partial_{v_2}\hat{f}_1 - 2\hat{\mathcal{Y}}_1\hat{f}_1\right)\end{array}}{\hat{f}_0}. \tag{5.46}$$

Note that for both equations (5.44), (5.45) there is a certain self-similarity: the left-hand-sides are the same, while the right-hand-sides are different. This property remains valid for all equations which define the higher order corrections, see below.

We arrive at the following equation for the nth correction,

$$2v_2\partial_{v_2}\hat{\mathcal{Y}}_n + (1+2p)\hat{\mathcal{Y}}_n - 2v_2\hat{\mathcal{Y}}_n$$
$$-2\frac{2v_2\,\partial^2_{v_2}\hat{f}_n + (1+2p-2v_2)\partial_{v_2}\hat{f}_n + 2\,N\,\hat{f}_n - 2\,(v_2\partial_{v_2}\hat{f}_0)\,\hat{\mathcal{Y}}_n}{\hat{f}_0}$$
$$= \left(\varepsilon_n - \tilde{Q}_n\right), \tag{5.47}$$

where

$$\varepsilon_0 = (2K+1), \hat{\mathcal{Y}}_0 = 1, \hat{f}_0 = L_N^{(-1/2+p)}(v_2),$$

see (5.41), correspond to the $(2N+p)$th excited state of the harmonic oscillator, which is the solution of the unperturbed equation (5.18) at $\lambda^2 = 0$.

Here the $\tilde{Q}_n$'s in the r.h.s. of (5.47):

$$\tilde{Q}_1 = v_2^2, \quad \tilde{Q}_n = -v_2 \sum_{k=1}^{n-1} \hat{\mathcal{Y}}_k \hat{\mathcal{Y}}_{n-k} + \frac{v_2^2 \hat{f}_{n-1}}{\hat{f}_0}$$
$$+ \frac{1}{\hat{f}_0} \sum_{k=1}^{n-1} \Big((1+2p)\hat{\mathcal{Y}}_k \hat{f}_{n-k} - \varepsilon_k \hat{f}_{n-k} \Big)$$
$$+ \frac{v_2}{\hat{f}_0} \sum_{k=1}^{n-1} \left(2\hat{f}_{n-k}\, \partial_{v_2} \hat{\mathcal{Y}}_k + 4\, \hat{\mathcal{Y}}_k\, \partial_{v_2} \hat{f}_{n-k} - \hat{f}_{n-k} \sum_{j=0}^{k} \hat{\mathcal{Y}}_j\, \hat{\mathcal{Y}}_{k-j} \right),$$
$$n = 2, 3, \ldots, \tag{5.48}$$

play the role of the *(effective) perturbation potentials*. For $K = 0, 1$ the ratio in the l.h.s. of (5.47) disappears while the r.h.s. is dramatically simplified, $\tilde{Q}_n = Q_n$, see (5.22).

By making concrete calculations for $n = 1, 2, 3$ we arrive at the following, very compact expressions,

$$\begin{aligned}
\varepsilon_1^{(K)} &= \frac{3}{4}\left(2K^2 + 2K + 1\right), \\
\varepsilon_2^{(K)} &= -\frac{1}{16}\left(34K^3 + 51K^2 + 59K + 21\right), \\
\varepsilon_3^{(K)} &= \frac{3}{64}\left(125K^4 + 250K^3 + 472K^2 + 347K + 111\right),
\end{aligned} \tag{5.49}$$

for the energy corrections, which at $K = 0$ coincide with ones given in Table 5.1 and

$$\begin{aligned}
\hat{\mathcal{Y}}_1(v_2) &= \frac{1}{2}\left(v_2 + K + \frac{3}{2}\right), \\
\hat{\mathcal{Y}}_2(v_2) &= -\frac{1}{8}\left(v_2^2 + \frac{1}{2}(6K+11)v_2 + \frac{1}{2}(9K^2 + 25K + 21)\right), \\
\hat{\mathcal{Y}}_3(v_2) &= \frac{1}{16}\Big(v_2^3 + \frac{1}{2}\left(10K + 21\right) v_2^2 + \left(13K^2 + 45K + 45\right) v_2 \\
&\quad + \frac{1}{4}(86K^3 + 337K^2 + 549K + 333)\Big),
\end{aligned} \tag{5.50}$$

for the $\hat{\mathcal{Y}}$-corrections, which at $K = 0$ coincide with ones given in Table 5.2, and for the corrections to prefactor $\hat{f}(v_2)$:

$$\begin{aligned}
\hat{f}_0(v_2) &= \mathcal{L}_N^{(-\frac{1}{2}+p)}(v_2),\\
\hat{f}_1(v_2) &= -2K\mathcal{L}_{N-1}^{(-\frac{1}{2}+p)}(v_2) + \mathcal{L}_{N-2}^{(-\frac{1}{2}+p)}(v_2),\\
\hat{f}_2(v_2) &= \frac{1}{16}\left(85K^2 + 3K + 36\right)\mathcal{L}_{N-1}^{(-\frac{1}{2}+p)}(v_2)\\
&\quad + \frac{1}{8}(16K^2 - 46K + 17)\mathcal{L}_{N-2}^{(-\frac{1}{2}+p)}(v_2)\\
&\quad - \frac{1}{3}\left(6K - 7\right)\mathcal{L}_{N-3}^{(-\frac{1}{2}+p)}(v_2) + \frac{1}{2}\mathcal{L}_{N-4}^{(-\frac{1}{2}+p)}(v_2),\\
\hat{f}_3(v_2) &= -\frac{1}{32}(600K^3 + 59K^2 + 805K + 51)\mathcal{L}_{N-1}^{(-\frac{1}{2}+p)}(v_2)\\
&\quad - \frac{1}{32}(340K^3 - 954K^2 + 826K - 551)\mathcal{L}_{N-2}^{(-\frac{1}{2}+p)}(v_2)\\
&\quad - \frac{1}{48}(64K^3 - 807K^2 + 1403K - 912)\mathcal{L}_{N-3}^{(-\frac{1}{2}+p)}(v_2)\\
&\quad + \frac{1}{24}(48K^2 - 250K + 249)\mathcal{L}_{N-4}^{(-\frac{1}{2}+p)}(v_2)\\
&\quad - \frac{1}{3}(3K - 7)\mathcal{L}_{N-5}^{(-\frac{1}{2}+p)}(v_2) + \frac{1}{6}\mathcal{L}_{N-6}^{(-\frac{1}{2}+p)}(v_2),
\end{aligned}$$

where the normalization of the Laguerre polynomials is conveniently changed to

$$\mathcal{L}_{N-J}^{(-\frac{1}{2}+p)}(v_2) = (-1)^N \frac{2^{K-3J}(N-J)!\,K!}{(K-2J)!}\, L_{N-J}^{(-\frac{1}{2}+p)}(v_2),$$

cf. (5.41), to simplify formulas.

The analysis of equation (5.47) with the effective potentials (5.48) reveals the general structure of the corrections:

A. The nth energy correction is a polynomial of degree $(n+1)$ in the principal quantum number K,

$$\varepsilon_n = Pol_{n+1}(K),$$

with rational coefficients,

B. The nth correction to $\hat{\mathcal{Y}}(v_2)$ is a polynomial of degree n in v_2,

$$\hat{\mathcal{Y}}_n(v_2) = Pol_n(v_2; K) = c_0\, v_2^n + c_1(K)\, v_2 + \cdots + c_i(K)\, v_2^i + \cdots + c_n(K),$$

with coefficients c_i, which are polynomials in variable K of degree i

$$c_i(K) = Pol_i(K),$$

also with rational coefficients,

C. The nth correction to the prefactor $\hat{f}(v_2)$ is a polynomial of degree N in v_2 for any n, which can be represented as a linear superposition of the $(2n)$ associated Laguerre polynomials with the same index $(-\frac{1}{2}+p)$,

$$\hat{f}_n(v_2) = \sum_{J=1}^{2n} d_J(K)\, \mathcal{L}_{N-J}^{(-\frac{1}{2}+p)}(v_2),$$

with coefficients, which are polynomials in K of degree $(2n-J)$,

$$d_J(K) = Pol_{2n-J}(K),$$

also with rational coefficients.

5.3 Generalized Bloch Equation for Quartic Anharmonic Oscillator

Following the presentation in Chapter 3 let us introduce a new variable u, function $\mathcal{Z}$ and energy,

$$u = g\, x = \lambda v, \quad y = \frac{1}{g}\mathcal{Z}(u), \quad E = \hbar\, \varepsilon, \tag{5.51}$$

for equation (5.10), cf. (1.18), keeping the same effective coupling constant (5.12)

$$\lambda = \hbar^{1/2}\, g,$$

while assuming $g \neq 0$ as always. We arrive at *the Generalized Bloch equation* (GB) for the quartic anharmonic oscillator [Shuryak and Turbiner (2018)],

$$\lambda^2\, \partial_u \mathcal{Z}(u) - \mathcal{Z}^2(u) = \lambda^2\, \varepsilon(\lambda^2) - u^2 - u^4, \quad \partial_u \equiv \frac{d}{du}, \quad u \in (-\infty, \infty), \tag{5.52}$$

cf. (2.3),(5.13) and (3.3). Note that it requires a regularization at the limit $\lambda \to 0$, which will lead to the RB equation. To avoid the discussion of regularization, we consider $\lambda \neq 0$, hence, λ plays a role of a formal parameter.

This equation describes dynamics in a *classical* ($\hbar$-independent) coordinate $u = g\,x$.

Evidently, the equation (5.52) is $\mathbf{Z}_2$-invariant: $(u \to -u)$ and $\mathcal{Z}(-u) = -\mathcal{Z}(u)$. Hence, $\mathcal{Z}(u)$ can be represented as

$$\mathcal{Z}(u) = u\hat{\mathcal{Z}}(u_2), \quad u_2 = u^2.$$

By plugging this into equation (5.52), we arrive at the equation

$$2\lambda^2\, u_2 \partial_{u_2} \hat{\mathcal{Z}} + \lambda^2\, \hat{\mathcal{Z}} - u_2 \hat{\mathcal{Z}}^2 = \lambda^2\, \varepsilon(\lambda^2) - u_2 - u_2^2, u_2 \in [0, \infty), \tag{5.53}$$

where $\partial_{u_2} \equiv \frac{d}{du_2}$, $\hat{\mathcal{Z}}(0) = \varepsilon$, this equation is free of any symmetry. Equation (5.53) can be easily generalized to the case of any even potential: $V(-u) = V(u)$.

Now we develop a PT in powers of λ for (5.52). It is evident that the expansion of the energy ε in powers of λ (5.14) remains the same as for the RB equation (5.13),

$$\varepsilon = \sum_{n=0}^{\infty} \lambda^{2n} \varepsilon_n,$$

while the expansion for $\mathcal{Z}$ is of the form

$$\mathcal{Z} = \sum \lambda^{2n}\, \mathcal{Z}_n(u), \mathcal{Z}_0 = u\sqrt{1+u^2}, \mathcal{Z}_1 = \frac{\mathcal{Z}_0' - \varepsilon_0}{2\mathcal{Z}_0}, \ldots, \tag{5.54}$$

cf. (2.9), (5.15) and (3.11), where we have to put $\varepsilon_0 = 1$. Any finite number of corrections $\mathcal{Z}_{2,3,4,\ldots}$ can be calculated purely algebraically. It can be immediately recognized that $\mathcal{Z}_0$ is, in fact, the classical momentum at zero energy and $\lambda = 1$, while $\mathcal{Z}_1$ is related to the derivative of the logarithm of the determinant, see Chapter 3. The PT in powers of λ can be developed in (5.53) as well,

$$\begin{aligned} \hat{\mathcal{Z}} &= \sum \lambda^{2n}\, \hat{\mathcal{Z}}_n(u_2), \hat{\mathcal{Z}}_0 = \sqrt{1+u_2}, \\ 2\hat{\mathcal{Z}}_1 &= \frac{1}{u_2} + \frac{1}{1+u_2} - \frac{1}{u_2\sqrt{1+u_2}}, \ldots, \end{aligned} \tag{5.55}$$

In general, the expansion (5.54) is the true semiclassical expansion in powers of $\hbar$, as well as the expansion in powers of g^2. By integrating $\int \mathcal{Z}\, du$ (and putting $\hbar = 1$) we arrive at the true semiclassical expansion of the phase,

$$\phi = \frac{1}{3g^2}(1+g^2x^2)^{3/2}$$
$$+\frac{1}{4}\log(1+g^2x^2) + \frac{1}{2}(4N+2p+1)\log[1+(1+g^2x^2)^{1/2}] + \cdots, \tag{5.56}$$

where for the ground state $N = p = 0$[9]. Remarkably, the higher order corrections $\mathcal{Z}_{2+i}, i = 0, 1, 2, \ldots$ in (5.54) do not generate logarithmic terms. For the $(2N+p)$-th excited state with quantum number $2N = 0, 2, \ldots$ and parity $p = 0, 1$, the function $\mathcal{Z}$ contains $(2N+p)$ simple poles with residues (-1) at real x in addition to a non-singular part. In the case of excited states the coefficient in front of the second logarithmic term in (5.56) is the state-dependent: it depends on N, p. The expansion (5.56) mimics the asymptotic expansion at large $u = gx$ in Eq.(5.52),

$$\phi = \frac{g}{3}x^2|x| + \frac{1}{2g}|x| + \left(N + \frac{p}{2} + \frac{1}{2}\right)\log(x^2) + O\left(\frac{1}{|x|}\right). \tag{5.57}$$

The first three terms in (5.57) are universal: they grow as $|x| \to \infty$ and do not depend on the energy ε. Furthermore, the first two terms in (5.57) do not depend on the state being studied, they emerge from the expansion of the classical action $\int \mathcal{Z}_0\, du$, which is the first term of the semiclassical expansion (5.56).

5.4 Quartic Anharmonic Oscillator: Approximating Eigenfunctions

Without loss of generality we set $\hbar = 1$ from now on, hence, the variable v is equal to x: $v = x$ and the RB equation (5.13) coincides with the Riccati equation (5.10). Let us perform the procedure of matching the expansions (5.16) and (5.56) of the phase in the exponent of (5.1) using a single function. Then we multiply the result by a polynomial pre-factor which carries information about the nodes in a *minimal* manner: the number of different nodes is equal to the degree of the polynomial. As a result we arrive almost unambiguously at the following function for the $(2N+p)$-th excited state

[9]The first term in this expansion was calculated in [Turbiner (1984)] under the name *leading log approximation in quantum mechanics*.

with quantum numbers (N,p), where $N=0,1,2,\ldots,p=0,1$, for quartic AHO:

$$\Psi^{(N,p;4)}_{(approximation)}(x) = \frac{x^p P_N^{(p)}(x^2;g^2)}{(B^2+g^2\,x^2)^{\frac{1}{4}}\left(\alpha B+\sqrt{B^2+g^2\,x^2}\right)^{2N+p+\frac{1}{2}}} \exp\left(-\frac{A+(B^2+3)\,x^2/6+g^2\,x^4/3}{\sqrt{B^2+g^2\,x^2}}+\frac{A}{B}\right), \qquad (5.58)$$

here $P_N^{(p)}$ is a polynomial of degree N in x^2 with positive roots only and the parameter $\alpha=1$. Here $A=A_{N,p}(g^2), B=B_{N,p}(g^2)$ are two interpolation parameters. If $\alpha=0$ pure formally, the expression (5.58) becomes the one which was found in [Turbiner (2005, 2010)] by interpolating the asymptotic expansions (5.16) and (5.57). In the limit $g^2\to 0$, function (5.58) becomes the one of the harmonic oscillator,

$$\Psi^{(0)} = x^p L_N^{(-\frac{1}{2}+p)}(x^2)e^{-\frac{x^2}{2}},$$

where $L_N^{(-\frac{1}{2}+p)}$ is the associated Laguerre polynomial with index $(-\frac{1}{2}+p)$.[10]

Formula (5.58) is the *central* formula of this Chapter and one of the key formulas of the entire book. It provides a highly-accurate approximate solution for the spectra of the quantum one-dimensional quartic symmetric anharmonic oscillator in its full generality.

The simplest way to fix the parameters A, B is to consider (5.58) as a variational trial function moving subsequently up from one by one from the ground state to higher excited states. In order to explain in concrete terms how realize this program it is necessary to do the following:

- calculate the effective potential,

$$V^{(N,p;4)} - E_0 = \frac{\partial_x^2 \Psi^{(N,p;4)}_{(approximation)}}{\Psi^{(N,p;4)}_{(approximation)}}, \qquad (5.59)$$

[10]It is worth mentioning the existence of a well-known relation between Hermite and Laguerre polynomials with standard normalization [Bateman (1953); Abramowitz and Stegun (1964)],

$$H_{2N+p}(x) = (-)^N\, 2^{2N+p} N! x^p\, L_N^{(-\frac{1}{2}+p)}(x^2).$$

where the condition $V^{(N,p;4)}(x=0)=0$ allows us to choose the reference point for the energy E_0, then

- impose the orthogonality conditions

$$\langle\Psi^{(M,p;4)}_{(approximation)},\Psi^{(N,p;4)}_{(approximation)}\rangle = 0, \quad M = 0,1,2\ldots(N-1),$$

assuming that the functions $\Psi^{(M,p;4)}_{(approximation)}, M = 0,1,2\ldots(N-1)$ are already found and they are mutually orthogonal, and

- then to calculate the variational energy,

$$E^{(N,p)}_{var} = \min_{\{A,B\}} \times \left[E_0 + \langle\Psi^{(N,p;4)}_{(approximation)}, \left(V - V^{(N,p;4)}\right)\Psi^{(N,p;4)}_{(approximation)}\rangle\right], \tag{5.60}$$

which contains the expectation value of the deviation of the original potential (5.1) from the effective potential (5.59).

In Fig. 5.3 the variational energies $E^{(N,p)}_{var}$ *vs* g^2 for the first six low-lying states are shown. It turns out that their accuracies are smaller than the width of any line on the plot! Concrete numerical results for various g^2 are presented in six Tables in App. C.2.1 .

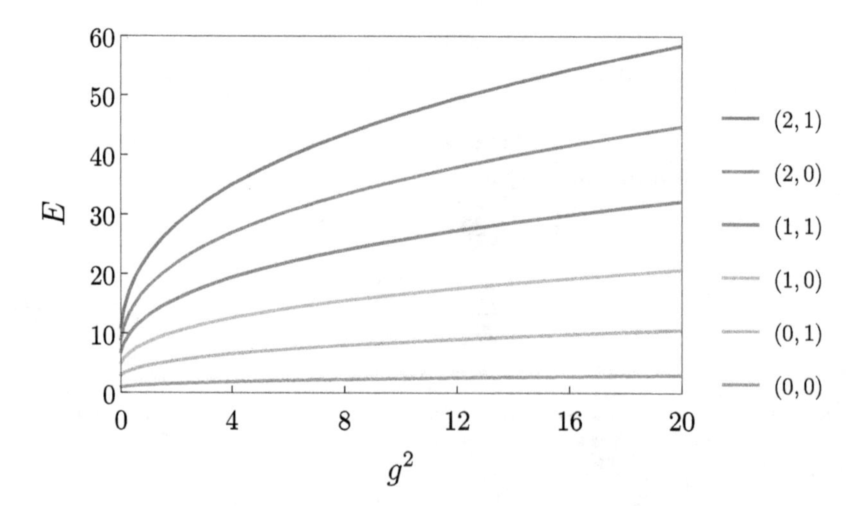

Fig. 5.3. Energies E of the first 6 low-lying states as function of g^2.

By introducing a new coupling constant

$$\tilde{\lambda}(g^2) = (0.008 + g^2)^{1/3}, \quad \tilde{\lambda}(0) = 0.2, \tag{5.61}$$

which mimic the asymptotic behavior of the energy at small and large couplings, one can easily interpolate the energies across the whole domain of variations of the coupling constant $g^4 \in [0, \infty)$ using two-point Padé approximants Padé$(N+1/N)(\tilde{\lambda})$. For example, the ground state energy is described like

$$\begin{aligned} E_{0,0} &= \text{Padé}(4/3)(\tilde{\lambda}) \\ &= \frac{0.0785 + 0.2581\tilde{\lambda} + 0.6650\tilde{\lambda}^2 + 0.5436\tilde{\lambda}^3 + 1.0604\tilde{\lambda}^4}{0.0775 + 0.2837\tilde{\lambda} + 0.5129\tilde{\lambda}^2 + \tilde{\lambda}^3}, \end{aligned} \tag{5.62}$$

cf. App. B, Eq. (C.7), with relative accuracy $\sim 10^{-5}$ (or better). In App. C.2.2 a fit of the energies of six low-lying states via Padé$(4/3)(\tilde{\lambda})$ in the whole domain $g^2 \geq 0$ is presented, (C.7)–(C.12). In general, by increasing N in the Padé approximant Padé$(N+1/N)(\tilde{\lambda})$ one can get better accuracy.

In the framework of the non-linearization procedure [Turbiner (1984)] one can develop a rapidly-convergent perturbation theory with respect to the deviation of the original potential V from the effective one (5.59): $(V - V^{(N,p)})$, see Chapter 2 for a general description. Usually, the rate of convergence is defined by the second correction E_2, see (2.50).

It can be shown that in the small and large g^2 limits, the parameters A, B behave like

$$A^{(N,p)} \to \frac{1}{3g^2}, \quad A^{(N,p)} \to -a^{(N,p)} g^{2/3},$$

while

$$B^{(N,p)} \to 1, \quad B^{(N,p)} \to b^{(N,p)}\, g^{2/3},$$

respectively. Note that the functions $a^{(N,p)}, b^{(N,p)} > 0$ are very smooth, they can easily be fitted with relative accuracy $\sim 10^{-3}$ by

$$a_{fit}^{(N,p)} = (8.869 + 23.120\, K + 7.856\, K^2 + 4.140 K^3 + 0.262\, K^4)^{1/3}, \tag{5.63}$$

$$b_{fit}^{(N,p)} = (10.040 + 3.255 K)^{1/3}, \tag{5.64}$$

where $K = 2N+p$, thus, realizing the dependence on the quantum numbers. The functions a_{fit}, b_{fit} are the interpolating parameters of the function

(5.58) applied to the power-like potential

$$V = g^2 x^4,$$

see below.

Returning to the quartic anharmonic oscillator, concrete numerical calculations are carried out for the six lowest states (N, p): $N = 0, 1, 2$ and $p = 0, 1$ with $g^2 = 0.1, 1, 10, 20, 100$ [Turbiner and del Valle (2021a)]. As a result the variational energy can be found with relative accuracy $\sim 10^{-10}$ — this was confirmed by making accurate calculations using the Lagrange Mesh Method (LMM) [Baye (2015)], see App. C for details, with 50 mesh points (giving relative accuracy $\sim 10^{-13}$), see also [Turbiner and del Valle (2021b)]. Comparison of the variationally calculated energies and the LMM results is presented in App. B. Plots of the parameters A, B are presented in Figs. 5.4–5.5. Surprisingly, in order to reach the above-mentioned accuracy it is sufficient to find the parameters A, B with 4 significant digits only. In Fig. 5.6 the behavior of the nodes x_{node} for the (N, p)-states with $N = 1$ and $p = 0, 1$ are shown. By introducing a new coupling constant

$$\tilde{\lambda}(g^2) = (0.008 + g^2)^{1/3}, \quad \tilde{\lambda}(0) = 0.2,$$

see (5.61), one can interpolate the parameters A, B for all the above-mentioned six states in a form,

$$A = \frac{1}{g^2 \tilde{\lambda}} P_5(\tilde{\lambda}), \quad B = \frac{Q_3(\tilde{\lambda})}{Q_2(\tilde{\lambda})},$$

where P_5, Q_3, Q_2 are polynomials of 5,3,2-th degrees, respectively, with different coefficients. In particular, for the ground state

$$A^{(0,0)} = \frac{-0.0171 + 0.4205\,\tilde{\lambda} - 0.1990\,\tilde{\lambda}^2 + 1.039\tilde{\lambda}^3 - 0.0567\,\tilde{\lambda}^4 - 1.797\,\tilde{\lambda}^5}{g^2\,\tilde{\lambda}},$$

$$B^{(0,0)} = \frac{0.3716 + 5.476\,\tilde{\lambda} + 2.231\,\tilde{\lambda}^2 + 33.51\,\tilde{\lambda}^3}{1 + 0.9981\,\tilde{\lambda} + 15.61\,\tilde{\lambda}^2},$$

cf. App. B, Eq. (C.1). Use of these interpolated parameters in (5.58) leads to a relative accuracy in the energy of $\sim 10^{-9}$ (or better). A similar accuracy is reached for the all six studied states, see App. C.1.

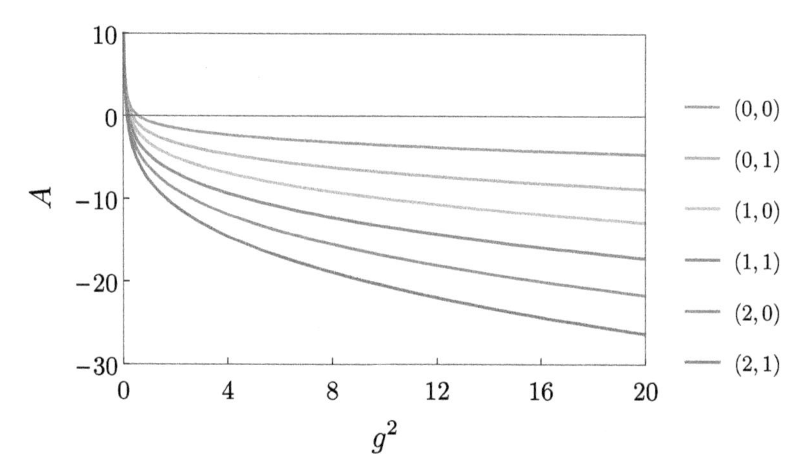

Fig. 5.4. Optimal parameter A as function of the coupling constant g^2 for the first six low-lying states, see App.B.

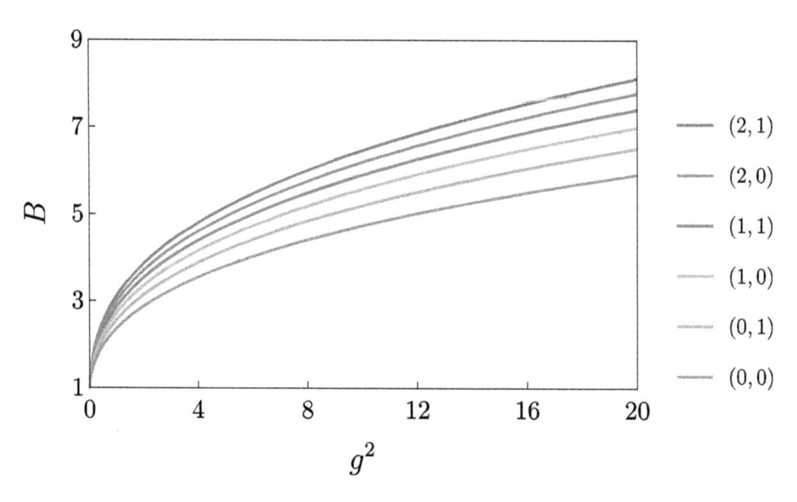

Fig. 5.5. Optimal parameter B as a function of the coupling constant g^2 for the first six low-lying states, see App.B.

It is interesting to investigate the polynomials $P_N^{(p)}$ in (5.58) which carry information about the non-zero nodes of the eigenfunction,

$$P_N^{(p)} = 1 - a_2^{(N,p)}\, x^2 + a_4^{(N,p)}\, x^4 + \cdots + a_{2N}^{(N,p)}\, (-x^2)^N .$$

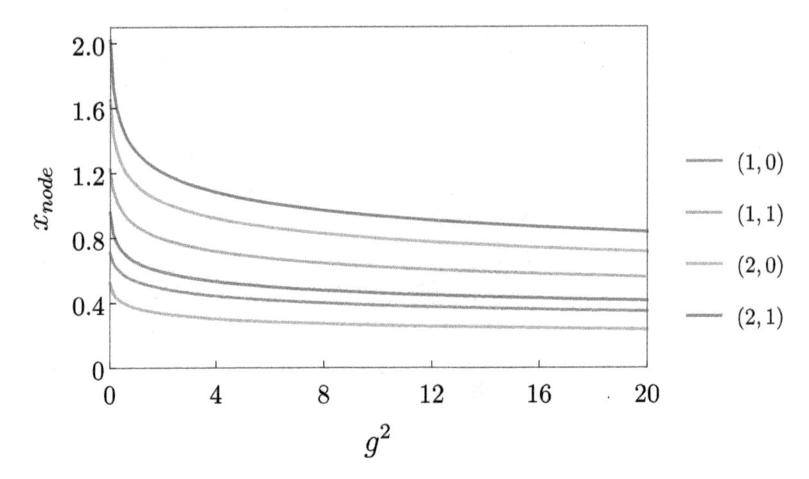

Fig. 5.6. Positive nodes $x_{node} > 0$ *vs* the coupling constant $g^2 \geq 0$ for the states $(1, 0)$, $(1, 1)$, $(2, 0)$, and $(2, 1)$. Note that for all states, there exist symmetric negative nodes at $x = -x_{node}$. Additionally, the negative parity states (1,1) and $(2, 1)$ have a node at $x_0 = 0$.

For the harmonic oscillator with $V = x^2$ (which corresponds to $g^2 \to 0$, the weak coupling regime)

$$a_2^{(N,p)} = \frac{4\,N}{(p+1)(p+2)},$$

$$a_4^{(N,p)} = \frac{16\,N(N-1)}{(p+1)(p+2)(p+3)(p+4)},$$

$$\cdots$$

$$a_{2N}^{(N,p)} = \frac{\Gamma(\frac{p+1}{2}+N)}{\Gamma(\frac{p+1}{2})},$$

which can be derived from the properties of the Laguerre polynomials. It can be easily shown that for large N,

$$a_{2k}^{(N,p)} \propto N^k.$$

Hence, for the ultra-strong coupling regime, $g^2 \to \infty$, with the potential $V = g^2\,x^4$, the first two coefficients can be found numerically any desired accuracy [Turbiner and del Valle (2021a)]. Then, the coefficients are fitted separately for positive and negative parity states as

- $p = 0$

$$a_{2,fit}^{(N,0)} = N\,(7.372 + 10.872\,N)^{1/3}\,,$$
$$a_{4,fit}^{(N,0)} = N\,(N-1)\,(0.830 + 1.420\,N)^{2/3}\,,$$

with relative error $\lesssim 0.5\%$ for any even $N \in [0, 40]$ and

- $p = 1$

$$a_{2,fit}^{(N,1)} = N\,(0.834 + 0.773\,N)^{1/3}\,,$$
$$a_{4,fit}^{(N,1)} = N(N-1)\,(0.167 + 0.127\,N)^{2/3}\,,$$

with relative error $\lesssim 0.12\%$ for any odd $N \in [0, 40]$.

Interested reader can calculate next coefficients $a_6, a_8, \ldots$. It is an open question to find the behavior of the coefficients as a function of the coupling constant g^2.

In the non-linearization procedure using multiplicative perturbation theory [Turbiner (1984)], see Chapter 2 one can analytically calculate the first corrections to (5.58) of the form:

$$\Psi = \Psi^{(N,p;4)}_{(approximation)}\,(1 - \phi_1 - \phi_2 \ldots).$$

It can be shown numerically that for all six studied states at any $g^2 \in [0, \infty)$, the first correction is extremely small

$$|\phi_1| \lesssim 10^{-6}, \forall x \in (-\infty, \infty). \tag{5.65}$$

Since for all six studied states the correction ϕ_1 can be found explicitly in closed analytic form and it is believed that the bound (5.65) can be verified analytically. Besides that in the non-linearization procedure the corrections E_2, E_3 to the energy (5.60), see (2.50), can be also calculated analytically and then verified numerically. They indicate an extremely high rate of convergence for the energy, 10^{-3} - 10^{-4} in agreement with the Lagrange Mesh Method calculations, see [Baye (2015)], [Turbiner and del Valle (2021b)] and App.F, with 75 mesh points for $g^2 = 0.1, 1, 10, 20, 100$, where a relative accuracy 10^{-20} in energy is easily reached.[11]

[11]Needless to say that the Lagrange Mesh Method in many occasions allows to reach rapidly the extremely high accuracy. Out of curiosity one can check that by using the pre-calculated positions of 1000 mesh points the energy of the states from $N = 0$ to $N = 80$ for $g^2 = 0.1, 1, 10, 20, 100$ can be found with more than 140 figures in ~ 250 seconds of CPU time on a single core 2.6 GHz processor for each value of g^2 and each concrete state, see [Turbiner and del Valle (2021b)].

5.5 Quartic (Anharmonic) Oscillator and Bohr-Sommerfeld Quantization Condition

As was previously mentioned for any eigenvalue of the quartic AHO (5.1) we can construct the strong coupling expansion (5.5),

$$E^{(N,p)} = g^{\frac{2}{3}} \sum_{0}^{\infty} b_k^{(N,p)} g^{-\frac{4k}{3}},$$

where the leading coefficient $b_0^{(N,p)}$ corresponds to the spectra of the quartic oscillator with potential,

$$V = g^2 x^4, \hat{V} = u^4,$$

which is the power-like potential, cf. (2.69), (3.9), (4.3). This potential realizes the ultra-strong coupling regime $g^2 \to \infty$ of the quartic AHO (5.1). We denote the eigenvalues as $E^{(4)} = \hbar\varepsilon^{(4)} \equiv b_0^{(N,p)}$.

Let us consider the (exact) Bohr-Sommerfeld quantization condition for the potential (3.9),

$$\int_{-\varepsilon^{1/4}}^{\varepsilon^{1/4}} \sqrt{\varepsilon - u^4} du = \pi \left(K + \frac{1}{2} + \gamma(K) \right), \tag{5.66}$$

where $K = 2N+p$ is the principal quantum number; sometimes, $(1/2+\gamma)$ is called the *quantization index* or, stated differently, $\gamma(K)$ is called the *WKB correction*. If $\gamma = 0$, this condition corresponds to the standard Bohr-Sommerfeld quantization condition, see e.g. [Landau and Lifshitz (1977)]. In this case we arrive at the so-called Bohr-Sommerfeld energy spectra $\varepsilon = \varepsilon_{BS}(K)$. Explicitly,

$$E_{BS}^{(4)} = \hbar\,(\hbar g^2)^{\frac{1}{3}} \left(\frac{3}{4}\, B\left(\tfrac{1}{2}, \tfrac{3}{4}\right) \left(K + \frac{1}{2} \right) \right)^{\frac{4}{3}}, \tag{5.67}$$

where $B(\alpha, \beta)$ is the Euler Beta function, see e.g. [Abramowitz and Stegun (1964)], [Bateman (1953)], This formula gives increasingly accurate energies for large quantum numbers K, while for small K it can differ significantly [Bender and Orszag (1978); Turbiner and del Valle (2021c)].[12]

The problem of the energy spectra of the power-like potential can be considered to be an inverse problem. If for some reason the exact spectra $\varepsilon = \varepsilon_{exact}$ is known (in one way or another) explicitly, by using (5.66) the

[12] At $K = 0$ the deviation is about 20%

function $\gamma(K)$ itself can be calculated explicitly. Vice versa if the function $\gamma(K)$ is known exactly, the exact spectra can be found solving the equation (5.66). In this case (5.66) is called the *exact WKB* quantization condition, see e.g. [Voros (1994)]. There are two cases where $\gamma(K)$ is known exactly: $\gamma = 0$, which corresponds to the harmonic oscillator $V = u^2$ and $\gamma = 1/2$, which corresponds to the square-well potential of width two, $u \in [-1, 1]$, with infinite wells. In general, γ is related to the boundary condition imposed at the turning point; it was claimed in [Landau and Lifshitz (1977)] that $(1/2 + \gamma)$ is of order one. For the quartic potential u^4, $\gamma(K)$ was calculated in [Turbiner and del Valle (2021c)], see Fig. 3 therein,

It is not clear how to calculate γ versus K in (5.66) theoretically in closed analytic form, it appears to be an open problem. In general, it is given by the expansion in the powers of $\hbar$. However, this function can be easily fitted with the requested in advance accuracy via the ratio of a polynomial to the square-root of another polynomial as the fitting function,

$$\gamma_n^{(4)}(K) = \frac{P_n(K)}{\sqrt{Q_{2n+2}(K)}}, n = 1, 2, 3, \ldots,$$

where P_n, Q_{2n+2} are polynomials of degrees $n, (2n + 2)$, respectively, with a condition $Q_{2n+2}(K) > 0$ for $K \geq 0$. The latter condition prevents the appearance in ratio of singularities for non-negative K and it is fulfilled for all concrete studied cases. In particular, the simplest approximation corresponds to $n = 1$,

$$\gamma_{n=1}^{(4)}(K) = \frac{0.026525\, K + 0.001203}{\sqrt{Q_4}}, \tag{5.68}$$

where

$$Q_4 = K^4 + 1.457179\, K^3 - 1.792243\, K^2 + 3.094978\, K + 0.000218.$$

When substituted into (5.66), it allows us to reproduce, at least, 8 figures in the energy for quantum numbers $0 \leq K \leq 100$, see [Turbiner and del Valle (2021c)]. However, further increase in the accuracy of the energy spectra going beyond the 8 significant digits requires us to take the next approximation, $n = 2$. This reveals a new feature: the WKB corrections $\gamma_{n=2}^{(4)}$ for positive and negative parity states are (slightly) different and they should be fitted separately. In particular, in order to get 10 correct significant digits in the energy the fit should be

$$\gamma^{(4)}_{n=2}(K)_+ = \frac{0.02652582\,K^2 - 0.03913219\,K + 0.08142236}{\sqrt{Q_6^{(+)}}}, \tag{5.69}$$

where

$$Q_6^{(+)} = K^6 - 1.95122872K^5 + 5.84890038\,K^4 - 2.38726524\,K^3$$
$$+5.55021093\,K^2 + 3.98746583\,K + 1.22015178,$$

for the positive parity states, $p = 0$, $K = 0, 2, 4, \ldots$, and

$$\gamma^{(4)}_{n=2}(K)_- = \frac{0.02652582\,K^2 + 0.01178769\,K - 0.00534128}{\sqrt{Q_6^{(-)}}}, \tag{5.70}$$

where

$$Q_6^{(-)} = K^6 + 1.88728273\,K^5 + 1.17139608\,K^4 - 0.12919918\,K^3$$
$$-0.17756539\,K^2 + 1.29044237\,K + 0.27714859,$$

for the negative parity states, $p = 1$, $K = 1, 3, 5, \ldots$, cf.(5.68). Eventually, the formula (5.66) with $\gamma^{(4)}_{n=2}(K)_\pm$ given by (5.69)-(5.70) provides no less than ten correct significant digits for the energy. It must be emphasized that γ is a small, positive, bounded function: for any $K \geq 0$,

$$\gamma^{(4)}(K) < 0.1.$$

The equality (5.66) has two outstanding properties: (i) the Bohr-Sommerfeld integral in the l.h.s. can be evaluated analytically and (ii) the resulting equation can be solved explicitly with respect to ε. Finally, the energy spectra is given by

$$E^{(4)}_{exact} = \hbar\,(\hbar g^2)^{\frac{1}{3}} \left(\frac{3}{4}\,B\left(\tfrac{1}{2}, \tfrac{3}{4}\right)\left(K + \frac{1}{2} + \gamma^{(4)}\right)\right)^{\frac{4}{3}}, \tag{5.71}$$

where $B(\alpha, \beta)$ is the Euler Beta function, see e.g. [Abramowitz and Stegun (1964)], [Bateman (1953)], $\gamma^{(4)}$ can be taken from the interpolations (5.68) or (5.69)–(5.70), see Fig. 5.7; they provided the accuracies in energy of 10 significant digits. In this case of 10 significant digits in the energy spectra the difference in γ's for positive and negative parity states occurs. However,

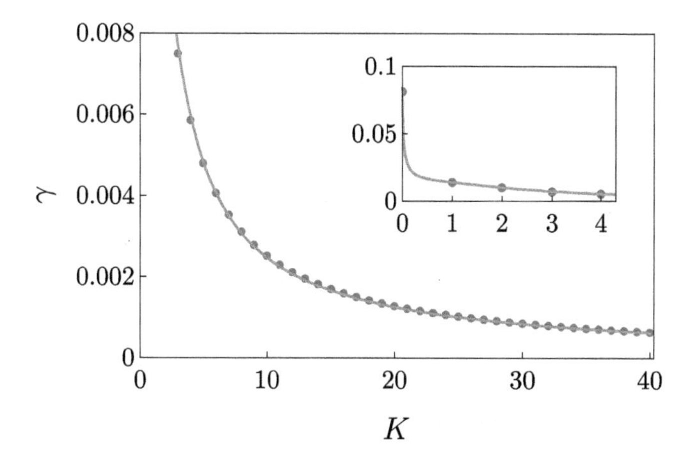

Fig. 5.7. WKB correction γ *vs* K for the quartic potential, see (5.66). Displayed points correspond to $K = 0, 1, 2, 3, \ldots, 40$. In (sub)Figure the domain $K = 0, 1, 2, 3, 4$ shown separately.

it can be checked that for accuracies up to 8 significant digits in the energy, these γ's (5.69)–(5.70) coincide in 8 significant digits.

As a result we arrive at the following approximate equality

$$E^{(4)}_{n=2} \equiv \hbar\,(\hbar g^2)^{\frac{1}{3}} \left(\frac{3}{4} B\left(\tfrac{1}{2}, \tfrac{3}{4}\right) \left(K + \frac{1}{2} + \gamma^{(4)}_{n=2,\pm}\right)\right)^{\frac{4}{3}}$$
$$\approx \; < \mathcal{H} > \Big|_{\Psi^{(N,p;4)}_{(approximation)}}, \tag{5.72}$$

which holds at the level of 9–10 significant digits for any $g^2 \in [0, \infty)$. Here $\gamma^{(4)}_{n=2,\pm}$ are given by (5.69)–(5.70). The expectation value of the Hamiltonian (5.6) (at $g^2 \to \infty$ of the quartic AHO oscillator and $a_2 = 1$) over the function

$$\Psi^{(N,p;4)}_{(approximation)}(x) = \frac{x^p P^{(p)}_N(x^2)}{(B^2 + x^2)^{\frac{1}{4}} \left(B + \sqrt{B^2 + x^2}\right)^{2N+p+\frac{1}{2}}}$$
$$\times \exp\left(-\frac{A + B^2 x^2/6 \; + \; x^4/3}{\sqrt{B^2 + x^2}} + \frac{A}{B}\right), \tag{5.73}$$

cf. (5.58) at $g = 1$, with parameters $A = a^{(N,p)}_{fit}, B = b^{(N,p)}_{fit}$, see (5.63), (5.64), respectively, is equal to the exact energy with 9–10 figures. In particular, for the ground state of the quartic oscillator, where $A_{0,0} = -1.8028, B_{0,0} = 2.1470$, the energy $E^{(4)}_{n=2} = 1.060362092$, while the accurate energy obtained using the LMM is $E_{LMM} = 1.060362090$, see App.F.

Hence, formula (5.72) presents the approximate energy spectra of the quartic oscillator in two different forms: as a function of γ obtained through fit (5.69)–(5.70) or as the expectation value of the Hamiltonian over the highly-accurate approximate eigenfunction (5.73). A similar formula for the quartic anharmonic oscillator at finite g^2 is unknown.

5.6 Quartic Double Well Potential

In this Section we briefly consider the quartic double-well potential with two symmetric harmonic wells characterized by reflection symmetry with respect to the point $x = x_s = \frac{1}{2g}$,

$$V_{dw}(x) = x^2(1 - gx)^2, \tag{5.74}$$

cf. Case **(v)** in Chapter 1, (1.10). Both the Riccati-Bloch (2.3) and the generalized Bloch (3.3) equations can be constructed for the potential (5.74). This will allow us to perform the asymptotic analysis of their solutions at $x \to 0$ and $x \to \pm\infty$ and to develop the perturbation theory (trans-series) in the effective coupling constant λ of the energy and of the logarithmic derivatives of the (ground state) wave function [Shuryak and Turbiner (2018)], see Section 3.2 for discussion. A semi-classical analysis in the path integral formalism is presented in [Shuryak and Turbiner (2018)].

Following the prescription, usually assigned in folklore to E.M. Lifschitz — one of the authors of the famous Course on Theoretical Physics by L.D. Landau and E.M. Lifschitz, see e.g. its Volume 3 [Landau and Lifshitz (1977)] cited in this book — if the wavefunction for a single well potential with minimum at $x = 0$ is known, $\Psi(x)$, the wavefunction for the double well symmetric potential with minima at $x = 0, 1/g$ can be written as $\Psi(x) \pm \Psi(x - 1/g)$. This prescription was successfully checked for the double-well potential (5.74) with a Gaussian function $\Psi(x)$ by E.M. Lifschitz and with a simplified version of the function (5.58) in [Turbiner (2010)].

By using the wavefunction (5.58) as entry we can form

$$\Psi^{(N,p;4)}_{(approximation, double-well)} = \frac{P_N(\tilde{x}^2; g^2)}{(B^2 + g^2\,\tilde{x}^2)^{\frac{1}{4}} \left(B + \sqrt{B^2 + g^2\,\tilde{x}^2}\right)^{2N+\frac{1}{2}}} \times \exp\left(-\frac{A + (B^2+3)\,\tilde{x}^2/6 + g^2\,\tilde{x}^4/3}{\sqrt{B^2 + g^2\,\tilde{x}^2}} + \frac{A}{B}\right) D^{(p)},$$

$$p = 0, 1, \tag{5.75}$$

where

$$D^{(0)} = \cosh\left(\frac{a_0\tilde{x} + b_0\tilde{x}^3}{\sqrt{B^2 + g^2\tilde{x}^2}}\right), \quad p = 0$$

$$D^{(1)} = \sinh\left(\frac{a_1\tilde{x} + b_1\tilde{x}^3}{\sqrt{B^2 + g^2\tilde{x}^2}}\right), \quad p = 1$$

and

$$\tilde{x} = x + \frac{1}{2g},$$

where the parameters $a_{0,1}, b_{0,1}$ depend on g^2 and N.

Expression (5.75) describes the states of positive parity ($p = 0$) and of negative parity ($p = 1$). By setting the polynomial $P_N = 1$ in (5.75) the resulting expression can be obtained by matching the expansion at small x (the Taylor expansion) and the two expansions at large positive/negative x (the semiclassical expansions at $x \to \pm\infty$), respectively. Here A, B and a, b are free parameters, which can be found through the minimization procedure of the variational energy. The polynomial $P_N(\tilde{x}^2; g^2)$ is fixed unambiguously by posing the orthogonality of the function $\Psi^{(N,p;4)}$ to the functions $\Psi^{(M,p;4)}, M = 0, 1, 2, \ldots, (N-1)$.

As an illustration, see [Turbiner and del Valle (2022)] we put $g = 1$ and consider the two lowest energy states of the single-well quartic AHO potential

$$V = u^2 + u^4, \tag{5.76}$$

and the double-well quartic AHO potential with degenerate minima

$$V = u^2(1-u)^2, \tag{5.77}$$

respectively. Both potentials (5.76), (5.77) are symmetric with respect to $u = 0$ and $u = 1/2$, respectively.

5.6.1 *Single-well Potential (5.76)*

Let us consider the single-well potential (5.76). Following the Section 5.4 by matching the small distance $u \to 0$ expansion and the large distance $u \to \infty$ expansion we can construct a function for the $K = (2N+p)$-th excited state with quantum numbers $N = 0, 1, 2, \ldots, p = 0, 1:$

$$\Psi^{(N,p;4)}_{(approximation)}(u) = \frac{u^p P_N^{(p)}(u^2)}{(B^2+u^2)^{\frac{1}{4}}\left(B+\sqrt{B^2\ +\ u^2}\right)^{2N+p+\frac{1}{2}}} \times \exp\left(-\frac{A+(B^2+3)\,u^2/6+u^4/3}{\sqrt{B^2\ +\ u^2}}+\frac{A}{B}\right), \tag{5.78}$$

cf. (5.58), where $P_N^{(p)}$ is a polynomial of degree N in u^2 with positive roots. Here $A = A_{N,p}, B = B_{N,p}$ are two parameters of interpolation. These parameters $(-A), B$ are growing slowly with quantum number N for fixed p. In particular, they are

$$A_{0,0} = -0.6244, B_{0,0} = 2.3667, \tag{5.79}$$

$$A_{0,1} = -1.9289, B_{0,1} = 2.5598, \tag{5.80}$$

for the ground state and the first excited state, respectively. This remarkably simple function (5.78), see Fig. 5.8 (top), provides 10–11 exact figures in the energies for the first 100 eigenstates which was checked with the LMM results, see App.F. Furthermore, the function (5.78) deviates from the exact function uniformly in domain $u \in (-\infty, +\infty)$ by $\sim 10^{-6}$. Taking (5.78) as the zeroth approximation for the eigenfunctions in the potential (5.76) one can develop the PT for non-linearization procedure (2.35)–(2.37). Even the first correction to the variational energy E_2, see (2.50), allows us to increase the number of the exact figures up to 13–14. In turn, the first correction to the phase $\phi(u)$ is of the order $\lesssim 10^{-6}$: it allows us to increase the number of exact figures (locally) up to 8–9.

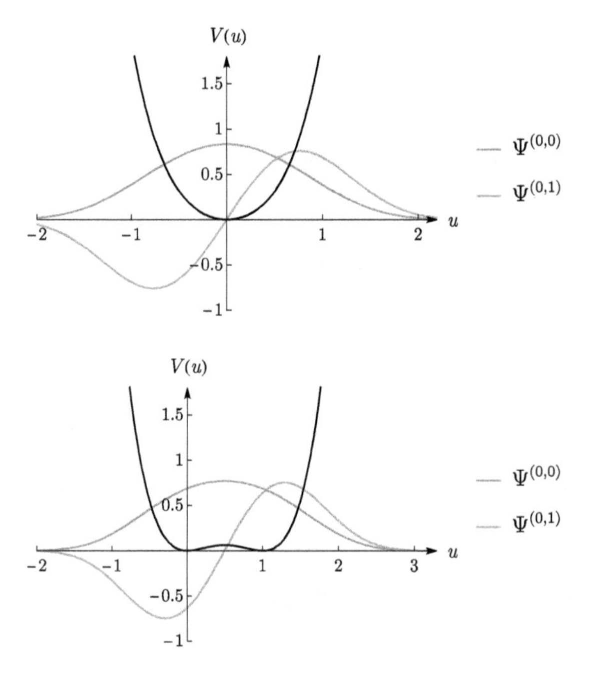

Fig. 5.8. The two lowest, normalized to one eigenfunctions of positive/negative parity: for the single-well potential (5.76), see (5.78) (top) and for the double-well potential (5.77), see (5.81)(bottom). The potentials shown by black lines. Both figures from [Turbiner and del Valle (2022)].

5.6.2 *Double-well Potential (5.77): Trial Function*

By using the wavefunction (5.78) one can construct a function similar to (5.75) by replacing $x \to \tilde{u}$ for the potential (5.77),

$$\Psi^{(N,p)}_{(approximation)}(\tilde{u}) = \frac{P_N(\tilde{u}^2)}{(B^2+\tilde{u}^2)^{\frac{1}{4}}\left(\alpha B + \sqrt{B^2+\tilde{u}^2}\right)^{2N+\frac{1}{2}}} \times \exp\left(-\frac{A+(B^2+3)\,\tilde{u}^2/6+\tilde{u}^4/3}{\sqrt{B^2+\tilde{u}^2}} + \frac{A}{B}\right) D^{(p)}, \tag{5.81}$$

where for $p = 0, 1$

$$D^{(0)} = \cosh\left(\frac{a_0\tilde{u}+b_0\tilde{u}^3}{\sqrt{B^2+\tilde{u}^2}}\right),$$

$$D^{(1)} = \sinh\left(\frac{a_1\tilde{u}+b_1\tilde{u}^3}{\sqrt{B^2+\tilde{u}^2}}\right).$$

Here

$$\tilde{u} = u - \frac{1}{2},$$

and $A, B, a_{0,1}, b_{0,1}$ are variational parameters. If $\alpha = 0$ as well as $b_{0,1} = 0$ the function (5.81) is reduced to the ones which were explored in [Turbiner (2010)] see Eqs. (10)–(11) therein. Note that the polynomial P_N is found unambiguously after imposing orthogonality conditions between $\Psi^{(N,p;4)}_{(approximation)}$ and $\Psi^{(M,p;4)}_{(approximation)}$ at $M = 0, 1, 2, \ldots, (N-1)$; here it is assumed that the polynomials P_M at $M = 0, 1, 2, \ldots, (N-1)$ are found beforehand.

5.6.3 *Double-well Potential (5.77): Results*

In this section we present concrete results for the energies of the ground state $(0,0)$ and of the first excited state $(0,1)$ obtained with the function (5.81) at $p = 0, 1$, respectively, see Fig. 5.8 (bottom). The obtained results are compared with the LMM results [Turbiner and del Valle (2021a)].

5.6.3.1 *Ground State (0,0)*

The ground state energy for the potential (5.77) obtained variationally using the function (5.81) at $p = 0$ and compared with the LMM results [Turbiner and del Valle (2021a)], where all printed digits (in the second line) are correct, is

$$E^{(0,0)}_{var} = 0.932\,517\,518\,401,$$
$$E^{(0,0)}_{mesh} = 0.932\,517\,518\,372.$$

Note that the first ten decimal digits in $E^{(0,0)}_{var}$ coincide with the ones in $E^{(0,0)}_{mesh}$ (after rounding). The variational parameters in (5.81) take the values,

$$A = 2.3237,$$
$$B = 3.2734,$$
$$a_0 = 2.3839,$$
$$b_0 = 0.0605,$$

cf. (5.79). Note that b_0 takes a very small value. Setting $b_0 = 0$ it reduces the accuracy of the energies to nine decimal digits.

5.6.3.2 *First Excited State (0,1)*

The energy of the first excited state $(0,1)$ of the potential (5.77) obtained variationally by using the function (5.81) at $p = 1$[13] and compared with LMM results [Turbiner and del Valle (2021a)], where all printed digits (see the second line below) are correct, is given by

$$E_{var}^{(0,1)} = 3.396\,279\,329\,936,$$
$$E_{mesh}^{(0,1)} = 3.396\,279\,329\,887.$$

Note that the first ten decimal digits of $E_{var}^{(0,1)}$ coincide with the ones in $E_{mesh}^{(0,1)}$. The variational parameters in (5.81) take the values,

$$A = -2.2957,$$
$$B = 3.6991,$$
$$a_1 = 4.7096,$$
$$b_1 = 0.0590,$$

cf. (5.80). Note that b_1 takes a very small value similar to b_0. Analogously to the ground state, setting $b_1 = 0$ leads to a reduction in accuracy for energy to nine decimal digits.

In general, the remarkably simple function (5.81), see Fig. 5.8 (bottom) for the two lowest states, provides 10–11 exact figures in energies for the first 100 eigenstates as was checked by making comparison with LMM results, see App.F. Furthermore, the function (5.81) deviates from the exact function uniformly in domain $u \in (-\infty, +\infty)$, its deviation is $\sim 10^{-6}$. It is worth noting that the accuracy of the simplified function, proposed in [Turbiner (2010)] where in (5.81) the parameters $\alpha = 0$ and $b_{0,1} = 0$, in the domain under the barrier $u \in (0.25, 0.75)$ is increased from 4 to 6 significant figures leaving the accuracy outside of this domain practically unchanged.

By taking (5.81) as the zeroth order approximation of the eigenfunctions in the potential (5.77) one can develop the PT for the non-linearization procedure (2.35)–(2.37). Even the first correction to the variational energy E_2, see (2.50), allows us to increase the number of exact figures from 10–11 up to 13–14. In turn, the first correction to the phase $\phi(u)$ is of order $\lesssim 10^{-6}$: it allows us to increase the number of the exact figures (locally) up to 8–9.

[13]The functions (5.81) at $p = 0$ and $p = 1$ have different parity: they are orthogonal by construction.

As for the general double-well potential (5.74) the trial function (5.75) can be explored for the first eigenstates for different coupling constants $g^2 = 0.1, 1, 10, 100$. The accuracy in energy is always $\sim 10^{-10}$ even in the semiclassical domain $g^2 \to 0$.

5.7 Quartic AHO: Concluding Remarks

5.7.1 *Asymmetric Quartic AHO*

Both single-well and double-well quartic oscillators are two important particular cases of the asymmetric (tilted) quartic double-well anharmonic oscillator potential

$$V_a = x^2\,(1 + agx + g^2x^2),$$

see e.g. [Brezin *et al.* (1977b); Escobar-Ruiz *et al.* (2017)], where a is the asymmetry parameter: $2 \geq a \geq -2$, see the Case **(iii)** (1.8). If $a = \pm 2, 0$, the potential V_a is symmetric in x-space. Quantum mechanics associated with potential V_a is essentially one-parametric: it depends on a combination $\hbar^{1/2}\,g$ *only*.

The formalism based on the Riccati-Bloch equation (see Chapter 2) and Generalized Bloch equation (see Chapter 3) can be successfully applied to the potential V_a. The perturbation theory in powers of g and the true semiclassical expansion in powers of $\hbar^{1/2}$ can be easily developed. As a result we arrive at the expansion of the phase at small x and at the two semiclassical expansions: one is at large positive and another one is at large negative distances. By matching these three expansions one can see that the general Approximant (4.1) for the arbitrary AHO is finally reduced to

$$\begin{aligned}\Psi^{(K)}_{(approximation)}(x) = {} & \frac{P_K(x;g)}{(B(K)^2 + \tilde{a}(K)gx + g^2\,x^2)^{\frac{1}{4}}} \\ & \times \frac{1}{\left(B(K) + \sqrt{B(K)^2 + \tilde{a}(K)gx + g^2\,x^2}\right)^{K+\frac{1}{2}}} \\ & \times \exp\left(-\frac{\begin{array}{c}A(K) + B(K)^2\,x^2/6 + \tilde{a}(K)gx^3 \\ +g^2\,x^4/3\end{array}}{\sqrt{B(K)^2 + \tilde{a}(K)gx + g^2\,x^2}} + \frac{A(K)}{B(K)}\right),\end{aligned} \tag{5.82}$$

cf. (5.58), where $A, B, \tilde{a}$ are free parameters which depend on the principal quantum number K. They can be fixed variationally. The coefficients of the polynomial $P_K(x;g)$ are found unambiguously by imposing the orthogonality conditions to the states with smaller principal quantum numbers $k = 0, 1, 2, \ldots (K-1)$.

5.7.2 *Radial Quartic AHO Oscillator*

The formalism developed in this Chapter for the one-dimensional quartic AHO can be extended to the case of the $O(d)$ spherically-symmetric, d-dimensional quartic anharmonic oscillator with potential

$$V = r^2 + g^2 r^4,$$

cf.(5.1), where the radial variable is given by $r^2 = \sum_{i=1}^{d} x_i^2$ and g^2 is the coupling constant. By separating out the angular variables in the Schrödinger equation we arrive at the familiar radial Schrödinger equation, which can be converted to a radial Riccati equation. In turn, by introducing the radial analogue of the quantum (5.11) and the classical (5.51) coordinates: $x \to r$ the radial Riccati equation can be transformed either to the Riccati-Bloch type or to the Generalized Bloch type equations, which are similar to equations (2.3) and (3.3), respectively. Importantly, these two radial equations share the same effective coupling constant λ as in one-dimensional case (2.2). Based on these two equations we can construct the PT and semiclassical expansions, respectively, in the same way as was done previously in this Chapter. In fact, these expansions look similar to (5.15) and (5.54) after the replacement $x \to r$ (and further replacements $N \to N_r, p \to \ell, 1 \to d$ for the degree of the second prefactor of the Approximant, see below). By matching the asymptotic expansion of the solution at small r of the radial RB equation with the semiclassical expansion of the radial GB equation for radial quartic AHO we arrive at

$$\Psi^{(n_r,\ell;4)}_{(approximation)}(r) = \frac{r^\ell P_{N_r,\ell}(r^2)}{(B^2+g^2\,r^2)^{\frac{1}{4}}\left(B+\sqrt{B^2+g^2\,r^2}\right)^{2N_r+\ell+\frac{d}{2}}} \times \exp\left(-\frac{A+(B^2+3)\,r^2/6+g^2\,r^4/3}{\sqrt{B^2+g^2\,r^2}}+\frac{A}{B}\right), \tag{5.83}$$

which we call the *Approximant* in Part II, Chapter 10, cf. (5.58). Here we set $\hbar = 1$ and $P_{N_r,\ell}$ is a polynomial of degree N_r in r^2 with positive

Table 5.4. Ground state energy of the quartic radial potential $V(r) = r^2 + g^2r^4$ in dimensions $d = 1, 2, 3, 6$ and coupling constant $g^2 = 0.1, 1, 10$, calculated using (5.83) in variational calculations. All printed digits exact.

g^2	$d=1$	$d=2$	$d=3$	$d=6$
0.1	1.065 285 509 54	2.168 597 211 3	3.306 872 013 2	6.908 332 111 2
1.0	1.392 351 642	2.952 050 092	4.648 812 704	10.390 627 296
10.0	2.449 174 07	5.349 352 82	8.599 003 46	19.936 900 4

roots; (N_r, ℓ) are radial and angular quantum numbers, $N_r = 0, 1, 2, \ldots$ and $\ell = 0, 1, 2, \ldots$, respectively. The parameters $A = A_{N_r,\ell}(g^2)$, $B = B_{N_r,\ell}(g^2)$ are two parameters of interpolation; they can be found variationally. In particular, for the ground state $(0, 0)$ this is done by making minimization of the expectation value of the radial Hamiltonian over the function (5.83). The results for the ground state for different d and g^2 are presented in Table 5.4. All printed digits for the energies in Table 5.4 are exact. They were checked in three different ways: (i) by making the Lagrange Mesh Method calculations for the radial Schrödinger equation, see App.F, (ii) by calculating the first correction to the variational energy E_2 and (iii) by comparing with the variational energies obtained with a certain generalization of the Approximant (5.83) containing more free parameters, see [Del Valle and Turbiner (2019, 2020)]. A detailed description of the quartic radial AHO will be presented in Part II, Chapter 6 including a generalization for the excited states.

5.7.3 *Conclusions*

By using the method of matching — a straightforward interpolation of the logarithm of the ground state wavefunction between the Taylor expansion at small distances and the true semiclassical expansion at large distances — the approximate ground state eigenfunction in the form of product of a three factors: two pre-factors and one exponential, is built for the quartic AHO with an arbitrary coupling constant. The modifications needed to obtain wave functions for the excited states involve a simple modification of the degree of one of the pre-factors and a further multiplication of the polynomial which carries the information about nodes. Finally, this brings us to formula (5.58), which should be applicable for an arbitrary coupling constant. Formula (5.58) leads to unprecedented accuracies of 9–10–11 significant figures for the eigenvalues for any strength of the quartic anharmonicity and of 6 significant figures for the eigenfunctions at any point in the coordinate space. Hence, formula (5.58) manifests the approximate (highly-accurate) solution to the problem of the quartic AHO. In the strong coupling regime $g \to \infty$, where the quartic AHO becomes the quartic oscillator, the energy spectra can be written in closed analytic form (5.71) in terms of a single function of the principal quantum number, which is bounded and small and which can be found with any desired accuracy.

By duplicating the function (5.58) of positive parity and center them at the points $x = 0$ and $x = \frac{1}{g}$,

$$\Psi^{(N,0)}_{(approximation)}(x) + (-)^p\, \Psi^{(N,0)}_{(approximation)}\left(x - \frac{1}{g}\right)$$

to follow the E.M. Lifschitz prescription, we arrive at a highly-accurate solution for quartic double well potential with harmonic wells situated at $x = 0, \frac{1}{g}$, respectively, for positive ($p = 0$) and negative ($p = 1$) parity states. Similarly to the solution for the single-well quartic AHO (5.58) this solution leads to unprecedented accuracies of 9–11 significant figures for the energies (and energy gaps) and of 6 significant figures for the eigenfunctions at any point in the coordinate space including the area under the barrier.

Formula (5.58) admits a straightforward generalization to (5.83) for the radial quartic AHO, see [Del Valle and Turbiner (2019, 2020)], which will be described in a greater details in Part II.

In the following Chapter 6 we will consider the sextic one-dimensional symmetric anharmonic oscillator.

Chapter 6

Sextic Anharmonic Oscillator

In this Chapter the knowledge presented in Chapters 2 and 3 for the general polynomial anharmonic oscillator and applied in Chapter 5 for the quartic anharmonic oscillator will be employed for the one-dimensional quantum single-well symmetric anharmonic oscillator with potential $V(x) = x^2 + g^4 x^6$. In this case the Perturbation Theory (PT) in powers of g^4 (weak coupling regime) and the Semiclassical Expansion (SE) in powers of $\hbar$ for the energies (eigenvalues) coincide. Again this is related to the fact that the dynamics in x-space and in (gx)-space correspond to the same energy spectrum with effective coupling constant $\hbar^2 g^4$. The two equations which govern the dynamics in those two spaces, the Riccati-Bloch (RB) and the Generalized Bloch (GB) equations, respectively, will be presented. The PT in g^4 for the logarithmic derivative of the wave function leads to the PT (with polynomial in x coefficients) of the RB equation and to the *true* SE in powers of $\hbar$ of the GB equation, which corresponds to a loop expansion of the density matrix in the path integral formalism. A 4-parametric interpolation of these two expansions leads to a uniform approximation of the wavefunction in x-space with an unprecedented accuracy of $\sim 10^{-6}$ (six significant figures) locally and an unprecedented accuracy of $\sim 10^{-10} - 10^{-12}$ (ten-twelve significant figures) for the associated energy eigenvalue, it remains true for any $g^2 > 0$, which will be studied. This Chapter is written in a brief manner and maximally self-contained manner.

6.1 Sextic Anharmonic Oscillator: Introduction and Generalities

The sextic symmetric anharmonic oscillator (pendulum), both classical and quantum, with potential

$$V_6(x) = x^2 + g^4x^6 = x^2(1+g^4x^4) = \frac{1}{g^2}\left((gx)^2 + (gx)^6\right) \equiv \frac{1}{g^2}\hat{V}_6(gx), \tag{6.1}$$

cf. (1.14), where

$$\hat{V}_6(u) = u^2 + u^6, \tag{6.2}$$

cf. (3.7), (5.2), together with the quartic AHO (5.1) is among the most fundamental problems in physics. It appears in many branches of physics and it is present in many textbooks on quantum mechanics. It reveals a great variety of non-trivial properties reflecting the complexity of Nature. A particular direction of interest for this problem is the field-theoretical treatment of (6.1) as a (0+1)-dimensional quantum field theory, which was considered to be a precursor for understanding the realistic three-dimensional quantum field theory $\lambda\phi^6$, see e.g. [Lipatov (1976)], this will be mentioned briefly.

We consider the real Hamiltonian

$$\mathcal{H} = -\frac{\hbar^2}{2m}\partial_x^2 + x^2 + g^4x^6, \quad x \in (-\infty, +\infty), \tag{6.3}$$

where the Planck constant $\hbar$, the mass m and the coupling constant g^4 are real parameters. We are looking for real eigenvalues (energies) and square-integrable, non-singular at real x eigenfunctions. According to the oscillation theorem, see e.g. [Landau and Lifshitz (1977)], the Kth eigenfunction of the excited state is characterized by K simple zeroes situated at real x; simultaneously, any eigenfunction has infinitely-many simple zeroes at complex x [Eremenko *et al.* (2008)]. The Hamiltonian is evidently a Hermitian operator with an infinite discrete spectra. Questions related to the analytic continuation to the complex x and g^4 planes are no going to be discussed in this Chapter. The Schrödinger equation in its final form reads,

$$-\frac{\hbar^2}{2m}\partial_x^2\Psi + (x^2 + g^4x^6)\Psi = E\Psi, \tag{6.4}$$

where the eigenstates are denoted as follows: the wavefunction $\Psi = \Psi(x; 1, g^4)$ and the energy $E = E(1, g^4)$, here 1 represents the value of the coefficient in front of the x^2 term in the potential. By rescaling of the x-variable: $x \to g^{-1/2}x$, we arrive at the Schrödinger equation for a similar potential with different parameters

$$-\frac{\hbar^2}{2m}\partial_x^2\Psi + (g^{-2}x^2 + x^6)\Psi = g^{-1}E\Psi. \tag{6.5}$$

This allows us to derive the following relations between eigenstates in these potentials:

$$\begin{aligned} E(1, g^4) &= gE(g^{-2}, 1), \\ \Psi(x; 1, g^4) &= \Psi(g^{1/2}x; g^{-2}, 1). \end{aligned} \tag{6.6}$$

These relations are called the Symanzik scaling relations.

Since the potential (6.1) is symmetric, $V_6(-x) = V_6(x)$, the eigenfunctions are characterized by a parity p,

$$\Psi(-x) = (-)^p\Psi(x),$$

where for $p = 0$ we have positive parity states and for $p = 1$ the states have negative parity. This implies the existence of the following representation for the wavefunction

$$\Psi = x^p\Psi_2(x^2),$$

where p takes value 0 or 1. Hence, the Hamiltonian (6.3) can be gauge rotated with the gauge factor x^p and then rewritten in the variable $x^2 \equiv x_2$,

$$\mathcal{H}_p = x^{-p}\mathcal{H}x^p|_{x_2=x^2} = -\frac{\hbar^2}{m}\left(2x_2\partial_{x_2}^2 + (1+2p)\partial_{x_2}\right) + x_2 + g^4x_2^3.$$

This simplifies the solution of the Schrödinger equation (6.4), which is now reduced to

$$\mathcal{H}_p\Psi_2 = E\Psi_2, \quad x_2 \in [0, \infty), \tag{6.7}$$

with the same energy spectra.

The theory for the weak coupling regime, where the coupling constant g^4 is small, can be created constructively. Equation (6.4) can be treated perturbatively by looking for a Taylor expansions in powers of g^4. In the remarkable paper by Bender and Wu [Bender and Wu (1969)], in a similar way to what was done for the quartic AHO, see Chapter 5, it was shown that in the weak coupling domain

- The problem is, in fact, *algebraic*: practically everything can be calculated by linear algebra means;
- The perturbation theory

$$\varepsilon = \sum_{k=0}^{\infty} \varepsilon_k g^{4k}, \tag{6.8}$$

for any eigenvalue has a zero radius of convergence, the coefficients ε_k are always rational numbers with alternating signs and their asymptotics at $k \to \infty$ is given by

$$\varepsilon_k = (-1)^{k+1} 2^{-1/2} \left(\frac{4}{\pi}\right)^{2(k+1)} \Gamma\left(2k + \frac{1}{2}\right) \left[1 - O\left(\frac{1}{k^2}\right)\right], \tag{6.9}$$

cf. (5.4), for the ground state,[1] this series is Borel summable, see for discussion [Graffi *et al.* (1970)];[2]
- There exist infinitely-many, complex-conjugated (thus symmetric with respect to real line in the m^2 complex plane), square-root branch points in the $m^2 = g^{-2}$ complex plane of a given eigenvalue, which are accumulated towards real m^2 axis when m^2 tends to minus infinity,[3] see Fig. 5.1; recently, their existence was proved rigorously in [Eremenko and Gabrielov (2009)]) (as the algebraic ramification points) as a reflection of the Landau-Zener theory of level crossings (in complex plane), see [Landau and Lifshitz (1977)].[4]

Each singularity occurs at a point of the level crossing of two eigenstates of the same parity.[5] This indicates a highly-complicated analytic structure for the eigenvalues in the coupling constant m^2 (or g^4) complex plane. The problem of summation of the asymptotic series (6.8) in order to get reliable results for $g^4 \neq 0$ remains, in general, unsolved.

---

[1]This formula is called the *Bender-Wu formula for sextic AHO.* It was derived by using quantum-mechanics methods and verified in [Brezin *et al.* (1977a,b)] by using quantum-field theory methods. It can be generalized for any excited state of the sextic AHO and for any two-term anharmonic oscillator.

[2]It implies that (6.8) does not contains the (hidden) exponentially-small terms, the Borel image coincides with the exact energy.

[3]In [Bender and Wu (1969)] in the g^4 complex plane this analytic structure was called the *horn structure* of the singularities.

[4]The exact location of singularities is still unknown.

[5]Each state is characterized by its parity, positive or negative. Singularity at the level crossing occurs for the eigenstates with same parity *only*, positive or negative.

In an attempt to overcome the asymptotic nature of the PT (6.8), it was proposed to study the strong coupling expansion in inverse (rational) powers of the coupling constant:

$$E = g \sum_{k=0}^{\infty} b_k g^{-2k}, \tag{6.10}$$

which is evidently convergent with a finite radius of convergence. However, there is a technical problem related to the question of finding the coefficients b_k constructively.

By taking an exponential representation for the wavefunction

$$\Psi = e^{-\frac{1}{\hbar}\phi},$$

cf. (1.17), and substituting it into (6.4) we arrive at the celebrated Riccati equation

$$\hbar y' - y^2 = 2m(E - x^2 - g^4 x^6), \quad y = \phi', \tag{6.11}$$

cf. (1.18), (5.10). This equation contains $\hbar$ in front of the leading derivative term and g^4 in front of the anharmonic term: this leads to a bifurcation in $\hbar$ and to the divergence of the PT in powers of g^4,[6] respectively.

Taking the generalized exponential representation (2.32),

$$\Psi = f(x) e^{-\frac{1}{\hbar}\Phi(x)},$$

and plugging it into the Schrödinger equation (1.16), see (6.4) we arrive at a generalized Riccati equation (2.33),

$$(\hbar y' - y^2) - \frac{\hbar^2 f'' - 2\hbar y f'}{f} = 2m(E - x^2 - g^4 x^6), \quad y = \Phi', \tag{6.12}$$

for the sextic anharmonic oscillator. At $f = 1$ this equation becomes (6.11).

Now let us get rid of the explicit $\hbar$ dependence in equations (6.11), (6.12). There are two ways to accomplish this: via the Riccati-Bloch equation and via the Generalized Bloch equation. These procedures correspond to preventing $\hbar$ to vanish. Surprisingly, those procedures allow us to get rid simultaneously from mass dependence, hence, prohibiting taking the limit $m \to 0$.

[6]Following the Dyson instability argument [Dyson (1952)].

6.2 Sextic AHO: Riccati-Bloch Equation

6.2.1 *Ground State*

Let us substitute into the Riccati equation (6.11) the new definitions for the variable $x \to v$, for the function $y \to \mathcal{Y}$ and for the energy $E \to \varepsilon$:

$$x = \left(\frac{\hbar^2}{m}\right)^{1/4} v, \quad y = (m\hbar^2)^{1/4}\mathcal{Y}(v), \quad E = \frac{\hbar}{m^{1/2}}\varepsilon, \tag{6.13}$$

respectively, cf. (5.11), and also the effective coupling constant

$$\lambda = \left(\frac{\hbar^2}{m}\right)^{1/4} g, \tag{6.14}$$

cf. (5.12). Note the existence of the relation

$$\lambda v = gx.$$

We arrive at the RB equation:

$$\partial_v \mathcal{Y} - \mathcal{Y}^2 = \varepsilon(\lambda^2) - v^2 - \lambda^4 v^6, \quad \partial_v \equiv \frac{d}{dv}, v \in (-\infty, \infty), \tag{6.15}$$

cf. (2.3). This equation has no $\hbar$-dependence: *it describes dynamics in a "quantum", $\hbar$-dependent coordinate* v (6.13) *instead of* x, *it is governed by the $\hbar$-dependent, effective coupling constant* λ (6.14). Note that

$$\partial_v^4 \left(\partial_v \mathcal{Y} - \mathcal{Y}^2\right)\Big|_{v=0} = 0, \tag{6.16}$$

cf. (2.5).

The asymptotic expansion of the ground state solution for Eq. (6.15) can be easily derived at small v,

$$\mathcal{Y} = \varepsilon v + \frac{\varepsilon^2 - 1}{3} v^3 + O(v^5). \tag{6.17}$$

Later this expansion will be used in the construction of the phase approximant $\Phi_{approximation}(x)$.

It is evident that the function $\mathcal{Y}$ in (6.15) should be antisymmetric, $\mathcal{Y}(-v) = -\mathcal{Y}(v)$. Hence, the function $\mathcal{Y}$ admits a natural representation

$$\mathcal{Y}(v) = v\hat{\mathcal{Y}}(v^2). \tag{6.18}$$

The RB equation (6.15), written for the function $\hat{\mathcal{Y}}$, is converted into the form

$$2v_2 \partial_{v_2} \hat{\mathcal{Y}}(v_2) + \hat{\mathcal{Y}}(v_2) - v_2 \hat{\mathcal{Y}}^2(v_2) = \varepsilon(\lambda) - v_2 - \lambda^4 v_2^3, \tag{6.19}$$

where

$$\partial_{v_2} \equiv \frac{d}{dv_2}, v_2 \equiv v^2,$$

cf. (5.18). This is the final form of the RB equation, where the PT in powers of λ^4 can be developed,

$$\varepsilon = \sum_{n=0}^{\infty} \lambda^{4n} \varepsilon_n, \tag{6.20}$$

$$\hat{\mathcal{Y}} = \sum_{n=0}^{\infty} \lambda^{4n} \hat{\mathcal{Y}}_n. \tag{6.21}$$

The first two terms in the expansions can be easily found

$$\varepsilon_0 = 1, \quad \hat{\mathcal{Y}}_0 = 1,$$

$$\varepsilon_1 = \frac{15}{8}, \quad \hat{\mathcal{Y}}_1 = \frac{1}{8}\left(4v_2^2 + 10v_2 + 15\right),$$

and, in general, the n-th correction $\hat{\mathcal{Y}}_n$ is a polynomial of degree $(2n)$,

$$\hat{\mathcal{Y}}_n = Pol_n(v_2) = a_0^{(n)} + a_1^{(n)} v_2 + a_2^{(n)} v_2^2 + \cdots + a_k^{(n)} v_2^k + \cdots + a_{2n}^{(n)} v_2^{2n}, \tag{6.22}$$

cf. (5.20).

By substituting the Taylor expansions for ε (6.20) and $\hat{\mathcal{Y}}$ (6.21) into equation (6.19) and collecting terms of order λ^{4n} we arrive at the equation to determine the nth correction,

$$2v_2\partial_{v_2}\hat{\mathcal{Y}}_n + \hat{\mathcal{Y}}_n - 2v_2\hat{\mathcal{Y}}_0\hat{\mathcal{Y}}_n = \varepsilon_n - Q_n. \tag{6.23}$$

This equation has the same functional form as one for the quartic AHO, cf. (5.21). Here

$$\varepsilon_0 = 1, \hat{\mathcal{Y}}_0 = 1,$$

correspond to the ground state of the harmonic oscillator, which is the solution to the unperturbed equation (6.19) at $\lambda^4 = 0$, and

$$Q_1 = v_2^2, \quad Q_n = -v_2 \sum_{i=1}^{n-1} \hat{\mathcal{Y}}_i \hat{\mathcal{Y}}_{n-i}, \; n = 2, 3, \ldots, \tag{6.24}$$

play the role of the *(effective) perturbation potentials*, which is the same role *exactly* as for the quartic AHO, cf. (5.22). The above-mentioned assumption about the polynomial nature of nth correction (6.22) can be immediately verified. Furthermore, the effective perturbation potential Q_n has the form of a polynomial of degree $(2n+1)$. As a result, the relation

$$\varepsilon_n = a_0^{(n)},$$

provides a connection between the energy correction ε_n and the normalization of the $\hat{\mathcal{Y}}_n$-correction at $v_2 = 0$: $\hat{\mathcal{Y}}_n(0)$. By inserting the polynomials $Pol_{2n-k}(v_2)$, $k = 0, 1, \ldots, 2n$, see (6.22), into equation (6.23) with Q_n given by (6.24) we obtain a system of recurrence relations, which can be solved iteratively, starting from the $(2n)$th coefficient $a_{2n}^{(n)}$ and moving down to the zeroth coefficient $a_0^{(n)} = \varepsilon_n$. The procedure of solving the recurrence relations can be easily programmed in MATHEMATICA, see Appendix B.2, where an explicitly written 8-line code is presented. On a reasonable CPU time one can get practically any finite number of corrections.[7] The only limitation is related to the memory needed to store the rational coefficients of $\hat{\mathcal{Y}}_k$, $k = 1, 2, \ldots, 2n$ and the RAM needed to carry out the large arithmetic operations. As an example, the first fifteen corrections ε_n are presented in Table 6.1, the first seven corrections $\hat{\mathcal{Y}}_n$ are presented in Table 6.2.

Interestingly, several leading coefficients (standing in front of the highest degrees) of the polynomial (6.22) can be found explicitly by using the

Table 6.1. Sextic AHO, the Ground State $(0, 0)$: First 15 corrections in λ^4 for ε, see (6.20).

Correction	Value	Correction	Value
ε_0	1	$-\varepsilon_8$	$\frac{60012267376487328104827 5}{2147483648}$
ε_1	$\frac{15}{8}$	ε_9	$\frac{225579055133599011315365 6625}{17179869184}$
$-\varepsilon_2$	$\frac{3495}{128}$	$-\varepsilon_{10}$	$\frac{21160722559334931139552067094465}{274877906944}$
ε_3	$\frac{1239675}{1024}$	ε_{11}	$\frac{1211704813993560160491091041923453 25}{2199023255552}$
$-\varepsilon_4$	$\frac{3342323355}{32768}$	$-\varepsilon_{12}$	$\frac{33279588623140635437373385248759373406 55}{70368744177664}$
ε_5	$\frac{3625381915125}{262144}$	ε_{13}	$\frac{269910219763912196748146318879785060322708 25}{562949953421312}$
$-\varepsilon_6$	$\frac{1156959285530359 5}{4194304}$	$-\varepsilon_{14}$	$\frac{5105855958102640324296351373326655993547051560 35}{9007199254740992}$
ε_7	$\frac{2558220350233785007 5}{33554432}$	ε_{15}	$\frac{5570677989498472438038350680373745491215204096114875}{72057594037927936}$

[7]This code was tested to obtain $n = 200$ corrections, an interested reader can repeat this calculation, see MATHEMATICA-12 code presented in App.B.2.

Table 6.2. Sextic AHO, the Ground State (0, 0): First 4 corrections for $\hat{Y}(v_2)$, see (6.21).

Correction	Value
$\hat{Y}_0$	1
$\hat{Y}_1$	$\frac{v_2^2}{2}+\frac{5v_2}{4}+\varepsilon_1$
$-\hat{Y}_2$	$\frac{v_2^4}{8}+\frac{19v_2^3}{16}+\frac{47v_2^2}{8}+\frac{545v_2}{32}-\varepsilon_2$
$\hat{Y}_3$	$\frac{v_2^6}{16}+\frac{37v_2^5}{32}+\frac{705v_2^4}{64}+\frac{2165v_2^3}{32}+\frac{72385v_2^2}{256}+\frac{197875v_2}{256}+\varepsilon_3$
$-\hat{Y}_4$	$\frac{5v_2^8}{128}+\frac{291v_2^7}{256}+\frac{4361v_2^6}{256}+\frac{86835v_2^5}{512}+\frac{624791v_2^4}{512}+\frac{13292713v_2^3}{2048}+\frac{102908501v_2^2}{4096}$ $+\frac{542621305v_2}{8192}-\varepsilon_4$

method of generating functions by starting again from the nth coefficient $a_{2n}^{(n)}$ and moving down, eventually, to the zeroth coefficient $a_0^{(n)}$. For example, the first two generating functions are

$$Z_0=\sum_{n=0}^{\infty}\lambda^{4n}v_2^{2n}a_{2n}^{(n)}=\sqrt{1+\lambda^4v_2^2}, \tag{6.25}$$

cf. (5.23),

$$Z_1=\sum_{n=1}^{\infty}\lambda^{4n}v_2^{2n-1}a_{2n-1}^{(n)}=\frac{1+2\lambda^4v_2^2-\sqrt{1+\lambda^4v_2^2}}{2v_2(1+\lambda^4v_2^2)}, \tag{6.26}$$

cf. (5.24), and the smallest one,

$$Z_\infty\equiv\sum_{i=0}^{\infty}\lambda^{2i}a_0^{(i)}=\varepsilon, \tag{6.27}$$

cf. (5.28), is the sum of the coefficients in front of $(v_2)^0$ in $\hat{\mathcal{Y}}_k$, $k=0,1,2,\ldots,\infty$. Asymptotically, at large n the coefficients behave like

$$a_{2n-k}^{(2n)}\sim\frac{2n!}{(2n-k)!}.$$

Hence, for finite fixed k, $k=0,1,2,\ldots$ the sum of coefficients is convergent, while for finite $(2n-k)=0,1,2,\ldots$ all these sums are divergent. This is similar to the quartic AHO and is reminiscent to the situation with the double-logarithmic approximation in Quantum Field Theory, see for discussion [Turbiner (1984)] and references therein.

6.2.2 *Padé-Borel Summation*

In a similar way to what was done for the quartic AHO, see Section 5.2.2, the Páde-Borel summation technique can be applied to the sextic AHO.

The divergent series (6.27) is of particular interest since it formally leads to the ground state energy ε. Since we can find a sufficiently large number of coefficients $\varepsilon_n = a_0^{(n)}$, the Páde-Borel summation technique can be used effectively to obtain an accurate value of the sum (6.27) for concrete values of λ^4). Let us again mention, that the series (6.27) is Borel summable [Graffi *et al.* (1970)]. The Páde-Borel summation technique is a two-step procedure:

(1) construct the Borel transform of (6.27) given by

$$\mathrm{BS}(t\lambda^2) = \sum_{n=0}^{\infty} \frac{\varepsilon_n}{(2n)!}(t^2\lambda^4)^n, \tag{6.28}$$

cf. (5.29), where t is an auxiliary variable. Since the $(2n)!$ in denominator cancels the factorial growth of the coefficients ε_n, see the asymptotic formula (6.9), the radius of convergence of the Borel transformed series (6.28) will be finite. The function BS is the Borel image of the sextic AHO.

(2) calculate the *inverse* Borel transform of $\mathrm{BS}(t^2, \lambda^4)$, defined by the following integral

$$\varepsilon(\lambda^2) = \int_0^\infty \mathrm{BS}(t^2\lambda^4)e^{-t}dt. \tag{6.29}$$

One can easily check that by inserting (6.28) into (6.29) and then changing the order of integration and summation, we arrive formally to (6.27).

In order for the equality (6.29)to hold the Borel transform (6.28) should be known with infinitely many PT coefficients (6.20). In reality, only a finite number of the PT coefficients can be found constructively. For the sextic AHO the first 15 PT coefficients in (6.27) are presented in Table 6.1 although a computational procedure presented in App.A allows us to obtain several hundred PT coefficients easily. In general, a closed analytic expression for the Borel image (6.28) is unknown, however, one can approximate it by using, for example, the diagonal Padé approximants:

$$\mathrm{BS}(t^2\lambda^4) \approx P[J/J](t\lambda^2) \equiv \frac{\sum_{j=0}^J A_j(t^2\lambda^4)^j}{\sum_{j=0}^J B_j(t^2\lambda^4)^j}.$$

see (6.2.2). Here A_j and B_j are coefficients that are determined by imposing the constraint that the first $(2J+1)$ Taylor coefficients in the expansion of the Padé approximant $P[J/J]$ coincide with the first $(2J+1)$ PT coefficients in (6.20) divided by $(2n)!$; schematically, at $(t^2\lambda^4) \to 0$,

$$\mathrm{BS}(t^2\lambda^4) - P[J/J](t^2\lambda^4) = O\left((t^2\lambda^4)^{2J+1}\right), \tag{6.30}$$

for a given J.[8] As a consequence, we can obtain the approximate value of the energy by taking the inverse Borel transform of the Padé approximant $P[J/J]$. As a result we arrive at the expression called the *Padé-Borel approximant.* This procedure carries the name the *Padé-Borel summation*, see e.g. [Brezin *et al.* (1977a)].

Calculations for the ground state energy with coupling constants $\lambda^4 = 0.1, 1, 10$ for different values of J with the Padé approximant $P[J/J]$ are presented in Table 6.3. In general, the accuracy improves with increasing J. However, the order of the approximant, which is needed to obtain a given accuracy, increases dramatically with growing λ^4. For example, for $\lambda^4 = 0.1$ in order to reach an accuracy of 19 significant figures, $J = 100$ is required (equivalently, 200 coefficients of the PT (6.20) should be known exactly), while for $\lambda^2 = 1$, the order $J = 150$ (or 300 PT coefficients) only yields 13 significant figures, and for $\lambda^2 = 10$ even with $J = 150$, the 7 significant

Table 6.3. Sextic AHO, the Ground State Energy: Padé-Borel Summation results for different λ^4 *versus* the order of the approximant $P[J/J] \equiv P_J^J$. E_{exact} are found with the LMM method, see App.F, all printed digits are correct.

	λ^4		
$P[J,J]$	0.1	1	10
$P[5,5]$	1.109 074 046 110 515 466	1.430 663 253 521 438 290	2.064 980 466 758 966 125
$P[10,10]$	1.109 087 031 132 705 922	1.435 296 258 945 946 349	2.167 885 828 824 732 683
$P[15,15]$	1.109 087 073 531 393 700	1.435 524 231 417 163 185	2.184 763 158 619 943 232
$P[25,25]$	1.109 087 078 436 674 144	1.435 617 619 451 880 454	2.200 348 346 073 633 744
$P[50,50]$	1.109 087 078 465 583 585	1.435 624 604 670 547 728	2.205 502 643 612 593 082
$P[100,100]$	1.109 087 078 465 583 703	1.435 624 619 004 637 852	2.205 721 558 672 596 907
$P[150,150]$	1.109 087 078 465 583 703	1.435 624 619 003 434 995	2.205 723 367 594 096 985
Exact	1.109 087 078 465 583 703	1.435 624 619 003 392 316	2.205 723 269 595 632 351

[8]It is known [Baker and Graves-Morris (2010)] that for convergent series the diagonal Pády approximants usually lead to fast convergence.

figures *only* can be found.[9] It does not seem feasible to get any reasonable accuracy for $\lambda^4 = 100$ with Padé-Borel summation.

6.2.3 *Excited States*

After introducing the new variable v, new energy ε and new function $\mathcal{Y}$, see (6.13), and also the effective coupling λ for equation (6.12), we arrive at the gRB equation for the sextic AHO,

$$\mathcal{Y}_v' - \mathcal{Y}^2 - \frac{f_v'' - 2\mathcal{Y} f_v'}{f} = \varepsilon - v^2 - \lambda^4 v^6, \tag{6.31}$$

where the subindex v indicates that the derivative is taken with respect to v, which will be dropped from now on. As a consequence of the $\mathbf{Z}_2$ invariance $(v \to -v)$ of the Hamiltonian $\mathcal{H}(v)$, see (6.3) with $(x \to v)$, any eigenstate is characterized by a definite parity $(-)^p, p = 0, 1$. Hence, for any eigenfunction

$$\mathcal{Y}(-v) = -\mathcal{Y}(v), f(-v) = (-)^p f(v),$$

where $p = 0$ corresponds to the positive parity states and $p = 1$ is for the negative parity states. This implies the existence of the representation

$$\mathcal{Y}(v) = v\hat{\mathcal{Y}}(v^2), f(v) = v^p \hat{f}(v^2). \tag{6.32}$$

After plugging (6.32) into the gRB equation (6.31) and introducing a new variable $v_2 = v^2$, we arrive at

$$\begin{aligned} &2v_2 \partial_{v_2} \hat{\mathcal{Y}} + (1 + 2p)\hat{\mathcal{Y}} - v_2 \hat{\mathcal{Y}}^2 \\ &\qquad - 2\frac{2v_2 \partial_{v_2}^2 \hat{f} + (1 + 2p)\partial_{v_2} \hat{f} - 2v_2 \hat{\mathcal{Y}} \partial_{v_2} \hat{f}}{\hat{f}} \\ &= \varepsilon - v_2 - \lambda^4 v_2^3, \end{aligned} \tag{6.33}$$

where

$$\partial_{v_2} \equiv \frac{d}{dv_2},$$

cf. (6.19) and (6.31). Namely this form of the gRB equation will be used to study the PT for the excited states of the sextic anharmonic oscillator.

[9] It must be emphasized that in order to avoid error accumulation the PT coefficients should be known with extremely high accuracy, better, with absolute accuracy, hence, in the form of rational numbers.

It must be emphasized that in this formalism the principal quantum number K of the excited state is equal to the number of symmetric nodes $(2N)$, where $N = 0, 1, 2, \ldots$, plus the parity $p = 0, 1$ of the state,

$$K = 2N + p.$$

This implies that if the prefactor $\hat{f}(v_2)$ is a polynomial with positive roots, it should be of degree N. For $N = 0$, the prefactor $\hat{f}(v_2) = 1$.

Now we are ready to develop the PT for the $(2N + p)$-th excited state as a Taylor expansion in powers of λ^2,

$$\varepsilon^{(K)}(\lambda) = \sum_{n=0}^{\infty} \varepsilon_n^{(K)} \lambda^{4n}, \tag{6.34}$$

and

$$\hat{\mathcal{Y}}^{(K)}(v_2) = \sum_{n=0}^{\infty} \hat{\mathcal{Y}}_n^{(K)}(v_2) \lambda^{4n}, \tag{6.35}$$

and

$$\hat{f}^{(K)}(v_2) = \sum_{n=0}^{\infty} \hat{f}_n^{(K)}(v_2) \lambda^{4n}. \tag{6.36}$$

cf. (2.35), (2.36), (2.37), respectively. We substitute the Taylor expansions (6.34), (6.35), (6.36) into equation (6.33), then we collect terms of order of λ^{4n} and then we set the sum of the coefficients in front of these terms equal to zero. It must be emphasized that for the ground states (the lowest energy states) of positive $(K = 0)$ and negative $(K = 1)$ parities, the prefactor is constant, $\hat{f}^{(0)} = \hat{f}^{(1)} = 1$; it is not deformed under the perturbation: all corrections can be set equal to zero, $\hat{f}_i^{(0)} = \hat{f}_i^{(1)} = 0$ for $i > 0$.

After quite cumbersome algebraic manipulations we arrive at the following equation for the nth correction,

$$\begin{aligned} &2v_2 \partial_{v_2} \hat{\mathcal{Y}}_n + (1 + 2p)\hat{\mathcal{Y}}_n - 2v_2 \hat{\mathcal{Y}}_n \\ &\qquad - 2\frac{2v_2 \partial_{v_2}^2 \hat{f}_n + (1 + 2p - 2v_2)\partial_{v_2} \hat{f}_n + 2N\hat{f}_n - 2(v_2 \partial_{v_2} \hat{f}_0)\hat{\mathcal{Y}}_n}{\hat{f}_0} \\ &= \left(\varepsilon_n - \tilde{Q}_n\right), \end{aligned} \tag{6.37}$$

where

$$\varepsilon_0^{(K)} = (2K + 1), \hat{\mathcal{Y}}_0 = 1, \hat{f}_0 = L_N^{(-1/2+p)}(v_2),$$

see (5.41), correspond to the $(2N+p)$-th excited state of the harmonic oscillator, which is the solution of the unperturbed equation (5.18) at $\lambda^2 = 0$. Here the $\tilde{Q}_n$'s in the r.h.s. of (6.37):

$$\tilde{Q}_1 = v_2^3, \quad \tilde{Q}_n = -v_2 \sum_{k=1}^{n-1} \hat{\mathcal{Y}}_k \hat{\mathcal{Y}}_{n-k} + \frac{v_2^2 \hat{f}_{n-1}}{\hat{f}_0}$$
$$+ \frac{1}{\hat{f}_0} \sum_{k=1}^{n-1} \left((1+2p)\hat{\mathcal{Y}}_k \hat{f}_{n-k} - \varepsilon_k \hat{f}_{n-k} \right)$$
$$+ \frac{v_2}{\hat{f}_0} \sum_{k=1}^{n-1} \left(2\hat{f}_{n-k} \partial_{v_2} \hat{\mathcal{Y}}_k + 4\hat{\mathcal{Y}}_k \partial_{v_2} \hat{f}_{n-k} - \hat{f}_{n-k} \sum_{j=0}^{k} \hat{\mathcal{Y}}_j \hat{\mathcal{Y}}_{k-j} \right),$$
$$n = 2, 3, \ldots, \tag{6.38}$$

cf. (5.48) play the role of the *(effective) perturbation potentials.* For $K = 0, 1$ the ratio in the l.h.s. of (6.37) disappears while the r.h.s. is dramatically simplified, $\tilde{Q}_n = Q_n$, see (6.24).

By making concrete calculations for $n = 1, 2$ we arrive at the following, very compact expressions,

$$\varepsilon_1^{(K)} = \frac{5}{8}(2K+1)\left(2K^2 + 2K + 3\right),$$
$$\varepsilon_2^{(K)} = -\frac{1}{128}(2K+1)\left(786K^4 + 1572K^3 + 5324K^2 + 4538K + 3495\right), \tag{6.39}$$

for the energy corrections, which at $K = 0$ coincide with ones given in Table 6.1 and

$$\hat{\mathcal{Y}}_1^{(K)}(v_2) = \frac{1}{2}\left(v_2^2 + \frac{1}{2}(2K+5)v_2 + \frac{1}{4}(6K^2 + 14K + 15) \right), \tag{6.40}$$

$$\hat{\mathcal{Y}}_2^{(K)}(v_2) = -\frac{1}{8}\left(v_2^4 + \frac{1}{2}(6K+19)v_2^3 + \frac{1}{2}(15K^2 + 63K + 94)v_2^2 \right.$$
$$+ \frac{1}{4}(50K^3 + 267K^2 + 587K + 545)v_2$$
$$\left. + \frac{1}{16}(390K^4 + 1644K^3 + 4416K^2 + 5770K + 3495) \right), \tag{6.41}$$

for the $\hat{\mathcal{Y}}$-corrections, which at $K=0$ coincide with ones given in Table 6.2, and for the corrections to prefactor $\hat{f}(v_2)$:

$$
\begin{aligned}
\hat{f}_0(v_2) &= \mathcal{L}_N^{(-\frac{1}{2}+p)}(v_2),\\
\hat{f}_1(v_2) &= -\frac{1}{4}\left(15K^2+K+15\right)\mathcal{L}_{N-1}^{(-\frac{1}{2}+p)}(v_2)\\
&\quad+\frac{1}{2}(6K-5)\mathcal{L}_{N-2}^{(-\frac{1}{2}+p)}(v_2)-\frac{4}{3}\mathcal{L}_{N-3}^{(-\frac{1}{2}+p)}(v_2),\\
\hat{f}_2(v_2) &= \frac{1}{64}\left(1834K^4+460K^3+8475K^2+975K+3735\right)\mathcal{L}_{N-1}^{(-\frac{1}{2}+p)}(v_2)\\
&\quad+\frac{1}{32}\left(225K^4-1358K^3+2561K^2-4682K+3030\right)\mathcal{L}_{N-2}^{(-\frac{1}{2}+p)}(v_2)\\
&\quad-\frac{1}{8}\left(90K^3-447K^2+839K-973\right)\mathcal{L}_{N-3}^{(-\frac{1}{2}+p)}(v_2)\\
&\quad+\frac{1}{24}\left(228K^2-880K+1377\right)\mathcal{L}_{N-4}^{(-\frac{1}{2}+p)}(v_2)\\
&\quad-\frac{2}{15}(30K-91)\mathcal{L}_{N-5}(v_2)^{(-\frac{1}{2}+p)}(v_2)\\
&\quad+\frac{8}{9}\mathcal{L}_{N-6}^{(-\frac{1}{2}+p)}(v_2),
\end{aligned}
\tag{6.42}
$$

where the normalization of the Laguerre polynomials is conveniently changed to

$$
\mathcal{L}_{N-J}^{(-\frac{1}{2}+p)}(v_2)=(-1)^N\frac{2^{K-3J}(N-J)!K!}{(K-2J)!}L_{N-J}^{(-\frac{1}{2}+p)}(v_2),
$$

cf. (5.41), to simplify formulas.

An analysis of equation (6.37) with the effective potentials (6.38) reveals the general structure of the corrections:

A. The nth energy correction ε_n is a polynomial of degree $(2n+1)$ in the principal quantum number K,

$$
\varepsilon_n=(2K+1)\,Pol_{2n}(K),
$$

with rational coefficients, note that at $K=-1/2$ all $\varepsilon_n=0$,

B. The nth correction to $\hat{\mathcal{Y}}(v_2)$ is a polynomial of degree $(2n)$ in v_2,

$$
\begin{aligned}
\hat{\mathcal{Y}}_n(v_2)=Pol_{2n}(v_2;K)&=c_0v_2^{2n}+c_1(K)v_2^{2n-1}\\
&\quad+\cdots+c_i(K)v_2^i+\cdots+c_n(K),
\end{aligned}
$$

with coefficients c_i, which are polynomials in K of degree i

$$c_i(K) = Pol_i(K),$$

also with rational coefficients,

C. The nth correction to the prefactor $\hat{f}(v_2)$ is a polynomial of degree $(N-1)$ in v_2 for any $n > 0$, which can be represented as a linear superposition of the $(3n)$ associated Laguerre polynomials of index $(-\frac{1}{2}+p)$,

$$\hat{f}_n(v_2) = \sum_{J=1}^{3n} d_J(K) \mathcal{L}_{N-J}^{(-\frac{1}{2}+p)}(v_2),$$

with coefficients d_J, which are polynomials in K of degree $(3n - J)$

$$d_J(K) = Pol_{3n-J}(K),$$

also with rational coefficients.

6.3 Generalized Bloch Equation for Sextic Anharmonic Oscillator

Following the presentation in Chapters 3,4 let us introduce a new variable u, function $\mathcal{Z}$ and energy ε,

$$u = gx = \lambda v, \quad y = \frac{(m)^{1/2}}{g} \mathcal{Z}(u), \quad E = \frac{\hbar}{m^{1/2}} \varepsilon, \tag{6.43}$$

for equation (6.11), cf. (1.18), keeping the same effective coupling constant (2.2), (5.12)

$$\lambda = \left(\frac{\hbar^2}{m}\right)^{1/4} g,$$

while assuming $g \neq 0$. We arrive at *the Generalized Bloch equation* (GB) for the sextic AHO

$$\lambda^4 \partial_u \mathcal{Z}(u) - \mathcal{Z}^2(u) = \lambda^4 \varepsilon(\lambda^4) - u^2 - u^6, \quad \partial_u \equiv \frac{d}{du}, \quad u \in (-\infty, \infty), \tag{6.44}$$

cf. (2.3), (5.13) and (3.3). For simplicity we consider $\lambda \neq 0$, hence, λ plays the role of a formal parameter. This equation describes dynamics in a *classical* ($\hbar$-independent) coordinate $u = gx$, see e.g. [Brezin *et al.* (1977b)].

Evidently, equation (6.44) is $\mathbf{Z}_2$-invariant: $(u \to -u)$ and $\mathcal{Z}(-u) = -\mathcal{Z}(u)$. Hence, $\mathcal{Z}(u)$ can be represented as

$$\mathcal{Z}(u) = u \hat{\mathcal{Z}}(u_2), \quad u_2 = u^2.$$

By plugging this into equation (6.44), we arrive at the equation

$$2\lambda^4 u_2 \partial_{u_2} \hat{\mathcal{Z}} + \lambda^4 \hat{\mathcal{Z}} - u_2 \hat{\mathcal{Z}}^2 = \lambda^4 \varepsilon(\lambda^4) - u_2 - u_2^3, u_2 \in [0, \infty), \tag{6.45}$$

where $\partial_{u_2} \equiv \frac{d}{du_2}$, $\hat{\mathcal{Z}}(0) = \varepsilon$, this equation has no symmetry constraints.

Now we develop a PT in powers of λ^4 for equation (6.45). It is evident that the expansion of the energy ε in powers of λ^4 (6.20) remains the same as for the RB equation (6.15),

$$\varepsilon = \sum_{n=0}^{\infty} \lambda^{4n} \varepsilon_n,$$

while the expansion for $\hat{\mathcal{Z}}$ is of the form

$$\begin{aligned} \hat{\mathcal{Z}} &= \sum \lambda^{4n} \hat{\mathcal{Z}}_n(u_2), \hat{\mathcal{Z}}_0 = \sqrt{1 + u_2^2}, \\ 2\hat{\mathcal{Z}}_1 &= \frac{1}{u_2} + \frac{2u_2}{1 + u_2^2} - \frac{\varepsilon_0}{u_2\sqrt{1 + u_2^2}}, \ldots, \end{aligned} \tag{6.46}$$

cf. (2.9), (6.21) and (3.11), where in order to find $\hat{\mathcal{Z}}_1$ explicitly we have to put $\varepsilon_0 = 1$. Any finite number of corrections $\hat{\mathcal{Z}}_{2,3,4,\ldots}$ can be calculated purely algebraically, without solving a differential equation. It can be immediately recognized that $u\mathcal{Z}_0$ is, in fact, the classical momentum at zero energy and $\lambda = 1$, while $u\mathcal{Z}_1$ is related to the derivative of the logarithm of the determinant, see Chapter 3.3.

In general, the expansion (6.46) is the *true* semiclassical expansion in powers of $\hbar^2$: the correction $\hat{\mathcal{Z}}_n(u_2)$ does not depend on the Planck constant $\hbar$. By integrating $\int u\hat{\mathcal{Z}}(u^2)du$ (and putting $\hbar = 1$) we arrive at the semiclassical expansion of the phase,

$$\begin{aligned} \phi &= \frac{x^2}{4}\sqrt{1 + g^4x^4} + \frac{1}{4g^2}\log(g^2x^2 + \sqrt{1 + g^4x^4}) \\ &\quad + \frac{1}{4}\log(1 + g^4x^4) + \frac{1}{4}\log(1 + \sqrt{1 + g^4x^4}) + \cdots, \end{aligned} \tag{6.47}$$

where the first two terms in the expansion (6.46) are taken into account; here for the ground state $N = p = 0$.[10] Remarkably, the higher order corrections $\hat{\mathcal{Z}}_{2+i}$, $i = 0, 1, 2, \ldots$ in (6.46) do *not* generate logarithmic terms. For the $(2N + p)$-th excited state with quantum number $N = 0, 1, \ldots$ and

[10] In the case of the quartic AHO the first term in this expansion was calculated in [Turbiner (1984)] under the name *leading log approximation in quantum mechanics*.

parity $p = 0, 1$, the function $\mathcal{Z}$ contains $(2N+p)$ simple poles with residues (-1) at real x in addition to a non-singular part. It can be shown that for excited states the factor $(1/4)$ in front of the third logarithmic term in (6.47) is modified to include a dependence on the quantum numbers N, p in a manner,

$$\frac{1}{4} \to \left(N + \frac{p}{2} + \frac{1}{4}\right),$$

while there are no other dependencies on quantum numbers occur. Expansion (6.47) becomes

$$\phi = \frac{x^2}{4}\sqrt{1+g^4x^4} + \frac{1}{4g^2}\log(g^2x^2 + \sqrt{1+g^4x^4}) + \frac{1}{4}\log(1+g^4x^4) + \left(N + \frac{p}{2} + \frac{1}{4}\right)\log(1+\sqrt{1+g^4x^4}) + \cdots. \quad (6.48)$$

Expansion (6.48) mimics the asymptotic expansion at large x of Eqs. (6.11), (6.12) and at large $u = gx$ of Eq. (6.44),

$$\phi = \frac{g^2}{4}x^4 + \left(N + \frac{p}{2} + \frac{3}{2} + \frac{1}{4g^2}\right)\log(x^2) + O(1), \quad (6.49)$$

where we set $\hbar = 1$ and $m = 1/2$ for simplicity, cf. (5.57) for the quartic oscillator. The first two terms in (6.49) are universal: they grow as $|x| \to \infty$ and they do not depend on the energy ε. Furthermore, the first term in (6.49) does not depend on the state being studied, it emerges from the expansion of the classical action $\int u\hat{\mathcal{Z}}_0 du$. This result is well-known under the name *the leading WKB asymptotics of the wave function.*

6.4 Sextic Anharmonic Oscillator: Approximating Eigenfunctions

Without loss of generality we set $\hbar = 1$ and $m = 1/2$, hence, the variable v is equal to x: $v = x$ and the RB equation (6.15) coincides with the Riccati equation (6.11). Let us perform the procedure of *matching* the expansions (6.17) and (6.48), cf. (2.11) and (5.16), of the phase in the exponent of (1.17) into a single function by introducing, if necessary, some extra parameters.[11] Then we multiply the result by a polynomial pre-factor

[11] In the particular case of the sextic AHO, it is enough to introduce four free parameters, which we denote A, B, C, D.

which carries the information about the nodes in a *minimal* manner: the number of different roots (nodes) is equal to the degree of the polynomial. As a result we arrive almost unambiguously at the following function for the $(2N+p)$-th excited state with quantum numbers (N,p), where $N=0,1,2,\ldots,p=0,1$, for sextic AHO,

$$\Psi^{(N,p;6)}_{(approximation)}(x)$$
$$= \frac{x^p P_N^{(p)}(x^2)}{(D^2+C^2g^2x^2+g^4x^4)^{\frac{1}{4}}(D+\sqrt{D^2+C^2g^2x^2+g^4x^4})^{N+\frac{p}{2}+\frac{1}{4}}}$$
$$\times \frac{\exp\left(-\frac{A+Bx^2+C^2g^2x^4/8+g^4x^6/4}{\sqrt{D^2+C^2g^2x^2+g^4x^4}}+\frac{A}{D}\right)}{\left(g^2x^2+\sqrt{D^2+g^4x^4}\right)^{\frac{1}{4g^2}}}, \qquad (6.50)$$

cf. (5.58); for details of the procedure see Chapter 4, (4.1) and Section 5.4. Here $P_N^{(p)}$ is a polynomial of degree N in x^2 with positive roots only. Here $A=A_{N,p}(g^4), B=B_{N,p}(g^4), C=C_{N,p}(g^4), D=D_{N,p}(g^4)$ are four interpolation parameters. In the limit $g^2\to 0$, function (6.50) becomes the eigenfunction of the harmonic oscillator,

$$\Psi^{(0)} = x^p L_N^{(-\frac{1}{2}+p)}(x^2)e^{-\frac{x^2}{2}},$$

where $L_N^{(-\frac{1}{2}+p)}$ is the associated Laguerre polynomial with index $(-\frac{1}{2}+p)$ in standard normalization, see [Bateman (1953); Abramowitz and Stegun (1964)].

Formula (6.50) is the *central* formula of this Chapter and one of the key formulas of the entire book. It provides a highly-accurate approximate solution for the eigenfunctions of the quantum one-dimensional sextic symmetric AHO (6.1) in its full generality.

The simplest way to fix the parameters A,B,C,D is to consider the function (6.50) as a variational trial function, see Chapter 4 and Section 5.4 for the details for the quartic AHO. For the ground state $(0,0)$ for chosen λ the function $\Psi^{(0,0;6)}_{(approximation)}$ is used as the entry in the energy functional $\equiv$ the expectation value of the Hamiltonian and then the minimization is performed with respect to the parameters A,B,C,D. The same procedure is used for the first excited state $(0,1)$ with the function $\Psi^{(0,1;6)}_{(approximation)}$ as the entry. In order to find the variational energy for further excited state (N,p) the function $\Psi^{(N,p;6)}_{(approximation)}$ is used as the entry with the orthogonality conditions imposed towards the functions of the all lower excited states,

$$\langle\Psi^{(N,p;6)}_{(approximation)}|\Psi^{(M,p;6)}_{(approximation)}\rangle = 0, \quad M = 0, 1, 2, \ldots, (N-1),$$

thus, moving subsequently up one by one from the ground state to higher excited states. The procedure of the orthogonalization allows to fix unambiguously the coefficients of the polynomial $P_N^{(p)}(x^2)$ in (6.50) in terms of four variational parameters $A = A_{N,p}(g^4)$, $B = B_{N,p}(g^4)$, $C = C_{N,p}(g^4)$, $D = D_{N,p}(g^4)$.

In Fig. 6.1 the variational energies $E_{var}^{(N,p)}$ *vs* g^4, obtained with 10–12 significant figures, for the first six low-lying states are shown. It turns out that their obtained accuracies are smaller than the width of any line on the plot! Concrete numerical results for various g^4 are presented in six Tables in App. C.2.1.

Let us introduce a new coupling constant

$$\tilde{\lambda}(g^4) = (0.3^4 + g^4)^{1/4}, \quad \tilde{\lambda}(0) = \frac{3}{10}, \tag{6.51}$$

which mimic the behavior of the energy at the asymptotics at small and large couplings g. By using two-point Páde approximants Páde$(N + 1/N)(\tilde{\lambda})$ the function $\tilde{\lambda}$ as the argument one can easily interpolate the eigenvalues across the whole domain of variations of the coupling constant

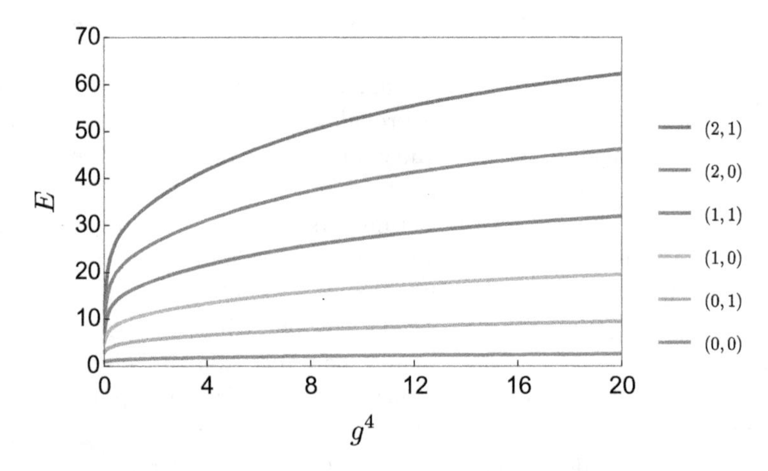

Fig. 6.1. Energies E of the first 6 low-lying states of the sextic AHO as function of g^4.

$g^4 \in [0, \infty)$. For example, the ground state energy is described by

$$E_{0,0} = \text{Páde}(6/5)(\tilde{\lambda}) = P_6/Q_5, \tag{6.52}$$

where

$$P_6 = 0.0721397 + 0.254139\tilde{\lambda} + 0.807002\tilde{\lambda}^2 + 1.25703\tilde{\lambda}^3 + 2.17277\tilde{\lambda}^4 + 1.60413\tilde{\lambda}^5 + 1.14482\tilde{\lambda}^6,$$

$$Q_5 = 0.0730737 + 0.269887\tilde{\lambda} + 0.728111\tilde{\lambda}^2 + 1.62709\tilde{\lambda}^3 + 1.4015\tilde{\lambda}^4 + \tilde{\lambda}^5,$$

with relative accuracy $\sim 10^{-5}$ (or better) for the whole domain of $g^4 \geq 0$, cf. App.D, Eq. (III.2). In App.D a fit of the energies of six low-lying states via Páde$(6/5)(\tilde{\lambda})$ in the whole domain $g^4 \geq 0$ is presented, (III.2)–(III.18), with 5-6 figures (or better). In general, by increasing N in the Páde approximant Páde$(N + 1/N)(\tilde{\lambda})$ one can get better accuracy. It must be emphasized that such high accuracies at small $g^4 > 0$ occur despite of the fact the exact energies have a type of the essential singularity at $g^4 = 0$ while the two-point Páde approximants do not.

In the framework of the non-linearization procedure [Turbiner (1984)] one can develop a rapidly-convergent perturbation theory with respect to the deviation of the original potential V from the effective one, see (5.59): $(V - V^{(N,p)})$, see also Chapter 2 for a general description. Usually, the rate of convergence is defined by the second correction E_2, see (2.50).

It can immediately be seen that the variational energy $E_{var}^{(N,p)}$ or, stated differently, the expectation value of the Hamiltonian (6.3) over the function (6.50) has a singularity at $g = 0$, independently on the values of the parameters A, B, C, D, with the branch cut along the negative $g^4 \in [0, -\infty)$. An interested reader can calculate the discontinuity

$$\delta E^{(N,p)} = \lim_{\epsilon \to 0} E_{var}^{(N,p)}(-g^4 + i\epsilon) - E_{var}^{(N,p)}(-g^4 - i\epsilon),$$

and find out that it vanishes exponentially as $|g^4| \to -0$. The behavior of this discontinuity defines the factorial growth of the PT coefficients E_n at large n, see e.g. [Bender and Wu (1969)].

It is interesting to compare the convergent expansion in powers of g^4 of the fitted variational energy, e.g. $E_{0,0}$ (6.52), and the corresponding divergent PT expansion of the exact energy. For the ground state $(0, 0)$ these expansions read

$$E_{fit}^{(0,0)} = 1 + 1.8742g^4 - 26.87g^8 + 1078.00g^{12} - 68231.8g^{16} + \cdots$$

$$E_{PT}^{(0,0)} = 1 + 1.8750g^4 - 27.30g^8 + 1210.62g^{12} - 102000g^{16} + \cdots$$

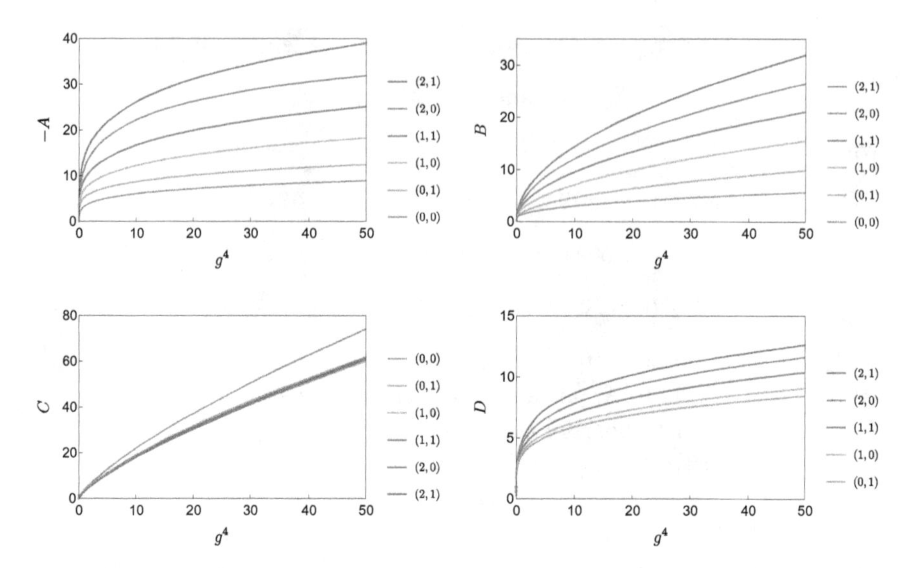

Fig. 6.2. Optimal parameters A, B, C, D for the first six low-lying states *versus* the coupling constant g^4.

see App. D.2.1.2. It is apparent that the disagreement in the coefficients in expansions grows with the order of term taken into account. For example, for g^{16} the coefficient in the fitted function differs from the exact PT coefficients in about two times. Next coefficients will differ much more. It occurs despite of the fact that the fitting function (6.52) provides sufficiently high accuracy for all positive values of the coupling constant g^4! This reflects the non-triviality of the summation procedure for divergent series. In App.D a comparison of the expansions is presented for the first six low-lying states. All of these expansions show a striking disagreement for the coefficients of the expansion in powers of g^4 of the accurately fitted variational energy and the exact perturbative coefficients. It is evident that this disagreement will be present for any excited state.

In the non-linearization procedure using multiplicative perturbation theory [Turbiner (1984)], see Chapter 2, one can analytically calculate the first corrections to (6.50) of the form:

$$\Psi = \Psi^{(N,p;6)}_{(approximation)}(1 - \phi_1 - \phi_2 \ldots).$$

It can be shown numerically that for all six studied states at any $g^4 \in [0, \infty)$, the first correction is extremely small

$$|\phi_1| \lesssim 10^{-6}, \ \forall x \in (-\infty, \infty). \tag{6.53}$$

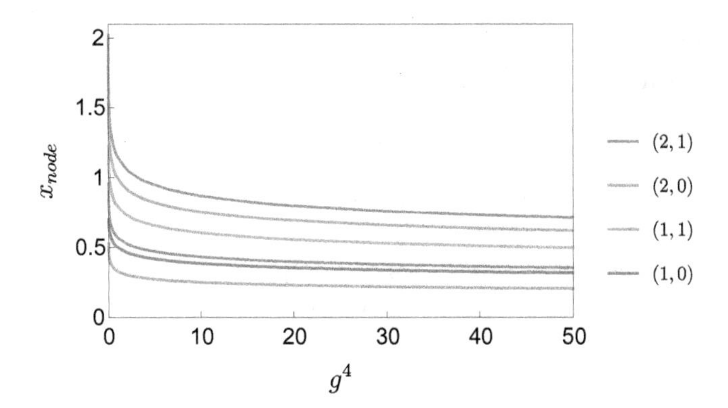

Fig. 6.3. Positive nodes $x_{node} > 0$ *vs* the coupling constant $g^4 \geq 0$ for the states $(1,0)$, $(1,1)$, $(2,0)$, and $(2,1)$. Note that for all excited states, there exist symmetric negative nodes at $x = -x_{node}$. Additionally, the negative parity states (1,1) and $(2,1)$ have a node at $x_0 = 0$.

Since for all six studied states the correction ϕ_1 can be found explicitly in closed analytic form, it is believed that the bound (6.53) can be verified analytically. In the non-linearization procedure the corrections E_2, E_3 to the energy (5.60), see (2.50), can be also calculated analytically and then verified numerically. They indicate an extremely high rate of convergence for the energy, $10^{-3} - 10^{-4}$, in agreement with the Lagrange Mesh Method calculations, see [Baye (2015)], [Turbiner and del Valle (2021b)] and App. F, with 75 mesh points for $g^4 = 0.1, 1, 10, 20, 100$, where a relative accuracy 10^{-20} in energy is easily reached.[12]

6.5 Sextic (Anharmonic) Oscillator and Bohr-Sommerfeld Quantization Condition

As was previously mentioned, for any eigenvalue of the sextic AHO (6.1) one can construct the strong coupling expansion (6.10),

$$E^{(N,p)} = g \sum_{0}^{\infty} b_k^{(N,p)} g^{-2k},$$

[12]The Lagrange Mesh Method, see Appendix F, in many occasions allows us to reach the extremely high accuracy rapidly. Out of curiosity it can be checked that by using the pre-calculated positions of 1000 mesh points $\{x_i\}$ and by choosing the global dilation parameter h in such a way that $h \times x_{1000} \sim 15$ (the optimized mesh), the energy of the states from $N = 0$ to $N = 80$ for $g^4 = 0.1, 1, 10, 20, 100$ can be found with at least 94 figures in ~660 seconds of CPU time on single core 2.6 GHz processor for each value of g^4 and each concrete state, see [Turbiner and del Valle (2021b)]. This is similar but slightly slower compared to what is achieved for the quartic AHO, see Chapter 5, footnote 8.

where the leading coefficient $b_0^{(N,p)}$ corresponds to the spectra of the sextic oscillator with potential,

$$V = g^2 x^6, \hat{V} = u^6,$$

which is a power-like potential, cf. (2.69), (3.9), (4.3). This potential realizes the ultra-strong coupling regime $g^4 \to \infty$ of the sextic AHO (6.1). We denote their eigenvalues as $E^{(6)} = \hbar\varepsilon^{(6)} \equiv b_0^{(N,p)}$.

Let us consider the (exact) Bohr-Sommerfeld quantization condition for the potential (3.9),

$$\int_{-\varepsilon^{1/6}}^{\varepsilon^{1/6}} \sqrt{\varepsilon - u^6}\, du = \pi \left(K + \frac{1}{2} + \gamma(K) \right), \tag{6.54}$$

cf. (5.66), where $K = 2N + p$ is the principal quantum number as always; sometimes, $(1/2 + \gamma)$ is called the *quantization index* or, stated differently, $\gamma(K)$ is called the *WKB correction.* If $\gamma = 0$, this condition corresponds to the standard Bohr-Sommerfeld quantization condition, see e.g. [Landau and Lifshitz (1977)]. In this case we arrive at the so-called Bohr-Sommerfeld energy spectra $\varepsilon = \varepsilon_{BS}(K)$. Explicitly,

$$E_{BS}^{(6)} = \hbar\varepsilon_{BS}^{(6)}(K) = \hbar(\hbar g^2)^{\frac{1}{2}} \left(\frac{2}{3} B\left(\tfrac{1}{2}, \tfrac{2}{3}\right) \left(K + \frac{1}{2} \right) \right)^{\frac{3}{2}}, \tag{6.55}$$

cf. (5.67), where $B(\alpha, \beta)$ is the Euler Beta function, see e.g. [Abramowitz and Stegun (1964)], [Bateman (1953)], This formula gives increasingly accurate energies for large quantum numbers K, while for small K it can differ significantly [Turbiner and del Valle (2021c)].[13]

The problem of the energy spectra of the power-like potential can be considered to be an inverse problem. If for some reason the exact spectra $\varepsilon = \varepsilon_{exact}$ is known (in one way or another) explicitly, by using (6.54) the function $\gamma(K)$ itself can be calculated explicitly. Vice versa if the function $\gamma(K)$ is known exactly, the exact spectra can be found by solving equation (6.54). In this case (6.54) is called the *exact WKB* quantization condition, see e.g.[Voros (1994)]. In general, γ is related to the boundary condition imposed at the turning point; it was claimed in [Landau and Lifshitz (1977)] that $(1/2 + \gamma)$ is of order one. For the sextic potential u^6, $\gamma(K)$ was calculated in [Turbiner and del Valle (2021c)], see Fig.3 therein. This is illustrated in Fig. 6.4.

[13] At $K = 0$ the deviation is about 20%.

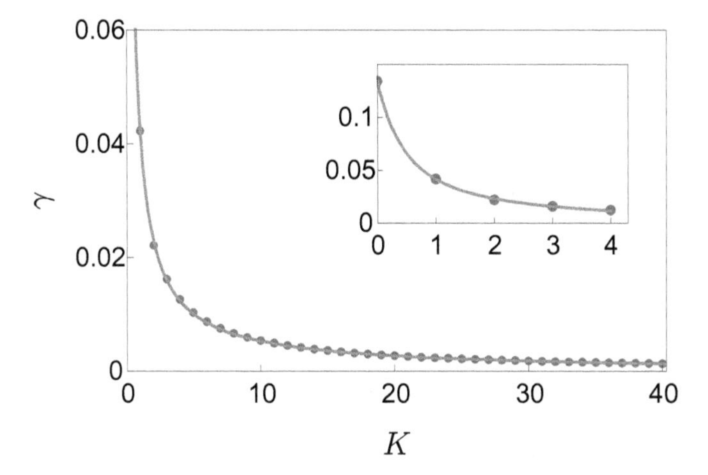

Fig. 6.4. WKB correction γ *vs* K for the sextic potential, see (6.54). Displayed points correspond to $K = 0, 1, 2, 3, \ldots, 40$. In (sub)Figure the domain $K = 0, 1, 2, 3, 4$ shown separately. Compare to Fig.5.7.

It is not clear how to calculate γ as a function of K in (6.54) theoretically, even approximately, in closed analytic form, so it appears to be an open problem. In general, it is given by an expansion in the powers of $\hbar$. However, this function can easily be fitted with a requested-in-advance accuracy via the ratio of a polynomial of degree n to the square-root of another polynomial of degrees $(2n + 2)$ as the fitting function,

$$\gamma_n^{(6)}(K) = \frac{P_n(K)}{\sqrt{Q_{2n+2}(K)}}, \quad n = 1, 2, 3, \ldots,$$

where P_n, Q_{2n+2} are polynomials, respectively, with the condition that $Q_{2n+2}(K) > 0$ for $K \geq 0$. The latter condition prevents the appearance of singularities in the fitting function for non-negative K and it is fulfilled for all concrete studied cases. In particular, the simplest approximation corresponds to $n = 1$,

$$\gamma_{n=1}^{(6)}(K) = \frac{0.0574K + 0.0520}{\sqrt{Q_4}}, \tag{6.56}$$

where

$$Q_4 = K^4 + 3.4958K^3 + 2.078K^2 - 0.0374K + 0.1497,$$

cf. (5.68). When (6.56) substituted into (6.54), it allows us to reproduce, at least, 8 significant figures in the energy for quantum numbers $0 \leq K \leq 100$, see [Turbiner and del Valle (2021c)]. However, further increase in the accuracy of the energy spectra beyond 8 significant digits requires us to take the next approximation, $n = 2$ in a similar way to the quartic oscillator. This reveals a new feature: the WKB corrections $\gamma^{(6)}_{n=2}$ for positive and negative parity states are given by (slightly) different curves: they should be fitted separately. In particular, in order to get at least 9 correct significant digits in the energy the fit should be

$$\gamma^{(6)}_{n=2}(K)_+ = \frac{0.05743009K^2 - 0.13783412K + 0.13449783}{\sqrt{Q_6^{(+)}}}, \tag{6.57}$$

where

$$Q_6^{(+)} = K^6 - 3.79952593K^5 + 5.98149752K^4 - 2.16026844K^3 - 3.75241407K^2 + 4.54042200K + 1.00000011,$$

cf. (5.69), for the positive parity states ($p = 0$): $K = 0, 2, 4, \ldots$, and

$$\gamma^{(6)}_{n=2}(K)_- = \frac{0.05743009K^2 - 0.05559154K + 0.03360701}{\sqrt{Q_6^{(-)}}}, \tag{6.58}$$

where

$$Q_6^{(-)} = K^6 - 0.94436250K^5 + 0.71874670K^4 - 1.06973081K^3 + 3.96675141K^2 - 2.97840199K + 0.00831615,$$

cf. (5.70), for the negative parity states ($p = 1$): $K = 1, 3, 5, \ldots$, cf. (5.68). Eventually, formula (6.54) with $\gamma^{(6)}_{n=2}(K)_\pm$ given by (6.57)–(6.58) provides 9-10 correct significant digits for the energy.

It must be emphasized that, in general, γ is a small, positive, bounded function: for any $K \geq 0$,

$$\gamma^{(6)}(K) < 0.1.$$

It is similar to the function $\gamma^{(4)}(K)$ of the quartic oscillator.[14]

[14] A similar property holds for any power-like potential $|u|^{2q}$ with $q \in [1, \infty)$ as well as for the logarithmic potential $\log |u|$: the WKB correction is a bounded (small) function $|\gamma^{(2q)}(K)| \leq 0.5$ changing smoothly with K, see [Turbiner and del Valle (2021c)]. It takes a minimal value for $q = 1$ (the harmonic oscillator), when $\gamma^{(2)}(K) = 0$ and a maximal value for $q = \infty$ (the square-well potential of the width equal to 2 with the infinite wells), when $\gamma^{(\infty)}(K) = 1/2$.

Equality (6.54) has two outstanding properties: (i) the Bohr-Sommerfeld integral in the l.h.s. can be evaluated analytically and (ii) the resulting equation can be solved explicitly with respect to ε. Finally, the energy spectra is given by

$$E^{(m=6)}_{exact} = \hbar(\hbar g^2)^{\frac{1}{2}} \left(\frac{2}{3} B\left(\tfrac{1}{2}, \tfrac{2}{3}\right) \left(K + \frac{1}{2} + \gamma^{(6)}\right)\right)^{\frac{3}{2}}, \qquad (6.59)$$

where $B(\alpha, \beta)$ is the Euler Beta function, see e.g. [Abramowitz and Stegun (1964); Bateman (1953)]. $\gamma^{(6)}$ can be taken from the interpolations (6.56) or (6.57)–(6.58), see Fig. 6.4; former one provides the accuracy in energy of 8-9 significant digits, latter ones provide the accuracies in energy of 9-10 significant digits. Note that in the second case of 9-10 significant digits in the energy spectra the difference in γ's for positive and negative parity states occurs. However, it can be checked that for accuracies up to 8 significant digits in the energy, these γ's (6.57)–(6.58) coincide with each other in 8 significant digits, they also coincide with (6.56).

As a result we arrive at the following approximate equality

$$\begin{aligned} E^{(6)}_{n=2} &\equiv \hbar(\hbar g^2)^{\frac{1}{2}} \left(\frac{2}{3} B\left(\tfrac{1}{2}, \tfrac{2}{3}\right) \left(K + \frac{1}{2} + \gamma^{(6)}\right)\right)^{\frac{3}{2}} \\ &\approx \langle \mathcal{H} \rangle \big|_{\Psi^{(N,p;6)}_{(approximation)}}, \end{aligned} \qquad (6.60)$$

which holds for 9-10 significant digits for any $g^2 \in [0, \infty)$. Here $\gamma^{(6)}_{n=2,\pm}$ are given by (6.57)–(6.58). The expectation value of the Hamiltonian (6.3) for $g^4 \to \infty$ (in the r.h.s. of (6.60)) taken over the function

$$\begin{aligned} \Psi^{(N,p;6)}_{(approximation)}(x) = & \frac{x^p P^{(p)}_N(x^2)}{(D^2 + C^2 x^2 + x^4)^{\frac{1}{4}}} \\ & \times \frac{1}{(D + \sqrt{D^2 + C^2 x^2 + x^4})^{N + \frac{p}{2} + \frac{1}{4}}} \\ & \times \exp\left(-\frac{A + Bx^2 + C^2 x^4/8 + x^6/4}{\sqrt{D^2 + x^4}} + \frac{A}{D}\right), \end{aligned} \qquad (6.61)$$

cf. (6.50) at $g = 1$, with four free parameters $A^{(N,p)}, B^{(N,p)}, C^{(N,p)}, D^{(N,p)}$, respectively, is equal to the exact energy within 9-10 significant figures.

Here $P_N^{(p)}(x^2)$ in (6.61) (and also in (6.50)) are some polynomials of degree N in x^2 with positive roots, which are found by imposing the orthogonality condition to the states (n,p) with quantum number n less than N, see above. We already showed that by taking (6.61) as a trial function, the variational energies are obtained with an unprecedented relative accuracy $\sim 10^{-12}$ (or better)[15] and the relative deviation of the trial function (6.61) from the exact one is $\sim 10^{-6}$ (or less) in any point in x-space. It is evident that the expectation values of the Hamiltonian (6.3) over $\Psi^{(N,p;6)}_{(approximation)}$ with variationally optimized parameters are equal to the modified Bohr-Sommerfeld energies $E^{(6)}_{exact}$ with appropriately accurate $\gamma^{(6)}$, with relative accuracy $\sim 10^{-10}$. Evidently, the WKB correction $\gamma^{(6)}$ can be expressed via the variational parameters in (6.61).

Hence, formula (6.60) presents the approximate energy spectra of the sextic oscillator in two different forms: as a function of $\gamma^{(6)}$ obtained through fit (6.57)–(6.58) or as the expectation value of the Hamiltonian over the highly-accurate approximate eigenfunction (6.61). A similar formula for the sextic AHO at finite g^4 is unknown.

6.6 Sextic AHO: Concluding Remarks

6.6.1 *Generalized Sextic Symmetric AHO*

The sextic AHO (6.1) is a particular case of the more general sextic symmetric oscillator with potential

$$V_6(x) = x^2 + 2ag^2x^4 + g^4x^6 = x^2(1 + 2ag^2x^2 + g^4x^4), \tag{6.62}$$

where a is parameter. At $a = 0$ the sextic AHO (6.1) is reproduced, if $a \to \infty$ we return to the quartic AHO (5.1). This oscillator has several widely used names: Generalized Sextic Symmetric AHO, Double AHO, Two-Parameter Sextic AHO. The potential always has a minimum at $x = 0$. However, depending on the parameter a the number of minima can vary from a single well to five-wells. The potential (6.62) is symmetric,

[15]In particular, for the sextic oscillator $V(u) = u^6$, see (3.9) by choosing $A_{0,0} = -3.2816$, $B_{0,0} = 0.5831$, $C_{0,0} = 4.0195$, $D_{0,0} = 3.0146$ the variational ground state energy is equal to $E_{var} = 1.144802453803$, while for the accurate energy obtained in LMM, see App.F, $E_{LMM} = 1.144802453797$, where all digits are exact: their absolute deviation is $\sim 6 \cdot 10^{-12}$.

$V_6(-x) = V_6(x)$, hence, the notion of parity can be introduced. The spectrum consists of positive and negative parity states. In the weak-coupling regime $g \to 0$ the algebraic perturbation theory in powers of g^2 can be developed. For a general coupling constant g and $-1 \le a < 1$ the expression (6.50) provides an accurate description of the spectra. At $a = -1$ the famous triple-well symmetric potential appears,

$$V_6(x) = x^2(1 - gx)^2(1 + gx)^2, \tag{6.63}$$

with three degenerate minima situated at $x = 0, \pm 1/g$, which correspond to three degenerate harmonic wells. Exponentially small terms at $g \to 0$ appear, and the Taylor expansion in powers of g becomes trans-series.

6.6.2 *Radial Quartic AHO Oscillator*

The formalism developed in this Chapter for the one-dimensional sextic AHO can be extended to the case of the $O(d)$ spherically-symmetric d-dimensional sextic anharmonic oscillator with potential

$$V = r^2 + g^4 r^6,$$

cf. (6.1), where the radial variable r is defined as $r^2 = \sum_{i=1}^{d} x_i^2$ and g^4 is the coupling constant. By separating out the angular variables in the d-dimensional Schrödinger equation we arrive at the familiar radial Schrödinger equation, which can be converted to a radial Riccati equation. In turn, by introducing the radial analogue ($x \to r$) of the quantum (6.13) and the classical (6.43) coordinates the radial Riccati equation can be transformed either to the (radial) Riccati-Bloch type or to the (radial) Generalized Bloch type equations, both of which are similar to equations (2.3), (5.13) and (3.3), (5.52), respectively. It must be emphasized that the same effective coupling constant λ (2.2) occurs in *all* these Riccati-Bloch and Generalized Bloch equations both radial and one-dimensional. Based on these radial Riccati-Bloch and Generalized Bloch equations one can construct the PT and semiclassical expansions, respectively, in the same way as was done in Part I. In fact, these expansions look similar to (6.21) and (6.46) after a replacement $x \to r$ (and replacement $N \to N_r, p \to \ell, 1 \to d$ in the degree of the second prefactor in the Approximant, see below). By matching the asymptotic expansion of the solution at small r of the radial RB equation with the semiclassical expansion of the radial GB equation we arrive at

$$\Psi^{(N_r,\ell;6)}_{(approximation)}(r)$$
$$= \frac{r^{\ell} P_{N_r}(r^2)}{(D^2+C^2g^2r^2+g^4r^4)^{\frac{1}{4}}(D+\sqrt{D^2+C^2g^2r^2+g^4r^4})^{N_r+\frac{\ell}{2}+\frac{d}{4}}}$$
$$\times \frac{\exp\left(-\frac{A+Br^2+C^2g^2r^4/8+g^4r^6/4}{\sqrt{D^2+C^2g^2r^2+g^4r^4}}+\frac{A}{D}\right)}{\left(g^2r^2+\sqrt{D^2+g^4r^4}\right)^{\frac{1}{4g^2}}}, \qquad (6.64)$$

cf. (5.83), which we call the *Approximant* in Part II, Chapter 5, cf. (6.50). Here we set $\hbar = 1$ and $P_{N_r,\ell}$ is a polynomial of degree N_r in r^2 with positive roots; (N_r, ℓ) are radial and angular quantum numbers, $N_r = 0, 1, 2, \ldots$ and $\ell = 0, 1, 2, \ldots$. The parameters $A = A_{N_r,\ell}(g^2), B = B_{N_r,\ell}(g^2), C = C_{N_r,\ell}(g^2), D = D_{N_r,\ell}(g^2)$ are four parameters of interpolation, they can be found variationally; in particular, for the ground state $(0, 0)$ this is done by minimizing the expectation value of the radial Hamiltonian over the function (6.64). A detailed description of the sextic radial AHO will be presented in Part II, Chapter 5 (including a generalization for the excited states).

6.6.3 *Conclusions*

As a results of matching for the logarithm of the ground state wavefunction of small distances expansion and the *true* semiclassical expansion at large distances, the approximate ground state eigenfunction in the form of a product of four factors: three pre-factors and one exponential, is built for the sextic AHO with an arbitrary coupling constant. The extension to the excited states is performed by a simple modification of the degree of one of the pre-factors and further multiplication by a polynomial, which carries the information about the nodes. Finally, it brings us to formula (6.50), which should be applicable for an arbitrary coupling constant. Formula (6.50) leads to unprecedented accuracies of 10-12 significant figures for the eigenvalues for any strength of the sextic anharmonicity and of 6 significant figures for the eigenfunctions for any point of the coordinate space similar to the case of the quartic AHO, cf. (5.58). The formula (6.50) manifests the approximate (highly-accurate) solution of the problem of the sextic AHO.

In the strong coupling regime $g \to \infty$, where the sextic AHO becomes the sextic oscillator, the energy spectra can be written in closed analytic form (6.59) in terms of a single function of the principal quantum number, which is bounded and small, and which can be found for any desired accuracy.

Formula (5.58) admits a straightforward generalization (6.64) for the radial sextic AHO, see [Del Valle and Turbiner (2019, 2020)], which will be described in Part II.

Part II

The Radial Anharmonic Oscillator

Chapter 7

Spherical Symmetrical Potentials: Generalities

7.1 Radial Schrödinger Equation

The Schrödinger equation for a spherical-symmetric potential $V(r) = V(|\mathbf{r}|)$ in a d-dimensional space has the form

$$\left[-\frac{\hbar^2}{2m}\nabla_d^2 + V(r)\right]\psi = E\psi, \tag{7.1}$$

where $d = 1, 2, \ldots$, $r = \sqrt{\sum_{k=1}^{d} x_k^2}$ is hyperradius and $\nabla_d^2 = \sum_{k=1}^{d} \partial_{x_k}^2$ is the d-dimensional Laplacian. This equation describes d-dimensional quantum particle of mass m. Following the spherical symmetry of the problem, we introduce in (7.1) the d-dimensional hyperspherical coordinates $\{r, \Omega\}$, where r is the hyperradius and Ω is solid angle parametrized by $(d-1)$ Euler angles, see e.g. [Efthimiou and Frye (2014)]. In these coordinates the Laplacian ∇_d^2 takes the form

$$\nabla_d^2 = \partial_r^2 + \frac{d-1}{r}\partial_r + \frac{\Delta_{S^{d-1}}}{r^2}, \quad \partial_r \equiv \frac{\partial}{\partial r}, \tag{7.2}$$

where $\Delta_{S^{d-1}}$ is the Laplacian on the S^{d-1} sphere. Indeed, $\Delta_{S^{d-1}} = -\hat{\mathbf{L}}^2$, where $\hat{\mathbf{L}}$ is the d-dimensional angular momentum operator. Two remarks are in order:

- for any spherically symmetric potential the d-dimensional angular momentum $\hat{\mathbf{L}}$ is conserved,
- in hyperspherical coordinates we can separate the hyperradial coordinate r from the angular coordinates $\{\Omega\}$.

Hence, for integer $d > 1$ the wave function ψ can be labeled by radial quantum number n_r, angular quantum momentum ℓ and $(d-2)$ magnetic quantum numbers $\{m_\ell\}$, it is represented as the product of two functions

$$\psi_{n_r,\ell,\{m_\ell\}}(r,\Omega) = \Psi_{n_r,\ell}(r)\,\xi_{\ell,\{m_\ell\}}(\Omega), \tag{7.3}$$

where $n_r = 0,1,2,\ldots$, $\ell = 0,1,2,\ldots$. For given ℓ the set of $(d-2)$ magnetic quantum numbers $\{m_\ell\}$, where each of them can take value from $-\ell$ to $+\ell$, in total, contains $\mathcal{N}(d,\ell)$ different states,

$$\mathcal{N}(d,\ell) = \frac{(2\ell+d-2)\,(\ell+d-3)!}{\ell!\,(d-2)!}, \tag{7.4}$$

and $\mathcal{N}(d,0) = 1$. For $d > 1$ the angular part of the wave function $\xi_{\ell,\{m_\ell\}}(\Omega)$ corresponds to a d-dimensional spherical harmonics, see e.g. [Müller (1966)], which satisfies the eigenvalue equation

$$-\Delta_{S^{d-1}}\,\xi_{\ell,\{m_\ell\}} = \hat{\mathbf{L}}^2\,\xi_{\ell,\{m_\ell\}} = \ell\,(\ell+d-2)\,\xi_{\ell,\{m_\ell\}}. \tag{7.5}$$

For given ℓ this equation is satisfied by $\mathcal{N}(d,\ell)$ different orthogonal spherical harmonics with the same eigenvalue $\ell\,(\ell+d-2)$. Its degeneracy (multiplicity) is equal to $\mathcal{N}(d,\ell)$ (7.4). Note that for two-dimensional case $d=2$ the multiplicity $\mathcal{N}(2,\ell) = 2$, while for three-dimensional case $d=3$ the degeneracy is given by the familiar formula

$$\mathcal{N}(3,\ell) = 2\ell+1,$$

see e.g. [Landau and Lifshitz (1977)]. Since the theory of d-dimensional spherical harmonics is well established, see e.g. [Efthimiou and Frye (2014)], we should focus solely on determining the radial part of the wave-function $\Psi_{n_r,\ell}(r)$, see (7.3).

After substituting (7.2), (7.3) and (7.5) to (7.1) we eventually arrive at the radial Schrödinger equation, which determines $\Psi_{n_r,\ell}(r)$,

$$\left[-\frac{\hbar^2}{2m}\left(\partial_r^2 + \frac{d-1}{r}\,\partial_r - \frac{\ell\,(\ell+d-2)}{r^2}\right) + V(r)\right]\,\Psi_{n_r,\ell}(r)$$
$$= E_{n_r,\ell}\,\Psi_{n_r,\ell}(r), \quad \partial_r \equiv \frac{d}{dr}. \tag{7.6}$$

For any radial potential, the energy $E_{n_r,\ell}$ of excited state is always degenerate with respect to the quantum numbers $\{m_\ell\}$. Specifically, for given n_r, ℓ and integer $d > 1$ we have $\mathcal{N}(d,\ell)$ (7.4) different wave functions with the same energy. For this reason since now on we omit the label $\{m_\ell\}$ in the notation of the energy and the wave function keeping only (n_r,ℓ) whenever necessary. Hence, we mark the eigenstate by radial and angular quantum numbers (n_r, ℓ). The operator in the r.h.s. of (7.6) is Hermitian on the half-line $r \in [0, \infty)$ with measure r^{d-1}.

The ground state is non-degenerate and its eigenfunction is nodeless, it has the zero angular momentum $\ell = 0$, thus, this corresponds to the S-state, and zero radial quantum number $n_r = 0$. The angular part corresponds to the zero harmonics (which is a constant for any d) with zero eigenvalue in (7.5). From (7.6) it can be seen that the ground wave function (as well as all S state eigenfunctions) depends only on the radial coordinate r that now on we denote by $\Psi_{(0,0)}(r)$ and then omit labels. The ground state will be the main object to study in the Part II. This corresponds to the lowest energy eigenfunction of the radial Schrödinger operator

$$\hat{h}_r = -\frac{\hbar^2}{2m}\left(\partial_r^2 + \frac{d-1}{r}\,\partial_r\right) + V(r), \tag{7.7}$$

for which the eigenvalue equation together with boundary conditions read

$$\hat{h}_r\,\Psi(r) = E\Psi(r), \qquad \int_0^\infty \Psi^2\, r^{d-1} dr < \infty, \quad \Psi(0) = 1,\ \Psi(\infty) = 0. \tag{7.8}$$

Here E denotes the ground state energy, assuming that $\hbar, m$ are dimensionless, it has dimension $[cm]^{-2}$ as well as the potential. From now on d can be considered as a continuous parameter. Interestingly, some non-trivial analytical properties of the ground state energy as a function of d appear at both non-physical dimensions $d \leq 0$ and in the physical ones, $d > 0$, see e.g. [Dolgov and Popov (1979)] and [Doren and Herschbach (1986)]. In general, the S-states are marked by $(n_r, 0)$, they are described by the eigenfunctions of the operator (7.7) and characterized by the n_r simple positive zeroes.

For arbitrary quantum numbers (n_r, ℓ) the differential part of the radial Schrödinger operator remains unchanged while the potential is modified by adding the centrifugal potential

$$V(r)\ \to\ V(r)\ +\ \frac{\hbar^2}{2m}\,\frac{\ell\,(\ell+d-2)}{r^2},$$

see (7.6). Making the gauge rotation of the operator in the l.h.s. of (7.6), the centrifugal potential is gauged away,

$$r^{-\ell}\left(\partial_r^2\ + \frac{d-1}{r}\,\partial_r - \frac{\ell\,(\ell+d-2)}{r^2}\right) r^{\ell} = \left(\partial_r^2\ + \frac{2\ell+d-1}{r}\,\partial_r\right).$$

As a result we arrive at the radial Schrödinger equation without the centrifugal potential,

$$\left[-\frac{\hbar^2}{2m}\left(\partial_r^2 + \frac{2\ell+d-1}{r}\,\partial_r\right) + V(r)\right]\,\Xi_{n_r,\ell}(r)$$
$$= E_{n_r,\ell}\,\Xi_{n_r,\ell}(r), \tag{7.9}$$

cf. (7.6), where

$$\Psi_{n_r,\ell}(r) = r^{\ell}\,\Xi_{n_r,\ell}(r)\ .$$

By taking the ground state wave function Ψ in the exponential representation (1.17),

$$\Psi = e^{-\frac{1}{\hbar}\,\Phi(r)},$$

and substituting it into the eigenvalue problem (7.6) with radial operator (7.7) we arrive at the radial Riccati equation for the ground state, see e.g. [Del Valle and Turbiner (2019)],

$$\hbar\,\partial_r y(r) - y(r)\,\left(y(r) - \frac{\hbar(d-1)}{r}\right) = 2m\,\Big[E - V(r)\Big], \tag{7.10}$$

where function $y(r)$ is the logarithmic derivative of $\Psi(r)$ or, equivalently, the derivative of the phase $\Phi(r)$,

$$y(r) = -\hbar\,\partial_r\,(\log\Psi(r)) = -\hbar\,\frac{\partial_r\Psi(r)}{\Psi(r)} = \partial_r\Phi(r). \tag{7.11}$$

We focus on the case when the potential $V(r)$ has the form of finite-degree polynomial in r,

$$V(r; a_2, a_3, \ldots, a_m) = \frac{1}{g^2}\sum_{k=2}^{p} a_k\,g^k\,r^k \equiv \frac{1}{g^2}\hat{V}(gr), \tag{7.12}$$

see below, which corresponds to the radial AnHarmonic Oscillator (rAHO). Hence, the final form of the radial Riccati equation for S-states reads [Del Valle and Turbiner (2019)]

$$\hbar\,\partial_r y(r) - y(r)\,\left(y(r) - \frac{\hbar(d-1)}{r}\right) = 2m\left[E - \frac{1}{g^2}\hat{V}(gr)\right], \tag{7.13}$$

c.f. (1.18). It should be emphasized that the domain for the radial Riccati equation is the half-line, $r \in [0,\infty)$, while the one-dimensional Riccati equation is defined on the entire line $(-\infty,\infty)$. Hence, the equation (7.13) at $d = 1$ corresponds to the one-dimensional Riccati equation with potential which has the infinite wall at $r = 0$ if $\Psi(0) = 0$.

By taking a generalized exponential representation,

$$\Psi = r^{\ell} e^{-\frac{1}{\hbar}\Phi(r)},$$

and substituting it into the eigenvalue problem (7.9) we arrive at the modified radial Riccati equation,

$$\hbar\,\partial_r y(r) - y(r)\left(y(r) - \frac{\hbar(2\ell + d - 1)}{r}\right) = 2m\left[E - V(r)\right], \tag{7.14}$$

cf. (7.13), where function $y(r)$ is the logarithmic derivative of $\Xi(r)$ or, equivalently, the derivative of the phase $\Phi(r)$,

$$y(r) = -\hbar\,\partial_r\left(\log\Xi(r)\right) = -\hbar\frac{\partial_r\Xi(r)}{\Xi(r)} = \partial_r\Phi(r). \tag{7.15}$$

For rAHO the potential V is given by (7.12). The equation (7.14) is highly suitable to study the eigenstates with $n_r = 0$.

By taking the mixed exponential representation,

$$\Psi = r^{\ell}\,P(r^2)\,e^{-\frac{1}{\hbar}\Phi(r)}, \tag{7.16}$$

and substituting it into the eigenvalue problem (7.9) with the potential (7.12) we arrive at the general radial Riccati equation,

$$\boxed{\begin{aligned}&\hbar\,\partial_r y(r) - y(r)\left(y(r) - \frac{\hbar(2\ell + d - 1)}{r}\right)\\ &\qquad - \frac{\hbar^2\,\partial_r^2 P + \hbar^2\,\frac{(2\ell+d-1)}{r}\partial_r P(r) - 2\,\hbar\, y\partial_r P}{P}\\ &= 2m\left[E - \frac{1}{g^2}\hat{V}(gr)\right],\end{aligned}} \tag{7.17}$$

cf. (2.33), where function $y(r)$ is the derivative of the phase $\Phi(r)$,

$$y(r) = \partial_r\Phi(r). \tag{7.18}$$

Evidently, equation (7.17) can be reduced to equations (7.13), (7.14).

Equation (7.17) as well as its degenerations (7.13), (7.14) are the *central equations* of Part II.

7.2 Radial Anharmonic Oscillators: Potentials

In d-dimensional coordinate space $x \in \mathbf{R^d}$ the radial quantum harmonic oscillator with $O(d)$ spherically symmetric potential has the form

$$V = \frac{1}{2} m\omega^2 r^2, \tag{7.19}$$

where $r = \left(\sum_{k=1}^{d} x_k^2\right)^{1/2}$ is the hyperradius, m is mass of the particle and ω is the characteristic frequency. It plays the important role in exploration of quantum world by describing small radial oscillations (vibrations) of diatomic molecules near equilibrium. Sometimes, this oscillator is called the linear radial oscillator. It is evident that the radial harmonic oscillator (7.19) is reduced to the sum of one-dimensional oscillators (1.1) once written in the Cartesian coordinates.

Since the potential (7.19) grows at $r \to \infty$, the particle can perform a finite motion in radial direction while moving freely in angular directions. Its spectra of energies is entirely discrete. Remarkable property of the radial harmonic oscillator is that it can be solved *exactly* in any dimension d: all eigenstates (eigenvalues and eigenfunctions) can be found in closed analytic form.

When oscillations are not small, the anharmonic effects can begin to play the role. They become significant by changing the energy spectra, expectation values and eigenfunctions, especially at large distances. Sometimes, the anharmonic effects are of the type that the spherical symmetry of the system is preserved. In this case it requires to modify the harmonic oscillator potential (7.19) by introducing radial, spherically symmetric anharmonic terms. There are four types of radial anharmonic potentials which, in general, are widely known and explored:

(i) Cubic radial anharmonic potential

$$V = r^2 + g r^3 \tag{7.20}$$

(ii) Quartic radial anharmonic potential

$$V = r^2 + g^2 r^4 \tag{7.21}$$

(iii) Sextic radial anharmonic potential

$$V = r^2 + g^4 r^6 \tag{7.22}$$

(iv) Quartic double-well radial anharmonic potential

$$V = r^2\,(1 - gr)^2, \tag{7.23}$$

sometimes, it is called the *Mexican hat* potential.

For all four potentials **(i)**–**(iv)** the parameter $g \geq 0$ is called the parameter of anharmonicity or the coupling constant. In the limit $g \to 0$ these potentials are reduced to the harmonic oscillator potential

$$V_{g=0} = r^2 .$$

In the opposite limit $g \to \infty$ (the so-called (ultra)-strong coupling regime) the potentials **(i)**–**(iv)** are reduced to a one-term radial oscillator potential

$$V_{g=\infty} = g^{p-2}\,r^p ,\ p = 3, 4, 6, \tag{7.24}$$

where g can be (naturally) placed equal to one, $g = 1$ (without loss of generality). For $p = 3, 4, 6$, this potential corresponds to the *cubic, quartic, sextic* radial oscillators, respectively. These are particular cases of the so-called *radial power-like* potentials,

$$V = r^p , \tag{7.25}$$

where $p > 0$ is a real number. The potential at $p = 1$ is called the *linear radial potential*: it has numerous applications.

All four potentials **(i)**–**(iv)** are particular cases of the general radial anharmonic oscillator potential

$$\boxed{\begin{aligned} V(r) &= \frac{1}{g^2}\,\hat{V}(g\,r) = \frac{1}{g^2}\sum_{k=2} a_k\,g^k\,r^k \\ &= a_2\,r^2 + a_3\,gr^3 + a_4\,g^2r^4 + \cdots + a_p\,g^{p-2}r^p + \cdots , \end{aligned} \tag{7.26}}$$

cf. [Brezin *et al.* (1977b)], eq. (8) therein, for $r \in [0, \infty)$, where $g \geq 0$ is the coupling constant and a_k, $k = 2, 3, \ldots, p, \ldots$ are dimensionless parameters; in particular, $a_2 = \frac{m\,\omega^2}{2}$, hence, it is proportional to the square of frequency ω, however, for most cases for the sake of convenience we set $a_2 = 1$. This potential is invariant (formally) with respect to a simultaneous change $r \to -r$ and $g \to -g$. It is assumed that like in the polynomial potentials of degree p, see e.g. **(i)**–**(iv)** the parameter a_p in front of the leading term r^p is positive, thus, the potential is confining. It is characterized by the infinite discrete spectra.

The general anharmonic potential (7.26) is assumed always *to be bounded from below.* For the sake of convenience the minimum of the potential at $r = 0$ is considered as the global minimum (or one among several global minima). For the potential **(iv)** there are two degenerate minima. The reference point for the potential (7.26) is chosen in such a way that $V(0) = 0$. It must be also emphasized the evident fact that *any* two-term radial anharmonic oscillator potential,

$$V(r) = a_2\, r^2 + \hat{\lambda}\, r^p,\ a_2 > 0,$$

where $\hat{\lambda}$ is a parameter of anharmonicity, can be *always* written in the form (7.26) by putting $a_k = 0$ for $k = 3, 4, \ldots, (p-1), (p+1), \ldots$ and redefining $\hat{\lambda} \equiv g^{p-2}$. The general form of two-term rAHO potential, which is going to be explored in Part II, reads as follows

$$V_p(r) = a_2\, r^2 + g^{p-2}\, r^p = \frac{1}{g^2}\left[(gr)^2 + (gr)^p\right] \equiv \frac{1}{g^2}\, \hat{V}_p(gr), \tag{7.27}$$

where one can naturally set $a_2 = 1$. Additionally, by also setting $a_2 = 0$ we can reproduce any one-term radial oscillator (power-like) potential, λr^p, see (7.24).

The goal of Part II is to study the energies (eigenvalues) and eigenfunctions of the d-dimensional radial anharmonic potential (7.26), (7.27), in particular, cubic ($p = 3$), quartic ($p = 4$) and sextic ($p = 6$) cases.

7.3 Radial Riccati-Bloch Equation

The radial Riccati equation (7.13) can be transformed into another non-linear equation without explicit dependence on the Planck constant $\hbar$ and the mass m by introducing a new variable

$$v = \left(\frac{2m}{\hbar^2}\right)^{\frac{1}{4}} r, \tag{7.28}$$

and making the replacements in y and E,

$$y = (2m\hbar^2)^{\frac{1}{4}}\ \mathcal{Y}(v)\,, \quad E = \frac{\hbar}{(2m)^{\frac{1}{2}}}\,\varepsilon. \tag{7.29}$$

Then, the equation (7.13) becomes [Del Valle and Turbiner (2019)],

$$\boxed{\partial_v \mathcal{Y}(v) - \mathcal{Y}(v)\left(\mathcal{Y}(v) - \frac{d-1}{v}\right) = \varepsilon\,(\lambda) - \frac{1}{\lambda^2}\,\hat{V}\,(\lambda v)\,, \quad \partial_v \equiv \frac{d}{dv},} \tag{7.30}$$

with the potential given by (7.26); here

$$\lambda = \left(\frac{\hbar^2}{2m}\right)^{\frac{1}{4}} g, \qquad (7.31)$$

plays the role of the *effective* coupling constant replacing g. It is evident that

$$\lambda v = gr.$$

The boundary conditions those we impose in equation (7.30) are

$$\partial_v \mathcal{Y}(v)\Big|_{v=0} = \frac{\varepsilon}{d}, \quad \mathcal{Y}(+\infty) = +\infty. \qquad (7.32)$$

Evidently, if $\hbar = 1$ and $m = 1/2$, the coupling constant $\lambda = g$ and the equation (7.30) coincides with (7.13). We call the equation (7.30) the *radial Riccati-Bloch* (rRB) equation for the ground state. At $d = 1$ the rRB equation coincides with one-dimensional RB equation (2.3), where the extra boundary condition $\mathcal{Y}(0) = 0$ should be imposed if the domain remains $v \in [0, \infty)$.

By making substitutions (7.28), (7.29), (7.31) in the general radial Riccati equation (7.17) we arrive at

$$\boxed{\begin{aligned} &\hbar\,\partial_r \mathcal{Y} - \mathcal{Y}\left(\mathcal{Y} - \frac{2\ell + d - 1}{v}\right) \\ &\quad - \frac{\partial_v^2 P + \frac{(2\ell+d-1)}{v}\partial_v P - 2\,\mathcal{Y}\,\partial_v P}{P} = \varepsilon - \frac{1}{\lambda^2}\,\hat{V}\,(\lambda v), \qquad (7.33) \end{aligned}}$$

cf.(2.34). We call the equation (7.33) the *general radial Riccati-Bloch* (grRB) equation. This remarkably simple equation would allow us to develop highly effectively the weak coupling expansions for excited states for $\varepsilon, \mathcal{Y}$ and, especially, P, which is beyond of the scope of the Part II. In fact, the MATHEMATICA codes, presented in Appendices B, E for one-dimensional case, can be easily modified to tackle the radial case. With a few exceptions the grRB equation is not going to be explored in the Part II unlike the rRB equation. We will be focused most of the time on the ground state.

The rRB equation (7.30) allows us to construct the asymptotic expansions of $\mathcal{Y}(v)$, and ultimately of $y(r)$, at small and large v for the polynomial potential of the degree p (7.26). For fixed $\lambda \neq 0$ and small v we have the Taylor series expansion

$$\mathcal{Y}(v) = \frac{\varepsilon}{d}\,v + \frac{(\varepsilon^2 - a_2 d^2)}{d^2(d+2)}\,v^3 - \frac{a_3\lambda}{d+3}\,v^4 - \frac{a_4 d^3(d+2)\lambda^2 - 2\,\varepsilon\,(\varepsilon^2 - a_2 d^2)}{d^3(d+2)(d+4)}\,v^5 + \cdots\,, \tag{7.34}$$

while for large v,

$$\mathcal{Y}(v) = a_p^{1/2}\lambda^{(p-2)/2}v^{p/2} + \frac{a_{p-1}\lambda^{(p-4)/2}}{2a_p^{1/2}}v^{(p-2)/2} + \frac{(4a_p\,a_{p-2} - a_{p-1}^2)\lambda^{(p-6)/2}}{8a_p^{3/2}}v^{(p-4)/2} + \cdots\,. \tag{7.35}$$

It is easy to check that for $\lambda = 0$, hence, for the radial harmonic oscillator $V = a_2\,r^2$, see (7.19), the equation (7.30) has the exact solution for the ground state

$$\mathcal{Y}_0(v) = \frac{\varepsilon}{d}\,v, \varepsilon_0 = a_2^{1/2}\,d. \tag{7.36}$$

Three remarks are in a row:

- the expansion (7.34) at $\lambda = 0$ is terminated and consists of the first term alone, leading to $\mathcal{Y}(v) = (\varepsilon/d)\,v$, in agreement with (7.36),
- in general, any coefficient in the expansion (7.34) depends on the energy ε explicitly,
- the coefficients in the first $[p/2]$ terms in the expansion (7.35) growing with v do not depend on the energy ε, they can be found explicitly; a few low order terms (among the first $[p/2]$ terms growing with v) can depend explicitly on dimension d (and the quantum numbers ℓ and n_r for the excited states).

It is worth mentioning that for fixed mass m, the behavior at large v can be obtained in two situations: at large r and fixed $\hbar$, or at fixed r and at small $\hbar$, see (7.28). The expansion (7.35) describes the semiclassical limit $\hbar \to 0$.

Integrating (7.34) and (7.35) in v, both expansions are converted into expansions for the phase Φ, namely,

$$\frac{1}{\hbar}\Phi = \frac{\varepsilon}{2d}v^2 + \frac{(\varepsilon^2 - a_2 d^2)}{4d^2(d+2)}v^4$$
$$-\frac{a_3\lambda}{5(d+3)}v^5 - \frac{a_4\, d^3(d+2)\,\lambda^2 - 2\varepsilon(\varepsilon^2 - a_2 d^2)}{6d^3\,(d+2)(d+4)}v^6 + \cdots, \quad (7.37)$$

and

$$\frac{1}{\hbar}\Phi = \frac{2a_p^{1/2}}{p+2}\lambda^{\frac{p-2}{2}}\, v^{\frac{p+2}{2}} + \frac{a_{p-1}\lambda^{\frac{p-4}{2}}}{p\, a_p^{1/2}}\, v^{\frac{p}{2}}$$
$$+\frac{(4\,a_p\,a_{p-2} - a_{p-1}^2)\lambda^{\frac{p-6}{2}}}{4(p-2)a_p^{3/2}}\, v^{\frac{p-2}{2}} + \cdots, \quad (7.38)$$

for small and large v, respectively. Note that in the expansion (7.37) the linear in v term is absent. Let us emphasize that in the expansion (7.38) the coefficients in front of the all terms growing at large v (or radius r) including the logarithmic term $\log r$ do *not* depend on the energy ε. All of them can be found explicitly. In particular, for $p = 3, 4, 6$ (cubic, quartic, sextic rAHO, see (7.20), (7.21), (7.22)) the three growing at large v terms do *not* depend on the energy ε and also the dimension d. These three coefficients remain the same for any excited state. Later the expansions (7.37) and (7.38) will be used to make interpolation between large and small v to construct via the procedure of matching a (uniform) approximation for the phase Φ in (7.16).

7.3.1 *Radial Generalized Bloch Equation*

Similar to the one-dimensional case there exists an alternative way to transform the original radial Riccati equation (7.13) into a non-linear equation without the explicit dependence on the Planck constant $\hbar$ and mass m. This is achieved by using the change of variable

$$u = gr, \quad (7.39)$$

and by introducing a new unknown function

$$\mathcal{Z}\,(u) = \frac{g}{(2m)^{1/2}}\, y. \quad (7.40)$$

It can be shown that $\mathcal{Z}(u)$ satisfies the non-linear differential equation

$$\boxed{\lambda^2\,\partial_u \mathcal{Z}(u) - \mathcal{Z}(u)\left(\mathcal{Z}(u) - \frac{\lambda^2(d-1)}{u}\right) = \lambda^2\,\varepsilon(\lambda) - \hat{V}(u), \quad \partial_u \equiv \frac{d}{du},} \tag{7.41}$$

c.f. (7.30), with boundary conditions

$$\partial_u \mathcal{Z}(u)\Big|_{u=0} = \frac{\varepsilon}{d}, \quad \mathcal{Z}(+\infty) = +\infty. \tag{7.42}$$

Here the energy ε and the effective coupling constant λ are the same as in (7.29) and (7.31). Thus, they play the role of the energy and the effective coupling constant in both v-space (7.28) and u-space (7.39) dynamics.

At $d = 1$ the equation (7.41) simplifies and coincides with one-dimensional GB equation (3.3), which was called in [Escobar-Ruiz *et al.* (2016, 2017); Shuryak and Turbiner (2018)] the *(one-dimensional) GB Equation.*[1] Here, the equation (7.41) is a natural extension of one-dimensional GB equation to d-dimensional case with a d-dimensional radial potential. For this reason we continue to call it the *(radial) Generalized Bloch* (rGB) equation. Naturally, (7.41) can also be used to construct the asymptotic expansions for small u,

$$\mathcal{Z}(u) = \frac{\varepsilon}{d}\,u + \frac{(\varepsilon^2 - a_2\,d^2)}{d^2(d+2)\lambda^2}\,u^3 - \frac{a_3}{(d+3)\lambda^2}\,u^4 - \frac{a_4 d^3(d+2)\lambda^2 - 2\,\varepsilon\,(\varepsilon^2 - a_2 d^2)}{d^3(d+2)(d+4)\lambda^4}\,u^5 + \cdots, \tag{7.43}$$

cf. (7.34), and large u,

$$\mathcal{Z}(u) = a_p^{1/2}\,u^{p/2} + \frac{a_{p-1}}{2a_p^{1/2}}\,u^{(p-2)/2} + \frac{(4a_p a_{p-2} - a_{p-1}^2)}{8a_p^{3/2}}\,u^{(p-4)/2} + \cdots, \tag{7.44}$$

cf. (7.35), respectively. The coefficients in front of all $[p/2]$ growing with u terms do not depend on the energy ε and can be found explicitly.

[1] In this case the extra boundary condition $\mathcal{Z}(0) = 0$ should be imposed, if the domain remain $u \in [0, \infty)$

7.4 The Weak Coupling Regime

The coupling constant g defines the strength of anharmonic effects in the rAHO potential (7.26),

$$V(r) = a_2\, r^2 + a_3\, gr^3 + a_4\, g^2 r^4 + \cdots + a_p\, g^{p-2} r^p + \cdots .$$

When the coupling constant g is small, the anharmonic effects are supposedly small and the radial quantum harmonic oscillator with potential (7.19) describes the main features of quantum dynamics. We call this regime *the weak coupling regime.* In this regime the anharmonic effects can be studied in the PT in powers of the coupling constant g.

In the formalism of rRB and rGB equations the coupling constant g is replaced by the effective coupling constant

$$\lambda = \left(\frac{\hbar^2}{2m}\right)^{\frac{1}{4}} g,$$

which is proportional to g. It allows us to study the weak coupling regime developing the PT in the effective coupling constant λ. The energy can be studied as the Taylor expansion

$$\varepsilon(\lambda) = \sum_{n=0}^{\infty} \varepsilon_n\, \lambda^n, \tag{7.45}$$

which, being divergent for any radial AHO, can provide some useful information.

In this section we describe how to solve both the rRB equation and the rGB equation using PT in powers of the effective coupling constant λ, developing in addition to (7.45) the expansions

$$\mathcal{Y}(v) = \sum_{n=0}^{\infty} \mathcal{Y}_n(v)\, \lambda^n, \tag{7.46}$$

and

$$\mathcal{Z}(u) = \sum_{n=0}^{\infty} \mathcal{Z}_n(u)\, \lambda^n, \tag{7.47}$$

respectively.

For the rRB equation the perturbative solution can be realized by applying the Non-Linearization Procedure [Turbiner (1984)] for the radial Schrödinger equation. Due to the importance of this procedure, we begin this section by giving its brief description.

7.4.1 The Non-Linearization Procedure

Let us take the RB equation (7.30)

$$\partial_v \mathcal{Y}(v) - \mathcal{Y}(v)\left(\mathcal{Y}(v) - \frac{d-1}{v}\right) = \varepsilon\,(\lambda_f) - V(v;\lambda_f), \tag{7.48}$$

for the potential $V(v;\lambda_f)$, which admits a Taylor expansion

$$V(v;\lambda_f) = \sum_{n=0}^{\infty} V_n(v)\,\lambda_f^n. \tag{7.49}$$

Here λ_f is a formal parameter and the coefficient functions $V_n(v)$, $n = 0, 1, \ldots$, are real functions in v. Let us now assume that the unperturbed equation at $\lambda_f = 0$,

$$\partial_v \mathcal{Y}_0(v) - \mathcal{Y}_0(v)\left(\mathcal{Y}_0(v) - \frac{d-1}{v}\right) = \varepsilon_0 - V_0(v), \tag{7.50}$$

can be solved explicitly. It can always be achieved via the *inverse problem*: we take some function $\mathcal{Y}_0(v)$ and then calculate r.h.s.: the potential $V_0(v)$ and ε_0. Evidently, once we know $\mathcal{Y}_0(v)$, the wave function can be found,

$$\Psi_0(v) = \exp\left(-\int^v \mathcal{Y}_0(s)\,ds\right). \tag{7.51}$$

It leads to a constraint on the choice of $\mathcal{Y}_0(s)$: the function $\Psi_0(v)$ should be normalizable. Now, we can develop PT (7.45),

$$\varepsilon(\lambda_f) = \sum_{n=0}^{\infty} \varepsilon_n\,\lambda_f^n$$

where the sum $(\varepsilon_0 + \varepsilon_1)$ makes sense of the variational energy with $\Psi_0(v)$ (7.51) as the trial function, and (7.46)

$$\mathcal{Y}(v) = \sum_{n=0}^{\infty} \mathcal{Y}_n(v)\,\lambda_f^n.$$

Substituting (7.45) and (7.46) into the rRB equation (7.48) it is easy to see that the nth correction $\mathcal{Y}_n(v)$ satisfies the first order linear differential equation

$$\partial_v \left(v^{d-1}\Psi_0^2\, \mathcal{Y}_n(v)\right) = \left(\varepsilon_n - Q_n(v)\right) v^{d-1}\Psi_0^2, \tag{7.52}$$

where

$$Q_1(v) = V_1(v), \quad Q_n(v) = V_n(v) - \sum_{k=1}^{n-1} \mathcal{Y}_k(v)\, \mathcal{Y}_{n-k}(v), \quad n = 2, 3, \ldots, \tag{7.53}$$

cf. (2.25), (2.26). Evidently, the solution for $\mathcal{Y}_n(v)$ is given by

$$\mathcal{Y}_n(v) = \frac{1}{v^{d-1}\,\Psi_0^2}\left(\int_0^v \left(\varepsilon_n - Q_n\right)\Psi_0^2\, s^{d-1}\, ds\right), \tag{7.54}$$

where ε_n in (7.52) remains unknown. Since we are interested in finding bound states we have to impose the condition of the absence of current of particles for both $v \to 0$ and $v \to \infty$ as the boundary condition [Turbiner (1984)],

$$\mathcal{Y}_n(v)\, v^{d-1}\, \Psi_0^2\Big|_{v=\{0,\infty\}} \to 0. \tag{7.55}$$

It can be easily checked that if v tends to zero the condition (7.55) is satisfied automatically while at $v = \infty$ the correction ε_n should be chosen accordingly

$$\varepsilon_n = \frac{\int_0^\infty Q_n\, \Psi_0^2\, v^{d-1}\, dv}{\int_0^\infty \Psi_0^2\, v^{d-1}\, dv}, \tag{7.56}$$

to satisfy (7.55).

For $d = 1$, this perturbative approach was called the *Non-Linearization Procedure*, see Chapter I.2. Here we have given the straightforward extension of it to the radial potentials in arbitrary dimension d. We will continue to call it the *Non-Linearization Procedure.* In contrast with the Rayleigh-Schrödinger PT, the knowledge of entire spectrum of the unperturbed problem is not required to find constructively perturbative corrections in (7.45) and (7.46). It is sufficient to know the unperturbed ground state wave function of the unperturbed problem to which we are looking for corrections. This approach gives the closed analytic expression for both corrections ε_n and $\mathcal{Y}_n(v)$ in form of nested integrals. Therefore, this procedure is the efficient method to calculate several orders in PT, see e.g. [Turbiner (1984)] and [Del Valle and Turbiner (2019)].

Interestingly, in this framework the convergence of the perturbation series (7.45) and (7.46) is guaranteed as long as the first correction $\mathcal{Y}_1(v)$ is bounded,

$$|\mathcal{Y}_1(v)| \ \leq \ Const. \tag{7.57}$$

which is true for any d, cf.(2.31) for the case of $d = 1$.

We have presented a brief review of the Non-Linearization Procedure applied to the ground state in d-dimensional general radial potential case. This approach can be modified to study the excited states by admitting a number of simple poles in v with residues equal to (-1) in $\mathcal{Y}_n(v)$. The pole positions are found in PT in the form of the Taylor expansion in powers of λ_f. Explicit formulas are presented in Chapter I.2 for one-dimensional case.

7.4.2 *Radial Anharmonic Oscillator: The Weak Coupling Expansion from Radial Riccati-Bloch Equation*

Now, let us consider the d-dimensional radial AHO potential (7.26) in weak coupling regime assuming that the effective coupling constant $\lambda = \left(\frac{\hbar^2}{2m}\right)^{\frac{1}{4}} g$ is *small.* In order to apply the Non-Linearization Procedure of Subsection 7.4.1 to construct the weak coupling expansion we choose in (7.49)

$$\begin{aligned} \lambda = \lambda_f &= \left(\frac{\hbar^2}{2m}\right)^{\frac{1}{4}} g, \\ V_k &= a_{k+2}\, v^{k+2} \quad \text{for} \quad k = 0,\, 1, \ldots, m-2, \ \text{and} \\ V_k &= 0 \quad \text{for} \quad k > m-2, \end{aligned} \tag{7.58}$$

hence, our potential is a finite degree polynomial in both λ and v. The corresponding unperturbed equation (7.50) describes the d-dimensional spherical harmonic oscillator: an exactly solvable problem in any $d > 0$. Their unperturbed ground state wave function is given by

$$\Psi_0 = e^{-\sqrt{\frac{m}{2\hbar^2}}\, a_2^{1/2} r^2} = e^{-\frac{1}{2} a_2^{1/2} v^2}. \tag{7.59}$$

In general, the explicit calculation of the perturbative corrections $\mathcal{Y}_n(v)$ and ε_n can be carried out analytically by using (7.54). The final expression involves the incomplete gamma function. Let us take the nth correction $\mathcal{Y}_n(v)$ and find the asymptotic expansion in two limits, $v \to 0$ and $v \to$

∞, respectively. For small v the correction $\mathcal{Y}_n(v)$ is given by the Taylor expansion

$$\mathcal{Y}_n(v) = v \sum_{k=0}^{\infty} b_k^{(n)} \, v^k , \tag{7.60}$$

where $b_k^{(n)}$ are some coefficients. It can be easily shown that the first coefficient $b_0^{(n)}$ is related to the energy correction ε_n,

$$b_0^{(n)} = \frac{\varepsilon_n}{d} , \tag{7.61}$$

while the next coefficient always vanishes,

$$b_1^{(n)} = 0, \tag{7.62}$$

due to the absence of the linear term in the potential (7.26). From other side, for large v at $n > 0$ we have a Laurent series expansion,

$$\mathcal{Y}_n(v) = v \sum_{k=0}^{\infty} c_k^{(n)} \, v^{n-k} , \tag{7.63}$$

where $c_k^{(n)}$ are some coefficients. Interestingly, the leading coefficient $c_0^{(n)}$ does not depend on d while the next-to-leading one always vanishes,

$$c_1^{(n)} = 0. \tag{7.64}$$

At $d = 0$ all energy corrections ε_n vanish,

$$\varepsilon_n = 0.$$

It can be demonstrated by using the expression (7.56) for ε_n, and the asymptotic expansions (7.60) and (7.63). When $d \to 0$ the numerator of (7.56) is bounded,

$$\int_0^{\infty} Q_n \, \Psi_0^2 \, v^{d-1} \, dv \; < \; \infty , \tag{7.65}$$

while the denominator has the asymptotic series expansion

$$\int_0^{\infty} \Psi_0^2 \, v^{d-1} \, dv = \frac{a_2^{-\frac{d}{4}}}{d} + O(d^0). \tag{7.66}$$

Consequently, when $d \to 0$ the correction ε_n vanishes linearly, the coefficient $b_0^{(n)}$ remains finite, see (7.61). If ε_n vanishes for all n, then their formal sum (7.45) results in $\varepsilon = 0$ and, ultimately, $E = 0$. In general, for $d \neq 0$

it can be shown that expansion for ε is asymptotic: ε_n grows factorially as $n \to \infty$, see in particular, [Bender and Wu (1969); Brezin *et al.* (1977a,b)] and references into them. Therefore, series (7.45) and (7.46) are divergent in λ.

An interesting situation occurs when all odd monomial terms in r in potential (7.26) are absent, i.e. the potential is (formally) an even function, $V(r) = V(-r)$. In this case, all odd corrections $\mathcal{Y}_{2n+1}(v)$ and ε_{2n+1} vanish. The even correction $\mathcal{Y}_{2n}(v)$ has the form of a polynomial of finite degree

$$\mathcal{Y}_{2n}(v) = v \sum_{k=0}^{n} c_{2k}^{(2n)} v^{2(n-k)}, \quad \mathcal{Y}_{2n}(-v) = -\mathcal{Y}_{2n}(v), \tag{7.67}$$

where

$$c_{2n}^{(2n)} = \frac{\varepsilon_{2n}}{d}. \tag{7.68}$$

This implies that ε_{2n} and $\mathcal{Y}_{2n}(v)$ can be calculated by linear algebra means. In turn, the energy correction ε_{2n} is a finite-degree polynomial in d,

$$\varepsilon_{2n} = \sum_{k=1}^{n+1} d_k^{(2n)} d^k, \tag{7.69}$$

where $d_k^{(2n)}$ are real coefficients. However, it is enough to have a single odd monomial term $a_{2q+1} v^{2q+1}$ to be present in the potential (7.26) (naturally, q is integer) to break the features (7.67)–(7.69). Hence, all coefficients in front of the singular terms in the expansion (7.63), as well as all higher order terms in the expansion (7.60) at $k > n$, are proportional to a_{2q+1}.

7.4.3 *Radial Anharmonic Oscillator: The Weak Coupling Expansion from Radial Generalized Bloch Equation*

Taking in the rGB equation (7.41) the effective coupling constant $\lambda = \left(\frac{\hbar^2}{2m}\right)^{\frac{1}{4}} g$ as a formal parameter we can develop PT in powers of λ. The expansion of the eigenvalue ε in powers of λ remains the same as for the RB equation, see (7.45),

$$\varepsilon(\lambda) = \sum_{n=0}^{\infty} \varepsilon_n \lambda^n,$$

while $\mathcal{Z}(u)$ is of the form of the Taylor series

$$\mathcal{Z}(u) = \sum_{n=0}^{\infty} \mathcal{Z}_n(u)\,\lambda^n. \tag{7.70}$$

We assume that the perturbative corrections ε_n are already known: they can be found using the Non-Linearization Procedure, or the standard Rayleigh-Schrödinger PT, or any other suitable method. It is immediate to see that the correction $\mathcal{Z}_n(u)$ is calculated by algebraic means and this depends on the corrections of smaller order,

$$\begin{aligned}
\mathcal{Z}_0(u) &= \pm\sqrt{\hat{V}(u)},\\
\mathcal{Z}_1(u) &= 0,\\
\mathcal{Z}_2(u) &= \frac{u\,\partial_u \mathcal{Z}_0(u) + (d-1)\,\mathcal{Z}_0(u) - u\,\varepsilon_0}{2\,u\,\mathcal{Z}_0(u)},\\
&\ \vdots\\
\mathcal{Z}_n(u) &= \frac{u\,\partial_u \mathcal{Z}_{n-2}(u) + (d-1)\,\mathcal{Z}_{n-2}(u) - u^2\sum_{i=2}^{n-2}\mathcal{Z}_i(u)\mathcal{Z}_{n-i}(u) - u\,\varepsilon_{n-2}}{2\,u\,\mathcal{Z}_0(u)}\\
&\ \vdots
\end{aligned} \tag{7.71}$$

for $n > 2$. Note that the boundary condition $\mathcal{Z}(\infty) = +\infty$ implies the positive sign in the expression for $\mathcal{Z}_0(u)$ should be chosen.

By making the analysis of the correction $\mathcal{Z}_n(u)$ in (7.71) one can see that the boundary condition at $u = 0$, see (7.55), can *not* be fulfilled in the case of arbitrary anharmonic radial potential (7.26), where at least one odd term is present, once the expansion (7.70) for $\mathcal{Z}(u)$ is used. In general, any correction $\mathcal{Z}_n(u)$ (and its derivative) at $n > 1$ diverges at small u. For instance, for the cubic potential (7.20), where $a_3 = 1$, the nth correction $\mathcal{Z}_n(u)$ with $n > 1$ behaves like

$$\mathcal{Z}_n(u) \sim u^{-n+2} \tag{7.72}$$

when $u \to 0$. In turn, for the quintic potential with $a_3 = 0$ but $a_5 > 0$ the nth correction at $n > 3$ behaves like

$$\mathcal{Z}_n(u) \sim u^{-n+4} \tag{7.73}$$

when u tends to zero. For the polynomial in u potential of degree $(2k+1)$, if all other odd terms are absent $a_3 = a_5 = \cdots = a_{2k-1} = 0$ but $a_{2k+1} > 0$, the function $\mathcal{Z}_n(u)$ behaves like

$$\mathcal{Z}_n(u) \sim u^{-n+2k} \tag{7.74}$$

for small u as long as $n > 2k-1$. It is the clear indication that the radius of convergence of the expansion of $\mathcal{Z}(u)$ (see (7.70)) in $1/u$ is finite. However, if the anharmonic potential $V(r)$ is even, $V(r) = V(-r)$ the boundary condition at $u = 0$ can be satisfied. It might be considered as an indication that the radius of convergence in $1/u$ is infinite in this case. It allows us to determine the correction ε_n by imposing the boundary condition

$$\partial_u \mathcal{Z}_n(u)\Big|_{u=0} = \frac{\varepsilon_n}{d}. \tag{7.75}$$

In general, at large u the semiclassical expansion (7.70) with corrections given by (7.71) mimics the asymptotic expansion (7.44). It can be seen explicitly by taking the expansion of the correction $\mathcal{Z}_n(u)$ in $1/u$ for different n. In particular, the expansion in $1/u$ of $\mathcal{Z}_0(u) = \sqrt{\hat{V}(u)}$, which is the classical radial momentum at zero energy, reproduces functionally the asymptotic expansion (7.44) but with different coefficients. This expansion (7.70) can be considered as the alternative to (7.44). Namely, the expansion (7.70) will be used in the matching procedure to construct an approximation of the eigenfunctions of the general radial AHO in the next Chapter.

At this point, it is worth emphasizing once again that the expansion in powers of λ (7.70), as well as (7.46), is divergent for any anharmonic oscillator due to the so-called Dyson instability, see [Dyson (1952)] and as for the discussion, [Turbiner (1984)] and Chapter 2, Part I.[2] This fact can be seen in the partial sums of (7.70), see Figs. 7.1 and 7.2.

Let us also emphasize that the presence of a single odd degree monomial in u in the potential $V(u)$ breaks the feature (7.75): the expansion for $\mathcal{Z}(u)$ in powers of λ is only able to satisfy the single boundary condition at $u = \infty$ but fail to satisfy the boundary condition at $u = 0$, see 7.2.

---

[2]The Dyson instability is related with the fact the change of the sign in front of the leading monomial r^p in the potential (7.26) leads to tunneling from classically-permitted domain at small r to the domain at large r.

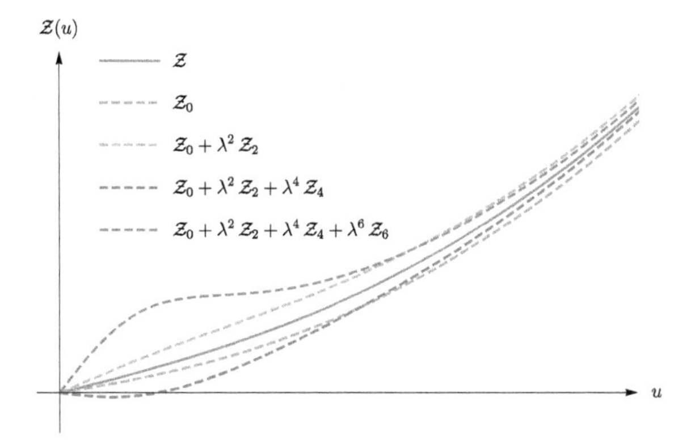

Fig. 7.1. Ground state: Partial sums of the expansion for $\mathcal{Z}(u)$ as functions of u for a general AHO potential with monomials of even degrees only. The function $\mathcal{Z}(u)$ (solid blue line), represents the accurate solution of RGB equation. Near $u = 0$, the deviation of the partial sums from the accurate solution is a consequence of the divergent nature of series (7.46).

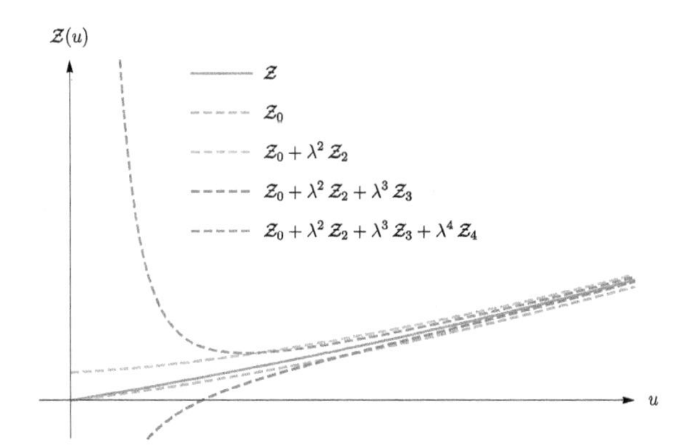

Fig. 7.2. Partial sums of the expansion for $\mathcal{Z}(u)$ as functions of u for a generic radial AHO potential which includes odd monomials. The function $\mathcal{Z}(u)$ (solid blue line) represents the accurate solution. The divergences in vicinity of $u = 0$ of the partial sums is the result of the impossibility of satisfying the boundary condition at $u = 0$.

7.4.4 *Connection between $\mathcal{Y}$ and $\mathcal{Z}$ Expansions*

Thus far we have constructed two different representations for the logarithmic derivative $y(r)$, see (7.11). From one side we have

$$y(r) = (2m\hbar^2)^{\frac{1}{4}}\, \mathcal{Y}(v(r)), \quad \mathcal{Y}(v) = \sum_{n=0}^{\infty} \mathcal{Y}_n(v)\, \lambda^n, \quad v(r) = \left(\frac{2m}{\hbar^2}\right)^{\frac{1}{4}} r, \tag{7.76}$$

from the other side

$$y(r) = \left(\frac{2m}{g^2}\right)^{\frac{1}{2}} \mathcal{Z}(u(r)), \quad \mathcal{Z}(u;\lambda) = \sum_{n=0}^{\infty} \mathcal{Z}_n(u)\,\lambda^n, \quad u(r) = g\,r, \quad (7.77)$$

where the coupling constant λ is defined in (7.31). The connection between them can be established if we use explicitly the expansion of $\mathcal{Y}_n(v)$ at large v. In this case $y(r)$ is given by

$$y = (2m\,\hbar^2)^{\frac{1}{4}}\, v \sum_{n=0}^{\infty} \lambda^n \sum_{k=0}^{\infty} c_k^{(n)}\, v^{n-k}, \quad (7.78)$$

or, equivalently, at large r,

$$y = \left(\frac{2m}{g^2}\right)^{\frac{1}{2}} \sum_{n=0}^{\infty} \lambda^n (gr) \sum_{k=1}^{\infty} c_n^{(k)} (gr)^{k-n}. \quad (7.79)$$

Comparing these expansions (7.78) and (7.79) we can draw a conclusion that

$$\mathcal{Z}_n(u) = u \sum_{k=1}^{\infty} c_n^{(k)} u^{k-n} \quad (7.80)$$

due to the uniqueness of the Taylor series expansion. Therefore, $\mathcal{Z}_n(u)$ is nothing but a generating function (!) for the coefficients $c_k^{(n)}$, $k = 0, 1, ...$, see for graphical illustration Fig. 7.3. Equation (7.80) reveals the reason why we cannot satisfy, in general, the boundary condition at $u = 0$: $\mathcal{Z}_n(u)$ is constructed from the expansion (7.63) which is valid in the limit $r \to \infty$ ($u \to \infty$ for fixed λ). In fact, some interesting properties of the corrections $\mathcal{Z}_n(u)$ occur when considering $r \to \infty$. For example, the first term in (7.77) has the expansion

$$\frac{\mathcal{Z}_0(gr)}{g} = a_p^{1/2}\, g^{(p-2)/2} r^{p/2} + \frac{a_{p-1}\, g^{(p-4)/2}}{2a_p^{1/2}} r^{(p-2)/2} + \frac{(4a_p\, a_{p-2} - a_{p-1}^2)\, g^{(p-6)/2}}{8a_p^{3/2}} r^{(p-4)/2} + \cdots . \quad (7.81)$$

This expansion reproduces exactly the dominant asymptotic behavior of the function $y(r)$ up to the term $O\left(r^{(p-4)/2}\right)$, c.f. (7.35). Note that the next generating function $\mathcal{Z}_1(u) = 0$ in agreement with (7.71). It is evident that the next generating functions in (7.70): $\mathcal{Z}_2(u)$, $\mathcal{Z}_3(u), \ldots$, also

$$
\begin{array}{rllllll}
y(r) = \frac{(2m)^{\frac{1}{2}}}{g} (& c_0^{(0)}(gr)^1 & & & & & \\
+ & c_0^{(1)}(gr)^2 & + \lambda^2 c_2^{(1)}(gr)^0 & + \lambda^3 c_3^{(1)}(gr)^{-1} & + \lambda^4 c_4^{(1)}(gr)^{-2} & + & \ldots \\
+ & c_0^{(2)}(gr)^3 & + \lambda^2 c_2^{(2)}(gr)^1 & + \lambda^3 c_3^{(2)}(gr)^0 & + \lambda^4 c_4^{(2)}(gr)^{-1} & + & \ldots \\
+ & c_0^{(3)}(gr)^4 & + \lambda^2 c_2^{(3)}(gr)^2 & + \lambda^3 c_3^{(3)}(gr)^1 & + \lambda^4 c_4^{(3)}(gr)^0 & + & \ldots \\
+ & c_0^{(4)}(gr)^5 & + \lambda^2 c_2^{(4)}(gr)^3 & + \lambda^3 c_3^{(4)}(gr)^2 & + \lambda^4 c_4^{(4)}(gr)^1 & + & \ldots \\
+ & \vdots & + \quad \vdots & + \quad \vdots & + \quad \vdots & + & \ldots) \\
\equiv \frac{(2m)^{\frac{1}{2}}}{g} (& \mathcal{Z}_0(gr) & + \lambda^2 \mathcal{Z}_2(gr) & + \lambda^3 \mathcal{Z}_3(gr) & + \lambda^4 \mathcal{Z}_4(gr) & + & \ldots)
\end{array}
$$

Fig. 7.3. Representation of the generating functions $\mathcal{Z}_n$ constructed from the perturbation series in powers of λ for the function $y(r)$.

contribute to the expansion of $y(r)$. It will be discussed in detail for the cubic anharmonic oscillator, see Chapter 9.

Let us conclude this Section by clarifying the perturbative approach for the rGB equation and its connection with the semiclassical WKB approximation. By integrating in r, the expansion (7.77) can be converted into an expansion for the phase

$$
\Phi(r;\lambda) = \sum_{n=0}^{\infty} \lambda^n G_n(r), \ G_1(r) = 0, \tag{7.82}
$$

where $G_n(r)$ is equal to

$$
G_n(r) = \left(\frac{2m}{g^2}\right)^{1/2} \int^r \mathcal{Z}_n(gr)\, dr. \tag{7.83}
$$

Note that keeping g and m fixed, the expansion (7.82) can be regarded as a semiclassical expansion of the phase in powers of $\hbar^{\frac{1}{2}}$, see (7.31). The leading order term is

$$
G_0(r) = \int^r \sqrt{2m\, V(r)}\, dr, \tag{7.84}
$$

for $V \geq 0$ it is the classical action at $E = 0$ and it corresponds to the zero order of the standard one-dimensional WKB method (developed in the classically forbidden region). This is the only term in expansion (7.82) which contains no dependence on the dimension d. The second order term in (7.82) is

$$\lambda^2 G_2(r) = \hbar \left(\frac{d-1}{2} \log[r] + \frac{1}{4} \log\left[2m\, V(r)\right] - \frac{\varepsilon_0}{2} \int^r \frac{dr}{\sqrt{V(r)}} \right). \quad (7.85)$$

Except for the integral — the third term — it looks like the first order correction in the standard WKB approach at $E = 0$ and $d = 1$. Thus, it defines the determinant in path integral formalism [Escobar-Ruiz *et al.* (2017)]. The appearance of the extra term in the form of integral can be explained as follows. Let us take the zero order term in the standard WKB method (in the classically-forbidden region) and expand it in powers of $\hbar^{1/2}$ using (7.29) and (7.31), thus

$$\int^r \sqrt{2m\,(V(r) - E)}\, dr = \int^r \sqrt{2m\, V(r)}\, dr - \frac{\hbar\, \varepsilon_0}{2} \int^r \frac{dr}{\sqrt{V(r)}} + O(\hbar^{3/2}). \quad (7.86)$$

Note that the first term is nothing but $G_0(r)$ while the second term is exactly the integral in (7.85). Consequently, one can see that (7.82) is the standard semiclassical WKB expansion in the classically-forbidden region, *re-expanded* in powers of $\hbar^{1/2}$. Needless to say, higher order generating functions $G_3(r), G_4(r), \ldots$ are related with higher order corrections in the WKB expansion in a similar way. They define two-, three-, etc loop contributions in the path integral formalism in Euclidian time for the density matrix at coinciding points, cf. Part I, Chapter 3 for the one-dimensional case [Escobar-Ruiz *et al.* (2016, 2017); Shuryak and Turbiner (2018)]. In these papers it was described how to obtain expansion (7.82) using a path integral formulation (in Euclidean time) considering a special classical solution at zero energy $E = 0$ called *flucton*. In a straightforward way this consideration can be extended to the d-dimensional radial case, where the flucton path appears in radial direction. Hence, as was explained earlier the perturbative approach developed for the rGB Equation is the *true* semiclassical approximation: both $\varepsilon(\lambda)$, and $\mathcal{Z}(u;\lambda)$ and $\Phi(r;\lambda)$ are expansions in powers $\hbar^{1/2}$.

Chapter 8

Radial AHO: Matching Perturbation Theory and Semiclassical Expansion

In this short Chapter we will present the main result of the Part II: the uniform approximation of the eigenfunctions of the radial AHO with a polynomial potential of degree p. As the first step we consider the ground state function and construct its approximation in the form of prefactor multiplied by exponential which depend on with several free parameters, then they will be fixed in variational procedure. Then, as the second step, we take the ground state function approximation which again depends on free parameters, modify the degrees of the prefactor accordingly, multiply it by r^ℓ and by the polynomial of degree n_r in r^2 with real coefficients, for which also all roots in r^2 are assumed positive and simple (no roots coincide). The coefficients of the polynomial are fixed by imposing the condition of the orthogonality to the approximations with lower-then-n_r-degree polynomials. In turn, the free parameters of this approximation are fixed by making the minimization of the variational energy.

Let us consider the polynomial, $O(d)$-symmetric, radial potential of degree p, see (1.26) for the general rAHO potential or (1.27) for the two-term rAHO potential. We will look for the radial eigenfunction of the n_r-th excited state at given ℓ in the general mixed exponential representation,

$$\Psi^{(n_r,\ell)} = r^\ell\, P_{n_r}^{(\ell)}(r^2) \times [\mathit{prefactor}] \times e^{-\frac{1}{\hbar}\Phi^{(n_r,\ell)}(r)},$$

$$P_{n_r}^{(\ell)}(r^2) = \prod_{i=1}^{n_r}(r^2 - r_i^2), \tag{8.1}$$

cf. (2.32) and (7.16), where all roots (r_i^2) of the polynomial $P_{n_r}^{(\ell)}(r^2)$ are assumed positive. Our main focus is to find the approximate expressions for $\Phi^{(n_r,\ell)}(r)$ and for [*prefactor*].

As the first step let us consider the ground state eigenfunction for which the first two factors in the representation (8.1) are absent. By taking the asymptotic expansion for the logarithmic derivative of the ground state wavefunction $\mathcal{Y}$ (1.11), (1.29) at small distances $v(r)$ (1.34) and by *matching* it with the semiclassical expansion of $\mathcal{Z}$ (1.40) at large distances $u(r)$ (or, said differently, the asymptotic expansion (1.44)) with emphasis to the exact reproduction of all $[p/2]$ growing terms at $r \to \infty$ we can arrive practically unambiguously(!) at an approximate expression for the ground state eigenfunction

$$\Psi^{(0,0)}_{approximation}(r) = [\text{prefactor}_{(0,0)}(r, g; \{\tilde{b}\})] \\ \times \exp\left(- \frac{A + \tilde{a}_1 r + \hat{V}(r, g; \{\tilde{a}\})}{\sqrt{\frac{1}{r^2} \hat{V}(r, g; \{\tilde{b}\})}} + \frac{A}{\tilde{b}_0^2} \right). \tag{8.2}$$

Here $\hat{V}(r, g;\ \{\tilde{a}\})$, $\hat{V}(r, g;\ \{\tilde{b}\})$ are two potentials, both of them were previously defined in (1.26) with the coefficients $a_2, a_3, \ldots, a_p$ (denoted here as $\{a\}$), but with a different sets of coefficients $\{\tilde{a}\}$ or $\{\tilde{b}\}$, respectively, instead. In general, $\{\tilde{a}\}, \{\tilde{b}\}$ and $\tilde{a}_1, A$ are $2p$ free (variational) parameters subject to $[p/2]$ constraints: for the phase $\Phi(r)$ (1.17) and its (reduced) logarithmic derivative $\mathcal{Z}$ (3.2) all $[p/2]$ growing terms at large distances $r(u)$ in (3.8) do not depend on energy and should be reproduced exactly. This leads to constraints on the coefficients $\{\tilde{a}\}, \{\tilde{b}\}$. The prefactor [$prefactor_{(0,0)}(r, g; \{\tilde{b}\})$] in (8.2) depends on the concrete potential under consideration, it is usually defined by the second correction $\mathcal{Z}_2$ (1.71) in the expansion (1.70) (hence, by the (logarithm of) determinant), see also (3.12), where $\{a\}$ are replaced by $\{\tilde{b}\}$. In total, the function depends on $p - 1 - [p/2]$ free parameters which are defined in variational calculation.

By multiplying it by n_r-th degree polynomial in r^2 we will end up with approximation for the n_r-th radial excited state at fixed ℓ. Finally, its explicit form is the following

$$\Psi^{(n_r,\ell)}_{approximation}(r) = r^{\ell}\ P^{(\ell)}_{n_r}(r^2)\ [\text{prefactor}_{(n_r,\ell)}\,(r,g;\{\tilde{b}\})] \\ \times \exp\left(\ -\ \frac{A+\tilde{a}_1 r+\hat{V}(r,g;\{\tilde{a}\})}{\sqrt{\frac{1}{r^2}\,\hat{V}(r,g;\ \{\tilde{b}\})}}+\frac{A}{\tilde{b}_0^2}\right). \tag{8.3}$$

The prefactor $[\text{prefactor}_{(n_r,\ell)}(r,g;\{\tilde{b}\})]$ in (8.3) depends on the concrete potential under consideration, it is usually defined by the second correction $\mathcal{Z}_2$ in (1.71) but for the case of excited state (hence, by the appropriate determinant), where $\{a\})$ are replaced by $\{\tilde{b}\}$. It is not very much different from $[\text{prefactor}_{(0,0)}(r,g;\{\tilde{b}\})]$ with sufficient simple dependence on the quantum numbers (n_r,ℓ). Here $P^{(\ell)}_{n_r}(r^2)$ is a polynomial in r^2 of degree n_r, whose coefficients are found unambiguously by imposing the orthogonality conditions to the functions $\Psi^{(m_r,\ell)}(r)$, see (8.3), of previous states with $m_r = 0,1,2,\ldots,(n_r-1)$.

It has to be emphasized that by taking the approximant $\Psi^{(n_r,\ell)}(r)$ (8.3) to be the zeroth order approximation for the non-linearization procedure, see Section 1.4.1 and also Part I, Chapter 2, Section 2.3,

$$\Psi_0^{(n_r,\ell)} = \Psi^{(n_r,\ell)}_{approximation}\,(r),$$

see SubSections 2.3.1 and 2.3.3, one can develop a perturbation theory explicitly with respect to the deviation of the approximant $\Psi_0^{(n_r,\ell)}$ from the exact eigenfunction. Mainly due to the correct asymptotic behavior of the wavefunction (8.3) at large r, this perturbation theory is rapidly convergent. In particular,

- the second correction to the energy

$$|E_2^{(n_r,\ell)}| \ll E_0^{(n_r,\ell)} + E_1^{(n_r,\ell)} \equiv E^{(n_r,\ell)}_{variational},$$

- the first correction to the (nodal) prefactor

$$|P_1^{(n_r,\ell)}(r)| \ll |P_0^{(n_r,\ell)}(r)|,$$

  which implies that the coefficients in $P_1^{(n_r,\ell)}(r)$ are much smaller than corresponding ones in $P_1^{(n_r,\ell)}(r)$, that the positions of nodes are almost unchanged,

- and the first correction to the phase is bounded

$$|\Phi_1^{(n_r,\ell)}| \ll |\Phi_0^{(n_r,\ell)}| \text{ for } \forall r \in \mathbf{R}_+,$$

  uniformly.

In all concrete calculations for the radial AHO

$$|E_2| \sim 10^{-9} - 10^{-10}$$

(or better) and the ratio

$$|\Phi_1/\Phi_0| \lesssim 10^{-6},$$

see e.g. Chapters 9, 11 below, similar to ones for one-dimensional case, see Chapters 5–6 of Part I.

Formula (8.3) is the **central** formula of Part II of this book. It is similar to the formula (4.1) of Part I for one-dimensional case. In the next three Chapters the formula (8.3) will be applied to cubic, quartic, sextic radial anharmonic oscillators, respectively.

Chapter 9

Radial Cubic Anharmonic Oscillator

The simplest radial anharmonic oscillator is characterized by a cubic anharmonicity (1.20),

$$V(r) = r^2 + gr^3. \tag{9.1}$$

The corresponding radial Schrödinger equation (without the centrifugal term) has the form

$$\left[-\frac{\hbar^2}{2m}\left(\partial_r^2 + \frac{2\ell+d-1}{r}\partial_r\right) + r^2 + gr^3\right]\Xi_{n_r,\ell}(r)$$
$$= E_{n_r,\ell}\Xi_{n_r,\ell}(r), \tag{9.2}$$

see (1.9) at $p = 3$, where the radial wavefunction is

$$\Psi_{n_r,\ell}(r) = r^{\ell}\,\Xi_{n_r,\ell}(r),$$

n_r is the radial quantum number and ℓ is the quantum angular momentum. For the logarithmic derivative of the radial function Ξ:

$$y = -\partial_r \log \Xi,$$

one can write the radial Riccati-Bloch (rRB) equation

$$\partial_v \mathcal{Y} - \mathcal{Y}\left(\mathcal{Y} - \frac{2\ell+d-1}{v}\right) = \varepsilon\,(\lambda) - v^2 - \lambda v^3, \quad \partial_v \equiv \frac{d}{dv}, \tag{9.3}$$

cf. (1.30), where

$$v = \left(\frac{2m}{\hbar^2}\right)^{\frac{1}{4}} r, \quad y = (2m\hbar^2)^{\frac{1}{4}}\,\mathcal{Y}\,(v)\,, \quad E = \frac{\hbar}{(2m)^{\frac{1}{2}}}\varepsilon,$$

cf. (7.28), (1.29), or one can write the radial Generalized Bloch (rGB) equation

$$\lambda^2 \partial_u \mathcal{Z} - \mathcal{Z}\left(\mathcal{Z} - \frac{\lambda^2(2\ell + d - 1)}{u}\right) = \lambda^2 \varepsilon(\lambda) - u^2 - u^3,\ \partial_u \equiv \frac{d}{du}, \quad (9.4)$$

cf. (1.41), where

$$u = gr, \quad \mathcal{Z}(u) = \frac{g}{(2m)^{1/2}} y,$$

cf. (1.39), (1.40). In both equations

$$\lambda = \left(\frac{\hbar^2}{2m}\right)^{\frac{1}{4}} g,$$

plays the role of the *effective* coupling constant.

The study of quantum dynamics in the cubic radial potential (9.1) is the main subject of this Chapter. It will be presented in maximally self-contained form.

9.1 PT in the Weak Coupling Regime

For the cubic anharmonic oscillator the perturbative expansions of ε and $\mathcal{Y}(v)$ (1.46), (1.45), respectively, remain the same as for the general oscillator,

$$\varepsilon = \varepsilon_0 + \varepsilon_1 \lambda + \varepsilon_2 \lambda^2 + \cdots$$

with $\varepsilon_0 = d$ and

$$\mathcal{Y}(v) = \mathcal{Y}_0 + \mathcal{Y}_1 \lambda + \mathcal{Y}_2 \lambda^2 + \cdots,$$

respectively, see (7.28), (1.29). Since potential (9.1) contains the odd degree monomial r^3, the correction $\mathcal{Y}_n(v)$ is characterized by the infinite series (1.60) and (1.63) at small and large v, respectively. The first three terms in the expansion of $\mathcal{Y}_n(v)$ at $n = 1, 2, 3$ for both $v \to 0$ and $v \to \infty$ asymptotics are presented in Appendix H. For the ground state the first corrections ε_1 and $\mathcal{Y}_1(v)$ can be calculated in closed analytic form in the

Non-Linearization Procedure, see Section 1.4.1,

$$\varepsilon_1 = \frac{\Gamma\left(\frac{d+3}{2}\right)}{\Gamma\left(\frac{d}{2}\right)}, \quad \mathcal{Y}_1(v) = \frac{e^{v^2}}{2v^{d-1}}\left\{\varepsilon_1 \gamma\left(\frac{d}{2}, v^2\right) - \gamma\left(\frac{d+3}{2}, v^2\right)\right\}, \tag{9.5}$$

where $\Gamma(a)$ is the Euler gamma function and $\gamma(a,b)$ denotes the incomplete gamma function, respectively, see [Bateman (1953)]. Note that the higher energy corrections ε_n can be computed only numerically. However, the behavior of ε_n at large d can be easily obtained in the $1/d$-expansion analytically. This expansion can be constructed via the Non-Linearization Procedure, it will not be covered in this Chapter.

9.2 Ground State: Generating Functions

We can determine the coefficients of $\mathcal{Y}_n(v)$, i.e. $c_k^{(n)}$ in (1.63), by algebraic means. In general, they are written in terms of the corrections ε_0, $\varepsilon_1, \ldots, \varepsilon_n$, see Appendix H for explicit formulas for $n = 1, 2, 3$.

It was pointed out in [Dolgov and Popov (1979)] as well as in [Dolgov and Popov (1978)], see also [Turbiner (1980, 1984)], that the general expression for several first coefficients $c_0^{(n)}$, $c_2^{(n)}, \ldots$ can be obtained by solving recurrence relations. For example, $c_0^{(n)}$ satisfies the non-linear recurrence relation

$$c_0^{(n)} = -\frac{1}{2}\sum_{k=1}^{n-1} c_0^{(k)} c_0^{(n-k)}, \quad c_0^{(1)} = \frac{1}{2}, \tag{9.6}$$

as for $c_2^{(n)}$ the recurrence relation is linear

$$c_2^{(n)} = \frac{1}{2}(2n+d)c_0^{(n)} - \sum_{k=1}^{n-1} c_0^{(k)} c_2^{(n-k)}, \quad c_2^{(1)} = \frac{1}{4}(d+2). \tag{9.7}$$

The solution of (9.7) requires the knowledge of the coefficients $c_0^{(n)}$. It is evident that in order to find $c_3^{(n)}$, the coefficients $c_0^{(n)}$, $c_2^{(n)}$ are needed, the recurrence relation remains linear, etc. The simplest way to solve these recurrence relations is by using generating functions. Surprisingly, we can calculate these generating functions straightforwardly via the algebraic, iterative procedure derived from the rGB equation, see (1.71) and (9.4): we demonstrated that the kth generating function is nothing but $\mathcal{Z}_k(u)$!

For example, the first two non-trivial generating functions are

$$\mathcal{Z}_0(u) = u\sqrt{1+u}, \tag{9.8}$$

$$\mathcal{Z}_2(u) = \frac{u + 2d\left(1 + u - \sqrt{1+u}\right)}{4u(1+u)}, \tag{9.9}$$

where $u = (gr)$. From these expressions, the explicit solutions of equations (9.6) and (9.7) can be derived,

$$c_0^{(n)} = \frac{(-1)^{n+1}\Gamma(2n+1)}{2^{2n-1}\Gamma(n)\Gamma(n+1)}, \quad c_2^{(n)} = \frac{(-1)^{n+1}}{2}\left(1 + \frac{\Gamma(2n+1)}{2^{2n}\Gamma(n+1)^2}\right). \tag{9.10}$$

Note that the coefficients $c_0^{(n)}$ do not depend on d, while $c_2^{(n)}$ depend on d linearly. It can be demonstrated by induction that $c_k^{(n)}$ is polynomial in d of degree $(k-1)$.

As previously mentioned, an important property of the generating functions $\mathcal{Z}_0(gr)$, $\mathcal{Z}_2(gr)$, $\mathcal{Z}_3(gr), \ldots$ is related to the asymptotic behavior of the function y at large v. From (1.35) we have

$$y = (2m\hbar^2)^{\frac{1}{4}} \times \left(\lambda^{1/2}v^{3/2} + \frac{1}{2\lambda^{1/2}}v^{1/2} - \frac{1}{8\lambda^{3/2}}v^{-1/2} + \frac{2d+1}{4}v^{-1} - \frac{(8\lambda^2\varepsilon - 1)}{16\lambda^{5/2}}v^{-3/2} + \cdots\right), \tag{9.11}$$

see (7.28). Note that the first four terms of this expansion are independent on the energy ε. By integrating y over r we arrive at the expansion of the phase at $r \to \infty$: the same four terms become growing at $r \to \infty$ terms, which can be found explicitly. Their respective coefficients will lead to four constraints which must be imposed on the *Approximant*, see below.

The expansion of $(2m)^{1/2}\mathcal{Z}_0(gr)$ at large r can be transformed into an expansion for large v, namely

$$(2m)^{1/2}\mathcal{Z}_0 = (2m\hbar^2)^{\frac{1}{4}} \times \left(\lambda^{1/2}v^{3/2} + \frac{1}{2\lambda^{1/2}}v^{1/2} - \frac{1}{8\lambda^{3/2}}v^{-1/2} + \frac{1}{16\lambda^{5/2}}v^{-3/2} + \cdots\right). \tag{9.12}$$

This reproduces exactly the expansion (9.11) up to $O(v^{-1/2})$. At large r, the next correction $(2m)^{1/2}\lambda^2\mathcal{Z}_2(gr)$ contributes to the expansion (9.11) in a similar way but starting from $O(v^{-1})$,

$$(2m)^{1/2}\lambda^2\mathcal{Z}_2 = (2m\hbar^2)^{\frac{1}{4}} \times \left(\frac{2d+1}{4}v^{-1} - \frac{d}{2\lambda^{1/2}}v^{-3/2} - \frac{1}{4\lambda}v^{-2} + \cdots\right). \tag{9.13}$$

Note that in the expansion of $(2m)^{1/2}\mathcal{Z}_0(gr)$ plus $(2m)^{1/2}\lambda^2\mathcal{Z}_2(gr)$ at large v reproduces exactly the coefficient in front of $O(v^{-1})$ in the expansion (9.11). The expansion of the next correction $(2m)^{1/2}\lambda^3\mathcal{Z}_3$ starts from the term $O(v^{-3/2})$. Combining this with the results for the first two corrections $\mathcal{Z}_{0,2}$ the sum of three these corrections reproduces the coefficient in front of the $O(v^{-3/2})$ term exactly. Similar "nested" structure occurs for the next corrections.

9.3 The Approximant and Variational Calculus

The first two generating functions, $G_0(r)$ and $G_2(r)$, in the expansion (1.82) are given by

$$G_0(r) = \frac{(2m)^{1/2}}{g^2}\left(\frac{2(-2+3gr)(1+gr)^{3/2}}{15}\right), \tag{9.14}$$

$$G_2(r) = \frac{(2m)^{1/2}}{g^2}\left(\frac{1}{4}\log[1+gr] + d\log\left[1+\sqrt{1+gr}\right]\right). \tag{9.15}$$

The next two functions $G_3(r)$ and $G_4(r)$ are presented explicitly in Appendix H, Part II. Note that the function $G_0(r)$ contains no logarithmic terms.

The generating functions $G_0(r)$ and $G_2(r)$ serve as the building blocks for the construction of the Approximant. Following the general prescription of Chapter 8, see Eq. (2.2) we write the Approximant for the ground state eigenfunction in the exponential representation

$$\Psi^{(0,0)}_{approximation} = e^{-\Phi^{(0,0)}_{approximation}},$$

with

$$\Phi^{(0,0)}_{approximation}(r) = \frac{\tilde{a}_0 + \tilde{a}_1 gr + \tilde{a}_2 r^2 + \tilde{a}_3 gr^3}{\sqrt{\tilde{b}_2^2 + \tilde{b}_3 gr}} - \frac{\tilde{a}_0}{\tilde{b}_2} + \frac{1}{4}\log[\tilde{b}_2^2 + \tilde{b}_3 gr]$$
$$+ d\log\left[\tilde{b}_2 + \sqrt{\tilde{b}_2^2 + \tilde{b}_3 gr}\right], \tag{9.16}$$

see [Del Valle and Turbiner (2019)]. Here $\tilde{a}_{0,1,2,3}, \tilde{b}_{2,3}$ are six free parameters. Without loss of generality one can choose $\tilde{b}_2(\tilde{b}_3) = 1$, which is a normalization of the ratio in the second line of (9.16). The constraint on the coefficient in front of the logarithmic term $(1/4+d/2)\log r$, see the 4th term in (9.11), is already realized.

9.3.1 *The First Approximate Solution*

By making preliminary variational calculations with trial function (9.16) with different d and g one can see that the optimal parameters $\tilde{a}_3$ and $\tilde{b}_3$ are related in a way that reproduces with high accuracy the leading semiclassical behavior at large r of the exact ground state eigenfunction $\sim\exp(-\frac{2}{5}g^{1/2}r^{5/2})$. Thus, we can set

$$\tilde{a}_3 = \frac{2}{5}\tilde{b}_3^{1/2}, \tag{9.17}$$

in order to reproduce *exactly* the leading term of the asymptotic behavior of the phase as $r \to \infty$, see the 1st term in (9.11). Additionally, let us impose the constraint

$$\tilde{a}_1 = \frac{\tilde{b}_3}{4}(2\tilde{a}_0 - d - 1), \tag{9.18}$$

in order to guarantee the vanishing of the derivative Φ'_r at $r = 0$, thus, the absence of a linear term in the expansion of (9.16). Eventually, the *first* Approximant for the ground state eigenfunction, which was presented for the first time in [Del Valle and Turbiner (2019)], depends on three free parameters $\{\tilde{a}_0, \tilde{a}_2, \tilde{b}_3\}$. It reads[1]

$$
\begin{aligned}
{}_1\Psi^{(0,0)}_{approximation}(r) &= \frac{1}{\left(1+\tilde{b}_3 g r\right)^{1/4}\left(1+\sqrt{1+\tilde{b}_3 g r}\right)^{d}} \\
&\quad \times \exp\left(-\frac{\tilde{a}_0 + \tilde{a}_1 g r + \tilde{a}_2 r^2 + \tilde{a}_3 g r^3}{\sqrt{1+\tilde{b}_3 g r}}\right),
\end{aligned} \tag{9.19}
$$

[1]Subindex 1 is reserved to denote the *first* Approximant.

where $\tilde{a}_{3,1}$ are defined by constraints (9.17), (9.18). This function can be used as a trial wavefunction in the variational calculus in order to fix the remaining 3 free parameters.

Now let us consider the excited state with quantum numbers (n_r, ℓ). By taking the mixed exponential representation (7.16),

$$\Psi = r^{\ell} P(r^2) e^{-\frac{1}{\hbar}\Phi(r)},$$

and the introducing $y(r) = \partial_r \Phi(r)$, see (7.18), one can show that as a result of the analysis of the general radial Riccati equation (7.17), the expansion (9.11) remains almost unchanged

$$y = (2m\hbar^2)^{\frac{1}{4}} \times \left(\lambda^{1/2} v^{3/2} + \frac{1}{2\lambda^{1/2}} v^{1/2} - \frac{1}{8\lambda^{3/2}} v^{-1/2} \right.$$
$$\left. + \frac{2d + 8n_r + 4\ell + 1}{4} v^{-1} + \cdots \right), \tag{9.20}$$

where v is given by (7.28). In particular, the first four terms of this expansion remain independent on the energy ε. By integrating y over r we arrive at the expansion of the phase $\Phi(r)$ at $r \to \infty$: the same four terms remain growing at $r \to \infty$, they can be found explicitly. The only change that occurs in those four terms is that the coefficient in front of the logarithmic term becomes

$$(1/4 + d/2 + 2n_r + \ell) \log r,$$

see the 4th term in (9.20). Eventually, this leads to a modification of the coefficient in front of the second logarithmic term in G_2, see (9.15),

$$d \to d + 4n_r + 2\ell,$$

while G_0 remains unchanged,

$$G_2(r) = \frac{(2m)^{1/2}}{g^2} \left(\frac{1}{4} \log[1 + gr] + (d + 4n_r + 2\ell) \log \left[1 + \sqrt{1 + gr}\right] \right). \tag{9.21}$$

By taking this fact into account the *first* Approximant for the excited states has the form

$$
{}_1\Psi^{(n_r,\ell)}_{approximation}(r) = \frac{r^\ell P^{(\ell)}_{n_r}(r^2)}{\left(1+\tilde{b}_3 gr\right)^{1/4}\left(1+\sqrt{1+\tilde{b}_3 gr}\right)^{d+4n_r+2\ell}} \times \exp\left(-\frac{\tilde{a}_0+\tilde{a}_1 gr+\tilde{a}_2 r^2+\tilde{a}_3 gr^3}{\sqrt{1+\tilde{b}_3 gr}}\right), \qquad (9.22)
$$

cf. (9.19), where $P^{(\ell)}_{n_r}(r^2)$ is a polynomial of degree n_r in r^2 with all real positive roots. The coefficients of this polynomial $P^{(\ell)}_{n_r}(r^2)$ are found by imposing the n_r orthogonality conditions:

$$
(\Psi^{(n_r,\ell)}, \Psi^{(k_r,\ell)}) = 0, \quad k_r = 0, \ldots (n_r - 1).
$$

After imposing the orthogonality conditions and constraints (9.17), (9.18) on $a_{3,1}$, the function (9.22) depends eventually on 3 free parameters $\{\tilde{a}_0, \tilde{a}_2, \tilde{b}_3\}$, which are found by making minimization of the variational energy. For all studied states the optimal parameters $\{\tilde{a}_0, \tilde{a}_2, \tilde{b}_3\}$ demonstrate smooth behavior versus g and d.

The computation of the variational energy with the trial function $\Psi^{(n_r,\ell)}_{approximation}$ (9.19) requires to carry out a numerical evaluation of the integrals standing in the numerator and the denominator of the expectation value of the Hamiltonian, the integrals defining the scalar products in the orthogonality constraints, and also a numerical minimization. For concrete calculations [Del Valle and Turbiner (2019)] the computational code was written in FORTRAN-90 with use of the integration routine D01FCF from the NAG-LIB. The optimization routine, which was used to find the optimal variational parameters, was performed by using the program MINUIT(ROOT) of CERN-LIB.

Without loss of generality we set for simplicity $\hbar = 1$ and $m = 1/2$, thus, $v = r$, $\varepsilon = E$, $\lambda = g$ and $\mathcal{Y} = y$, see (7.28), (1.29) and (1.31). The calculations of the variational energy for the four low-lying states with quantum numbers (0,0), (1,0), (0,1) and (0,2) for different values of d and g are presented in Tables 9.1–9.4. For example, for the ground state (0,0) the plots of the parameters $\tilde{a}_{0,2,3}$ *vs* g for $d = 2, 3, 6$ are shown in Fig. 9.1.

For some states the variational energy $E_{var} = E^{(1)} = E_0 + E_1$, the first correction E_2 to it as well as its corrected value $E^{(2)} = E_{var} + E_2$

Table 9.1. Ground state energy for the cubic rAHO potential (r^2+gr^3) for $d=1,2,3,6$ and $g=0.1,1,10$, labeled by quantum numbers (0,0) for $d>1$. Variational energy $E_{var}=E_0^{(1)}$, the first correction E_2 (rounded to 3 s.d.) found with ${}_1\Psi^{(0,0)}_{approximation}$, the corrected energy $E_0^{(2)}=E_{var}+E_2$ shown. $E_0^{(2)}$ coincides with LMM, see Appendix G.1 in 9 displayed decimal digits, all printed digits are exact.

	$d=1$			$d=2$		
g	$E_0^{(1)}$	$-E_2$	$E_0^{(2)}$	$E_0^{(1)}$	$-E_2$	$E_0^{(2)}$
0.1	1.053120300	5.39×10^{-7}	1.053119761	2.124027648	4.40×10^{-7}	2.124027208
1.0	1.387428891	4.00×10^{-8}	1.387428851	2.877490906	3.76×10^{-8}	2.877490868
10.0	2.729533139	6.56×10^{-7}	2.729532483	5.794213459	5.58×10^{-7}	5.794212901
	$d=3$			$d=6$		
g	$E_0^{(1)}$	$-E_2$	$E_0^{(2)}$	$E_0^{(1)}$	$-E_2$	$E_0^{(2)}$
0.1	3.208922743	4.00×10^{-7}	3.208922343	6.528432540	2.02×10^{-7}	6.528432338
1.0	4.442965260	3.15×10^{-8}	4.442965229	9.465319951	1.85×10^{-8}	9.465319933
10.0	9.094985589	4.23×10^{-7}	9.094985166	19.981458504	1.96×10^{-7}	19.981458308

Table 9.2. The first excited state energy for the cubic rAHO potential (r^2+gr^3) for different d and g. They labeled by quantum numbers (0,1) for $d>1$, as for $d=1$ it corresponds to 1st excited state (1,0). Variational energy $E_0^{(1)}$, the first correction E_2 found with $\Psi^{(1,0)}_{approximation}$ for $d=1$ and ${}_1\Psi^{(0,1)}_{approximation}$ for $d>1$, the corrected energy $E_0^{(2)}=E_0^{(1)}+E_2$ shown, the correction E_2 rounded to 3 s.d. All printed digits for $E_0^{(2)}$ are exact.

	$d=1$			$d=2$		
g	$E_0^{(1)}$	$-E_2$	$E_0^{(2)}$	$E_0^{(1)}$	$-E_2$	$E_0^{(2)}$
0.1	3.208922765	4.21×10^{-7}	3.208922343	4.305557665	3.55×10^{-7}	4.305557309
1.0	4.442965265	3.59×10^{-8}	4.442965229	6.068723537	2.92×10^{-8}	6.068723507
10.0	9.094985630	4.64×10^{-7}	9.094985166	12.579594377	3.48×10^{-7}	12.579594029
	$d=3$			$d=6$		
g	$E_0^{(1)}$	$-E_2$	$E_0^{(2)}$	$E_0^{(1)}$	$-E_2$	$E_0^{(2)}$
0.1	5.412425220	2.86×10^{-7}	5.412424933	8.784695351	1.21×10^{-7}	8.784695230
1.0	7.745092165	2.41×10^{-8}	7.745092141	13.018486318	1.49×10^{-8}	13.018486303
10.0	16.215748127	2.66×10^{-7}	16.215747861	27.841430199	1.37×10^{-7}	27.841430061

are shown. All states are studied for different values of dimension d and coupling constant g. For all cases the variational energy E_{var} is obtained with absolute accuracy 10^{-7}–10^{-8} (7–8 s.d.). This accuracy is found by calculating the second correction E_2 (the first correction to the variational

Table 9.3. The second excited state energy for the cubic rAHO potential $r^2 + gr^3$ for different d and g labeled by quantum numbers (0,2) for $d > 1$, as for $d = 1$ it corresponds to 2nd excited state $n_r = 2$. Variational energy $E_0^{(1)}$ found with $\Psi^{(2,0)}_{approximation}$ for $d = 1$ and ${}_1\Psi^{(0,2)}_{approximation}$ for $d > 1$, the first correction E_2 and the corrected energy $E_0^{(2)} = E_0^{(1)} + E_2$ shown. Correction E_2 rounded to 3 s.d. All 9 printed decimal digits in $E_0^{(2)}$ are correct.

g	$d = 1$	$d = 2$		
	$E_0^{(1)}$	$E_0^{(1)}$	$-E_2$	$E_0^{(2)}$
0.1	5.436849553	6.528432582	2.43×10^{-7}	6.52843233834
1.0	7.879141644	9.465319955	2.21×10^{-8}	9.46531993256
10.0	16.641305904	19.981458531	2.23×10^{-7}	19.98145830814

g	$d = 3$			$d = 6$		
	$E_0^{(1)}$	$-E_2$	$E_0^{(2)}$	$E_0^{(1)}$	$-E_2$	$E_0^{(2)}$
0.1	7.652743974	1.87×10^{-7}	7.652743787	11.069434802	6.73×10^{-8}	11.069434735
1.0	11.224406591	1.87×10^{-8}	11.224406573	16.699837135	1.22×10^{-8}	16.699837123
10.0	23.860743313	1.78×10^{-7}	23.860743135	36.070426676	1.01×10^{-7}	36.070426576

Table 9.4. Variational energy $E_0^{(1)}$ of the third excited state with quantum numbers (1,0) — the first radial excitation for the rAHO potential $(r^2 + gr^3)$ for $d = 2, 3, 6$ and at $g = 0.1, 1, 10$ and its node $r_0^{(0)}$, found with ${}_1\Psi^{(1,0)}_{approximation}$. Correction E_2 (not shown) contributes systematically to the 7th d.d.

g	$d = 2$		$d = 3$		$d = 6$	
	$E_0^{(1)}$	$r_0^{(0)}$	$E_0^{(1)}$	$r_0^{(0)}$	$E_0^{(1)}$	$r_0^{(0)}$
0.1	6.570942086	0.953377788	7.709696613	1.162457356	11.15814973	1.626236134
1.0	9.690374810	0.780305457	11.517370500	0.941956538	17.128462944	1.289494458
10.0	20.681623429	0.532055331	24.758598615	0.638726047	37.346045552	0.865045854

energy). Furthermore, making comparison of energies $E_0^{(2)}$ with the ones calculated by using the LMM, see Appendix G.1, one can see that an accuracy of 9 decimal digits is reached. This indicates that 3 s.d. taken from the correction E_2 added to 7–8–9th decimal digit in the variational energy $E_0^{(1)}$ allow us to obtain 9 decimal digits of the exact energy. This accuracy is confirmed independently by calculating the second correction E_3 to the variational energy: this correction is $\lesssim 10^{-9}$ for any d and g it has been studied. This indicates to a very fast rate of convergence $\sim 10^{-3}$ of the Non-Linearization Procedure with trial function (9.19) taken as the zero approximation.

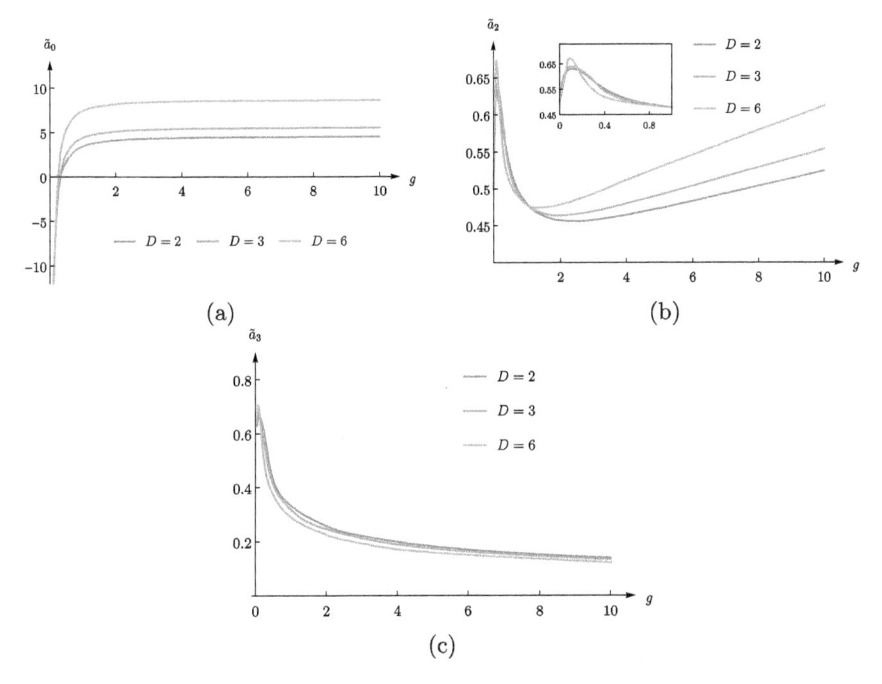

Fig. 9.1. Cubic rAHO: variational parameters $\tilde{a}_0$, $\tilde{a}_2$, $\tilde{a}_3$ for the ground state (0,0) *vs.* g for $D(\equiv d) = 2, 3, 6$ for the function (9.22).

These numerical calculations show that all digits presented for $E_0^{(2)}$ — the variational energy with first (second order in PT) correction E_2 taken into account — in Tables 9.1–9.4 are exact(!): we obtain not less than 9 decimal digits correctly within the Non-Linearization Procedure. Several next corrections $E_{3,4,...}$ can be easily calculated numerically: they increase systematically the number of correct decimal digits. In general, the energies grow with increase of d and/or g. Classifying states by radial quantum number n_r and angular momentum ℓ as (n_r, ℓ), we see a hierarchy of eigenstates: it holds for any fixed integer d and positive g:

$$(0,0), (0,1), (0,2), (1,0).$$

These results represent the first accurate variational calculations (and the corrections to them found in the Non-Linearization Procedure) of the energy of the low-lying states for the d-dimensional cubic AHO, see [Del Valle and Turbiner (2019)].

The Non-Linearization Procedure allows us to estimate a deviation of the Approximant $\Psi^{(n_r,\ell)}_{approximation}$ from the exact wave function $\Psi_{n_r,\ell}$. For the ground state the relative deviation is bounded and very small,

$$\left|\frac{\Psi_{0,0}(r) - \Psi^{(0,0)}_{approximation}(r)}{\Psi^{(0,0)}_{approximation}(r)}\right| \lesssim 10^{-4}, \tag{9.23}$$

in the range of $r \in [0,\infty)$ at any dimension d we explored and at any coupling constant $g \geq 0$ studied. Therefore, the first Approximant leads to a locally accurate approximation of the exact wave function $\Psi_{0,0}(r)$ once the optimal parameters for ${}_1\Psi^{(0,0)}_{approximation}$ are chosen. These optimal parameters are smooth, slow-changing functions in d and g. A similar situation appears for the excited states for different d and g.

The Approximant ${}_1\Psi^{(1,0)}_{approximation}$ also provides an accurate estimate of the position of the radial node of the exact wavefunction, see (9.22). The orthogonality condition imposed to ${}_1\Psi^{(0,0)}_{approximation}$ and ${}_1\Psi^{(1,0)}_{approximation}$ provides a simple analytic expression for the zero order approximation $r_0^{(0)}$ to r_0. Making comparison of $r_0^{(0)}$ with numerical estimates which come from the LMM with 50 optimized mesh points, see Appendix G of Part II, we can see the coincidence of r_0 and $r_0^{(0)}$ for, at least, 5 d.d. at integer dimension d at any coupling constant $g \geq 0$. Results are presented in Table 9.4. Note that at fixed d the node position decreases monotonously with growth of g, while at fixed g it increases with the growth of d.

In all studied cases the first correction y_1 to the logarithmic derivative y of the ground state function is not bounded: it grows at large r as $\sim r^{3/2}$. However, the ratio $|y_1/y_0|$ is bounded and small, thus, y_1 is a *small* function in comparison with y_0. In Figs. 9.2, 9.3, 9.4 it is presented y_0 and y_1 *vs.* r for $g = 1$ in physics dimensions $d = 1, 2, 3$ for the ground state. Similar plots appear for $d > 3$. Making analysis of these plots one see that in domain $0 < r \lesssim 3$, which gives dominant contribution to the integrals defining the energy corrections, the $|y_1|$ is extremely small comparing to $|y_0|$ being close to zero. It explains why the energy correction E_2 is small being the order of $\sim 10^{-7}$, or even $\sim 10^{-8}$ depending on g. In similar way one can show numerically that the higher corrections $y_2, y_3, \ldots$ drop down to zero in the domain $0 < r \lesssim 3$ even faster, indicating the expected fast convergence both y_n and $E^{(n)}$ as $n \to \infty$. Similar behavior of corrections occurs for the excited states.

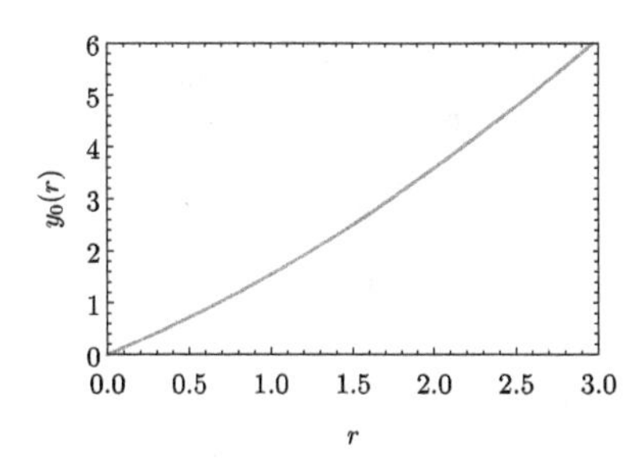

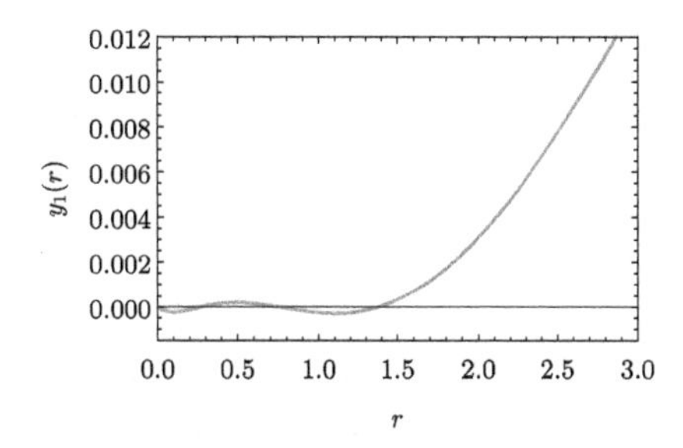

Fig. 9.2. Ground state: Function $y_0 = (\Phi^{(0,0)}_{approximation})'_r$ (9.16) (on left) and its first order correction y_1 (on right) as a function in r at $d = 1$ and $g = 1$.

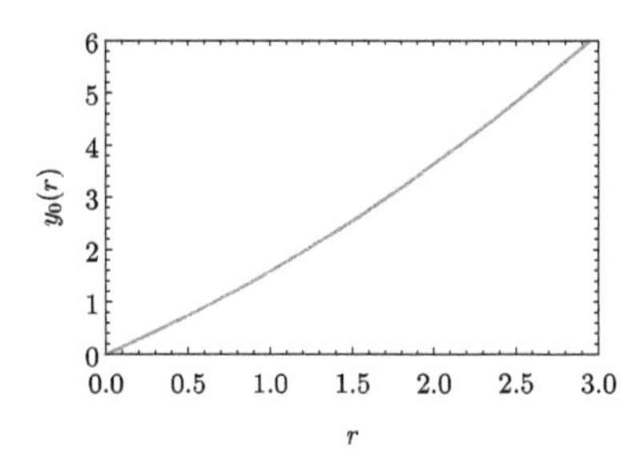

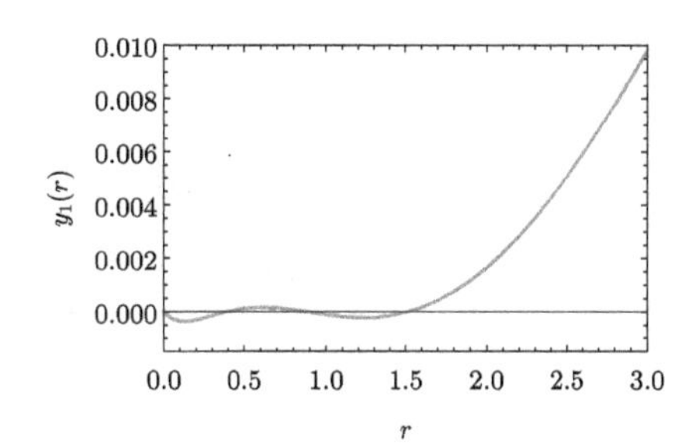

Fig. 9.3. Ground state: Function $y_0 = (\Phi^{(0,0)}_{approximation})'_r$ (9.16) (on left) and its first order correction y_1 (on right) as a function of r at $d = 2$ and $g = 1$.

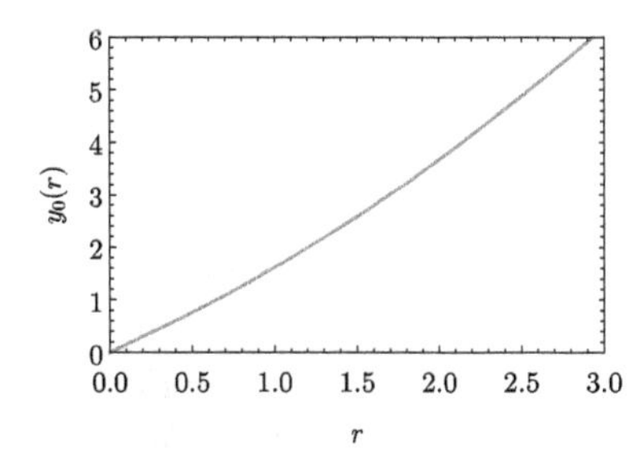

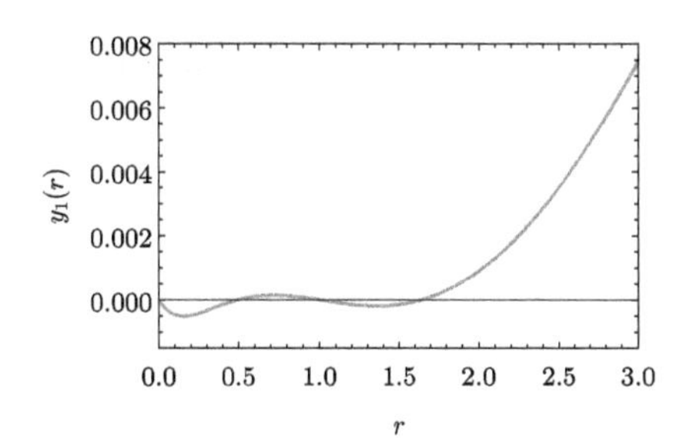

Fig. 9.4. Ground state: Function $y_0 = (\Phi^{(0,0)}_{approximation})'_r$ (9.16) (on left) and its first order correction y_1 (on right) as a function of r at $d = 3$ and $g = 1$.

9.3.2 *The Second Approximate Solution*

As was mentioned above the first four terms of the expansion of y (9.20) remain independent on the energy ε. Integrating y over r we will arrive at the expansion of the phase $\Phi(r)$ at $r \to \infty$: now the same four terms become growing at $r \to \infty$, all their coefficients can be found explicitly. Let us require that in the expansion of the approximate phase $\Phi^{(n_r,\ell)}_{approximation}(r)$ at $r \to \infty$ the same *four growing terms* are reproduced with the same coefficients as for the exact y. It leads to four constraints: three of them are on the parameters $\tilde{a}_{0,1,2,3,4}, \tilde{b}_{2,3}$ and the fourth one is on the degrees of the prefactors. The condition (9.18), in order to guarantee vanishing the derivative Φ'_r at $r = 0$, thus, the absence of linear term in the expansion of (9.16), is relaxed. By changing the normalization in the denominator in (9.16): $\tilde{b}_3 = 1$ and by renaming $\tilde{b}_2 \to B$, $\tilde{a}_0 \to A$ we arrive at the *second* approximate trial function,

$$
{}_2\Psi^{(n_r,\ell)}_{(approximation)} = \frac{r^\ell P_{n_r}(r^2)}{(B^2+gr)^{\frac{1}{4}}(B+\sqrt{B^2+gr})^{d+4n_r+2\ell}} \times \exp\left(-\frac{A-(3B^4-10B^2+15)\frac{r}{60g}+(3B^2+5)\frac{r^2}{15}+\frac{2g}{5}r^3}{\sqrt{B^2+gr}}+\frac{A}{B}\right), \tag{9.24}
$$

cf. (9.22). Optimal parameters A, B are easily found in the variational calculus. They are smooth, slow-changing functions of d and g. It can be checked that y_0 remains almost unchanged, see for the ground state Figs. 9.2, 9.3, 9.4 (on left), while the first correction $y_1(r)$ changes dramatically: it becomes bounded function, $|y_1(r)| \leq$ const, cf. (1.57) as well as all higher y_n-corrections at $n > 1$. This guarantees convergence of the Non-Linearization Procedure in rigorous mathematical sense. Similar property holds for all excited states. However, this remarkable property comes with a certain cost: the accuracy of the variational energies deteriorates with respect to ones obtained with the first approximate trial function ${}_1\Psi^{(n_r,\ell)}_{(approximation)}(r)$ (9.22).

Variational energies obtained with function (9.24) for the first 4 eigenstates at $g = 0.1, 1, 10$ for $d = 1, 2, 3, 6$ are presented in Tables 9.5–9.8, where they are compared with the exact results from LMM, see Appendix G, Part II. Accuracy of the variational energies remains on the level of 4–5 d.d. (before rounding). However, taking into account the first correction E_2 to the variational energy E_{var} allows us to get the 9 d.d. correctly in the corrected energy $E_0^{(2)} = E_{var} + E_2$. This indicates to much

Table 9.5. Ground state energy found in LMM for the cubic rAHO potential $V(r) = r^2 + gr^3$ for $d = 1, 2, 3, 6$. Variational energy, found with ${}_2\Psi^{(0,0)}_{approximation}$, given by underlined digits.

g	$d = 1$	$d = 2$	$d = 3$	$d = 6$
0.1	1.053119761	2.124027208	3.208922343	6.528432338
1.0	1.387428851	2.877490868	4.442965229	9.465319933
10.0	2.729532483	5.794212901	9.094985166	19.981458308

Table 9.6. First excited state energy found in LMM for the cubic rAHO potential $V(r) = r^2 + gr^3$ for $d = 1, 2, 3, 6$. Variational energy, found with ${}_2\Psi^{(0,1)}_{approximation}$ at $d > 1$, given by underlined digits.

g	$d = 1$	$d = 2$	$d = 3$	$d = 6$
0.1	3.208922343	4.305557309	5.41242493	8.784695230
1.0	4.442965229	6.068723507	7.745092141	13.018486303
10.0	9.094985166	12.579594029	16.215747861	27.841430061

Table 9.7. Second excited state energy found in LMM for the cubic rAHO potential $V(r) = r^2 + gr^3$ for $d = 1, 2, 3, 6$. Variational energy, found with ${}_2\Psi^{(0,2)}_{approximation}$ at $d > 1$, given by underlined digits.

g	$d = 1$	$d = 2$	$d = 3$	$d = 6$
0.1	5.436856575	6.528432338	7.652743787	11.069434735
1.0	7.879141595	9.465319933	11.224406573	16.699837123
10.0	16.641305277	19.981458308	23.860743135	36.070426576

Table 9.8. Third excited state energy found in LMM for the cubic rAHO potential $V(r) = r^2 + gr^3$ for $d = 1, 2, 3, 6$. Variational energy, found with ${}_2\Psi^{(1,0)}_{approximation}$ at $d > 1$, given by underlined digits. Node position is not shown.

g	$d = 2$	$d = 3$	$d = 6$
0.1	6.570941476	7.709696057	11.158149471
1.0	9.690376045	11.517370455	17.128462904
10.0	20.681622942	24.758598047	37.346045150

higher rate of convergence in the energy than in the case of the first approximate trial function ${}_1\Psi^{(n_r,\ell)}_{(approximation)}$ (9.22). It seems natural to investigate the case of the second approximate trial function with the constraint on the parameter A imposed to guarantee the absence of the linear in r term in the expansion of the phase at small r. It is not done.

9.4 The Strong Coupling Expansion

The strong coupling expansion for the energy of any excited state of cubic rAHO can be developed in the form,

$$E = g^{2/5}\left(\tilde{\varepsilon}_0 + \tilde{\varepsilon}_1 g^{-4/5} + \tilde{\varepsilon}_2 g^{-8/5} + \cdots\right), \tag{9.25}$$

assuming for simplicity $2m = \hbar = 1$. Evidently, it has a finite radius of convergence. This expansion corresponds to PT in powers of $\hat{\lambda} \equiv g^{-4/5}$ for the cubic potential

$$V(w) = w^3 + \hat{\lambda} w^2, \tag{9.26}$$

defined in $w \in [0, \infty)$. Later we present the explicit calculations of the first two terms in the strong coupling expansion (9.25) for the ground state energy.

As the first step we focus on calculation of the $\tilde{\varepsilon}_0$ in (9.25), which is, in fact, the ground state energy in the cubic radial oscillator potential $V = gr^3$ in the ultra-strong coupling regime $g \to \infty$, see Appendix G.1 of Part II. In order to do it we begin with one of the simplest *physically relevant trial function*,

$$\Psi_{0,0}^{(trial)} = e^{-\frac{2}{5} w^{5/2}}, \tag{9.27}$$

for the ground state in cubic AHO potential (9.26) at $\hat{\lambda} = 0$. Taking $\Psi_{0,0}^{(trial)}$ as zero approximation we calculate variational energy analytically and develop the PT in Non-Linearization procedure, described in Section 1.4.1, numerically. In this manner we obtain different PT corrections $\tilde{\varepsilon}_{0,k}$, $k = 0, 1, 2, 3, \ldots$ to the exact energy,

$$\tilde{\varepsilon}_0 = \tilde{\varepsilon}_{0,0} + \tilde{\varepsilon}_{0,1} + \tilde{\varepsilon}_{0,2} + \cdots, \tag{9.28}$$

and then form the partial sums. The first non-trivial partial sum $(\tilde{\varepsilon}_{0,0} + \tilde{\varepsilon}_{0,1})$ is equal to variational energy $\tilde{\varepsilon}_0^{(1)}$. Explicit results including partial sums up to sixth order are presented in Table 9.9 for the one-dimensional case, $d = 1$. We have to note a slow convergence to the exact result with increase of the number of perturbative terms in expansion (9.28) taken into account. Even the partial sum $\sum_{k=0}^{6} \tilde{\varepsilon}_{0,k}$ with seven terms included allows us to reproduce 7 s.d. *only* of the highly-accurate result obtained in LMM with 50 optimized mesh points, see Appendix G.1 of Part II. It is natural to try to accelerate convergence taking a more advanced trial function than (9.27) as the entry.

Table 9.9. Different partial sums (9.28) for the leading term $\tilde{\varepsilon}_0$ of the strong coupling expansion (9.25) for the cubic rAHO at $d = 1$ in PT with trial function (9.27) as zero approximation. Exact $\tilde{\varepsilon}_0$ obtained via LMM with 50 optimized mesh points, see Appendix G.1 where all printed digits are correct.

$\tilde{\varepsilon}_{0,0}$	$\sum_{k=0}^{1} \tilde{\varepsilon}_{0,k}$	$\sum_{k=0}^{2} \tilde{\varepsilon}_{0,k}$	$\sum_{k=0}^{3} \tilde{\varepsilon}_{0,k}$	$\sum_{k=0}^{4} \tilde{\varepsilon}_{0,k}$	$\sum_{k=0}^{5} \tilde{\varepsilon}_{0,k}$	$\sum_{k=0}^{6} \tilde{\varepsilon}_{0,k}$	Exact $\tilde{\varepsilon}_0$
0	1.053006976	1.021174929	1.022989568	1.022956899	1.022946414	1.022947763	1.022947875

As for the second term $\tilde{\varepsilon}_1$ in the expansion (9.25) it can be estimated by calculating the expectation value of w^2 with (9.27),

$$\tilde{\varepsilon}_1 = \frac{\langle \Psi_{0,0}^{(trial)} | w^2 | \Psi_{0,0}^{(trial)} \rangle}{\langle \Psi_{0,0}^{(trial)} | \Psi_{0,0}^{(trial)} \rangle} = 0.495, \tag{9.29}$$

This straightforward estimate with simplest trial function (9.27) provides $\tilde{\varepsilon}_1$ with accuracy $\sim$20%, see below. In similar way how it was done for $\tilde{\varepsilon}_0$ one can develop PT for $\tilde{\varepsilon}_1$ in (9.25). This PT is fast convergent: the first PT correction to (9.29) is equal to -0.086: it leads to $\tilde{\varepsilon}_1 = 0.409$ which differs from the exact value in $\sim$0.5%, see Table 9.11.

As an alternative to the trial function (9.27) let us use the first Approximant (9.16) in order to calculate the first two terms in the strong coupling expansion (9.25) for the ground state. In Table 9.10 we present for different d the leading coefficient of the strong coupling expansion $\tilde{\varepsilon}_0$ found variationally $\tilde{\varepsilon}_0^{(1)}$ and with the second PT correction $\tilde{\varepsilon}_2$ included, which was calculated via the Non-Linearization Procedure; here we introduce the partial sum $\tilde{\varepsilon}_0^{(2)} = \tilde{\varepsilon}_0^{(1)} + \tilde{\varepsilon}_2$. The final results are verified using the LMM with 50 optimized mesh points, see Appendix G.1. One can see that systematically the correction $\tilde{\varepsilon}_2$ is of order of $\sim 10^{-7}$. Hence, the first 6 d.d. in variational energy are correct: it defines the accuracy of variational calculations of $\tilde{\varepsilon}_0$ with the first Approximant (9.16) as the trial function. One can estimate the order of the third PT correction in the strong coupling expansion (9.25): for all studied d there exists a relation

$$\tilde{\varepsilon}_3 \sim 10^{-2} \tilde{\varepsilon}_2 .$$

It indicates to a high rate of convergence $\sim 10^{-2}$ of this PT. In Table 9.11 we present as well the first two approximations of the coefficient $\tilde{\varepsilon}_1$ in (9.25).

9.5 Cubic rAHO: Conclusions

Summarizing the studies of d-dimensional cubic rAHO for fixed d, one can imagine that the numerical results for given eigenvalue can be *described* via a simple analytical formula. Inspiration comes from the fact that for all studied eigenvalues *vs.* coupling constant g the obtained behavior is presented as very smooth, slow growing function. Basic idea is to interpolate the expansions in g for the weak and strong coupling regimes, see (1.45) and (9.25). It has been already shown that this approach is appropriate and is successful in description of various physical systems of different nature,

Table 9.10. Ground state energy $\tilde{\varepsilon}_0$ (see (9.25)) for the power-like potential $W = r^3$ for $d = 1, 2, 3, 6$ found in PT with the Approximant ${}_1\Psi^{(0,0)}_{approximation}$ as the entry: $\tilde{\varepsilon}_0^{(1)}$ is the variational energy, $\tilde{\varepsilon}_2$ is the second PT correction and $\tilde{\varepsilon}_0^{(2)} = \tilde{\varepsilon}_0^{(1)} + \tilde{\varepsilon}_2$ is corrected variational energy. Eight decimal digits in $\tilde{\varepsilon}_0^{(2)}$ are correct: they confirmed in LMM calculation, see Appendix G.1.

$d=1$			$d=2$		
$\tilde{\varepsilon}_0^{(1)}$	$-\tilde{\varepsilon}_2$	$\tilde{\varepsilon}_0^{(2)}$	$\tilde{\varepsilon}_0^{(1)}$	$-\tilde{\varepsilon}_2$	$\tilde{\varepsilon}_0^{(2)}$
1.022948250	3.75×10^{-7}	1.022947875	2.187461809	3.09×10^{-7}	2.187461499
$d=3$			$d=6$		
$\tilde{\varepsilon}_0^{(1)}$	$-\tilde{\varepsilon}_2$	$\tilde{\varepsilon}_0^{(2)}$	$\tilde{\varepsilon}_0^{(1)}$	$-\tilde{\varepsilon}_2$	$\tilde{\varepsilon}_0^{(2)}$
3.450562918	2.29×10^{-7}	3.450562689	7.647118254	1.01×10^{-7}	7.647118153

Table 9.11. Subdominant (next-to-leading) term $\tilde{\varepsilon}_1$ in the strong coupling expansion (9.25) of the ground state energy for the cubic rAHOpotenti al for $d = 1, 2, 3, 6$.

$d=1$			$d=2$		
$\tilde{\varepsilon}_1^{(1)}$	$\tilde{\varepsilon}_{1,1}$	$\tilde{\varepsilon}_1^{(2)}$	$\tilde{\varepsilon}_1^{(1)}$	$\tilde{\varepsilon}_{1,1}$	$\tilde{\varepsilon}_1^{(2)}$
0.410598524	6.78×10^{-7}	0.410599202	0.766573847	5.24×10^{-7}	0.766574371
$d=3$			$d=6$		
$\tilde{\varepsilon}_1^{(1)}$	$\tilde{\varepsilon}_{1,1}$	$\tilde{\varepsilon}_1^{(2)}$	$\tilde{\varepsilon}_1^{(1)}$	$\tilde{\varepsilon}_{1,1}$	$\tilde{\varepsilon}_1^{(2)}$
1.092125224	2.67×10^{-7}	1.092125491	1.967599668	1.42×10^{-7}	1.967599810

e.g. [Del Valle and Nader (2018)] and [Olivares-Pilón and Turbiner (2018)] as well as one-dimensional AHO, see Part I, Chapters 5–6.

For the ground state of the cubic rAHO potential, the simplest interpolation of the weak and strong coupling expansions has the form

$$E(g) \sim d(1 + ag + b^5 g^2)^{\frac{1}{5}}, \tag{9.30}$$

where a is a free parameter whilst we set $b = (\tilde{\varepsilon}_0/d)$. This interpolation reproduces exactly the leading terms in both expansions (1.45) and (9.25). Parameter a is fixed by fit of numerical data with requirement of minimal χ^2. Optimal parameter a and the parameter b are presented in Table 9.12. For $d = 1, 2, 3, 6$ the above simple formula describes the ground state energy

Table 9.12. Ground state energy of cubic AHO: Parameters of fit (9.30) for different d.

Parameter	$d=1$	$d=2$	$d=3$	$d=6$
a	3.281	3.922	4.823	5.994
b	1.023	1.094	1.150	1.275

E for any $g \in [0, \infty)$ with accuracy $\lesssim 2\%$. Straightforwardly modified formula (9.30) with different a, b works very well for the energies of excited states leading to a similar accuracy $\sim 2\%$.

For one-dimensional quartic AHO $V = x^2 + gx^4$, see Chapter 5 of Part I, it was discovered long ago by Bender and Wu [Bender and Wu (1969)] and recently was proved rigorously by Eremenko and Gabrielov [Eremenko and Gabrielov (2009)] that even (odd) parity eigenvalues form infinitely-sheeted Riemann surface in space of coupling constant g. This surface is characterized by infinitely many square-root branch points situated in the complex g-plane having a meaning of the points of level crossings at complex g. Each Riemann sheet contains infinitely many square-root branch points manifesting that every two energy levels intersect. Seemingly, the same phenomenon holds for any one-dimensional AHO with two term potential $V = x^2 + gx^{2m}$ defined in the whole line $x \in (-\infty, +\infty)$. One can guess that for cubic rAHO $V = r^2 + gr^3$ at $r \in [0, \infty)$ for fixed integer dimension $d > 1$ and fixed value of the angular momentum $\ell = 0, 1, \ldots$ the energies with quantum numbers $(n_r, \ell), n_r = 0, 1, 2, \ldots$ form infinitely-sheeted Riemann surface in variable $\tilde{\lambda} = g^{-4/5}$ with square-root branch points. It is evident that due to the Hermiticity of the original radial Hamiltonian in (9.2) each branch point should have its complex-conjugate. It could be a definite challenge to calculate the distance $\tilde{\lambda}^{\star}$ from the origin to the nearest square-root branch point in the complex plane $\tilde{\lambda}$ and find its d dependence, since it defines the radius of convergence of strong coupling expansion for two intersecting energy levels (n_r, ℓ) and $(\tilde{n}_r, \ell)$. The most important particular case is to find $\tilde{\lambda}^{\star}$ for the ground state $(0, 0)$ and the first excited state $(1, 0)$ (if exists): it probably defines the radius of convergence of the strong coupling expansion (9.25) for the respective states. It is not clear what intersection of the ground state eigenvalue with one of the excited state is closest to the origin in $\tilde{\lambda}$. How many square-root branch points in $\tilde{\lambda}$ exist for a given sheet of the Riemann surface it is also an open question.

Next Chapter will be devoted to quartic radial anharmonic oscillator.

Chapter 10

Radial Quartic Anharmonic Oscillator

In this Chapter we consider the radial anharmonic oscillator (rAHO) with quartic anharmonicity (1.21),

$$V(r) = r^2 + g^2 r^4. \tag{10.1}$$

The corresponding radial Schrödinger equation (without the centrifugal term) has the form

$$\left[-\frac{\hbar^2}{2m}\left(\partial_r^2 + \frac{2\ell + d - 1}{r}\partial_r\right) + r^2 + g^2 r^4\right] \Xi_{n_r,\ell}(r) = E_{n_r,\ell}\Xi_{n_r,\ell}(r), \tag{10.2}$$

see (1.9) at $p = 4$, where the radial wavefunction is defined as

$$\Psi_{n_r,\ell}(r) = r^\ell \Xi_{n_r,\ell}(r),$$

with n_r as the radial quantum number and ℓ as quantum angular momentum. For the logarithmic derivative of the radial function Ξ:

$$y = -\partial_r \log \Xi,$$

one can write the radial Riccati-Bloch (rRB) equation

$$\partial_v \mathcal{Y} - \mathcal{Y}\left(\mathcal{Y} - \frac{2\ell + d - 1}{v}\right) = \varepsilon\,(\lambda) - v^2 - \lambda^2 v^4,\ \partial_v \equiv \frac{d}{dv}, \tag{10.3}$$

cf. (1.30), where

$$v = \left(\frac{2m}{\hbar^2}\right)^{\frac{1}{4}} r, \quad y = (2m\hbar^2)^{\frac{1}{4}}\mathcal{Y}\,(v)\,, \quad E = \frac{\hbar}{(2m)^{\frac{1}{2}}}\varepsilon,$$

cf. (7.28), (1.29), or one can write the radial Generalized Bloch (rGB) equation

$$\lambda^2 \partial_u \mathcal{Z} - \mathcal{Z}\left(\mathcal{Z} - \frac{\lambda^2(2\ell + d - 1)}{u}\right) = \lambda^2 \varepsilon(\lambda) - u^2 - u^4, \partial_u \equiv \frac{d}{du}, \tag{10.4}$$

cf. (1.41), where

$$u = gr, \quad \mathcal{Z}\,(u) = \frac{g}{(2m)^{1/2}} y,$$

cf. (1.39), (1.40). In both equations

$$\lambda = \left(\frac{\hbar^2}{2m}\right)^{\frac{1}{4}} g,$$

plays the role of the *effective* coupling constant.

Study of quantum dynamics in the quartic rAHO potential (10.1) is the main subject of this Chapter. It will be presented in the maximally self-contained form.

10.1 PT in the Weak Coupling Regime

For the quartic rAHO the perturbative expansions of ε and $\mathcal{Y}(v)$, derived from rRB equation (10.3), where v and λ are defined in (7.28) and (1.31), correspondingly, are of the form

$$\varepsilon = \varepsilon_0 + \varepsilon_2 \lambda^2 + \varepsilon_4 \lambda^4 + \varepsilon_6 \lambda^6 + \cdots, \quad \varepsilon_0 = d, \tag{10.5}$$

and

$$\mathcal{Y}(v) = \mathcal{Y}_0(v) + \mathcal{Y}_2(v)\lambda^2 + \mathcal{Y}_4(v)\lambda^4 + \mathcal{Y}_6(v)\lambda^6 + \cdots, \quad \mathcal{Y}_0(v) = v, \tag{10.6}$$

respectively. All terms with odd powers in λ vanish in both expansions. In general, a finite number of corrections can be calculated by linear algebra means. In particular, the first non-vanishing corrections for the ground state are

$$\varepsilon_2 = \frac{1}{4} d(d+2), \quad \mathcal{Y}_2(v) = \frac{1}{2} v^3 + \frac{1}{4}(d+2)v, \tag{10.7}$$

while the next two corrections $\varepsilon_{4,6}$ and $\mathcal{Y}_{4,6}(v)$ are presented in Appendix II. I. In principle, the algebraic procedure of finding PT corrections holds for all even anharmonic potentials $V(r) = V(-r)$, however, it is enough to have a single odd monomial term to occur in the potential this property breaks down. In this situation, the calculation of correction ε_n becomes numerical procedure like it happens for the cubic case, see Chapter 9. Eventually, in contrast to the cubic rAHO, it can be shown that for all even potentials any correction ε_{2n} is a polynomial in d.

In general, all corrections $\mathcal{Y}_{2n}(v)$ are odd-degree polynomials in v of the form,

$$\mathcal{Y}_{2n}(v) = v \sum_{k=0}^{n} c_{2k}^{(2n)} v^{2(n-k)}, \tag{10.8}$$

where any coefficient $c_{2k}^{(2n)}$ is a polynomial in d of degree k,

$$c_{2k}^{(2n)} = P_k^{(2n)}(d), \quad c_{2n}^{(2n)} = \frac{\varepsilon_{2n}}{d}. \tag{10.9}$$

Due to invariance $v \to -v$ of the original equation (10.3) and the formal property that $\mathcal{Y}$ should be antisymmetric, $\mathcal{Y}(-v) = -\mathcal{Y}(v)$, it is convenient to simplify the equation by introducing a new unknown function and changing v-variable,

$$\mathcal{Y} = v\tilde{\mathcal{Y}}(v_2) \quad \text{and} \quad v_2 = v^2,$$

cf. (5.17). As a result, (10.3) is reduced to

$$2v_2\partial_{v_2}\tilde{\mathcal{Y}} - \tilde{\mathcal{Y}}\left(v_2\tilde{\mathcal{Y}} - d\right) = \varepsilon(\lambda) - v_2 - \lambda^2 v_2^2, \quad \partial_{v_2} \equiv \frac{d}{dv_2}. \tag{10.10}$$

This is a convenient form of the rRB equation to carry out the PT consideration. In particular, the first correction (10.7) to $\tilde{\mathcal{Y}}$ becomes a linear function,

$$\tilde{\mathcal{Y}}_2 = \frac{1}{4}\left[2v_2 + (d+2)\right].$$

In general, $\tilde{\mathcal{Y}}_{2n}$ is a polynomial in v_2 of degree n, cf. (10.8),

$$\tilde{\mathcal{Y}}_{2n}(v_2) = \sum_{k=0}^{n} c_{2k}^{(2n)} v_2^{(n-k)}, \tag{10.11}$$

like it is for one-dimensional quartic AHO (5.20). Corrections $\tilde{\mathcal{Y}}_{4,6}$ in the explicit form are presented in Appendix II. I.

The energy corrections ε_{2n} are of the form [Dolgov and Popov (1978)]

$$\varepsilon_{2n}(d) = d(d+2)R_{n-1}(d), \tag{10.12}$$

where $R_{n-1}(d)$ is a polynomial of degree $(n-1)$ in d, in particular, $R_0 = \frac{1}{4}$.

From (10.12) one can see that any energy correction ε_{2n} vanishes when $d = 0$. Consequently, their formal sum results in $\varepsilon(d = 0) = 0$ and ultimately in $E(d = 0) = 0$. Thus, at $d = 0$ the radial Schrödinger equation is reduced to

$$-\frac{\hbar^2}{2m}\left(\frac{d^2\Psi(r)}{dr^2} - \frac{1}{r}\frac{d\Psi(r)}{dr}\right) + (r^2 + g^2r^4)\Psi(r) = 0. \tag{10.13}$$

Needless to say that this equation may define the zero mode of the Schrödinger operator. This can be solved exactly in terms of Airy functions [Dolgov and Popov (1979)],

$$\Psi = C_1 \mathrm{Ai}\left(\frac{1+(\lambda v)^2}{\lambda^{4/3}}\right) + C_2 \mathrm{Bi}\left(\frac{1+(\lambda v)^2}{\lambda^{4/3}}\right), \tag{10.14}$$

where $C_{1,2}$ are arbitrary constants. However, this linear combination can not be made square-integrable at $d = 0$ by any choice of constants C_1 and C_2. Hence, the original assumption that the lowest energy eigenvalue vanishes, $\varepsilon = 0$, is incorrect; it opens the possibility for non-perturbative (exponentially-small in the coupling constant λ) contributions at $d = 0$ in order to have $E \neq 0$. Interestingly, at another non-physical dimension $d = -2$, all corrections ε_{2n} at $n > 1$ also vanish, thus, the formal sum of corrections results in $\varepsilon = \varepsilon_0 = -2$, see (10.12). In this case, no exact solution for the corresponding radial Schrödinger equation is found. It is an open question whether the Schrödinger equation has the solution in the Hilbert space at $\varepsilon = -2$.

10.1.1 *Generating Functions*

As it was mentioned above, one can determine the coefficients in the polynomial correction $\mathcal{Y}_{2n}(v)$, i.e. $c_{2k}^{(2n)}$, $k = 0, 1, \ldots, n$, see (10.8), by algebraic means. However, there is a more efficient procedure to calculate them through constructing their generating functions in $(u = gr)$-space. It will be shown that the correction $\mathcal{Z}_{2n}(u)$ is, in fact, a generating function of the coefficients $c_{2k}^{(2n)}$, $k = 0, 1, 2, \ldots$, see below.

For quartic rAHO the function $\mathcal{Z}(u)$, derived from rGB equation (10.4), can be written as the Taylor expansion in powers of λ namely,

$$\mathcal{Z}(u) = \mathcal{Z}_0(u) + \mathcal{Z}_2(u)\lambda^2 + \mathcal{Z}_4(u)\lambda^4 + \cdots, \tag{10.15}$$

where the coefficient in front of λ^{2k} is, in fact, the generating function for the coefficients $c_{2k}^{(2n)}$,

$$\mathcal{Z}_{2k}(u) = u \sum_{n=k}^{\infty} c_{2k}^{(2n)} u^{2(n-k)}, \quad k = 0, 1, \ldots, \tag{10.16}$$

Here the expansion of ε in powers of λ is given by (10.5). Note that all generating functions $\mathcal{Z}_{2k+1}(u)$, $k = 1, 2, \ldots$ of odd order λ^{2k+1} are absent in the expansion (10.15) as well as in (10.5).

Due to invariance $u \to -u$ it is convenient to simplify (10.4) by introducing

$$\mathcal{Z} = u\tilde{\mathcal{Z}}(u_2) \quad \text{and} \quad u_2 = u^2.$$

Finally, (10.4) is reduced to

$$2\lambda^2 u_2 \partial_{u_2} \tilde{\mathcal{Z}} - \tilde{\mathcal{Z}}\left(u_2 \tilde{\mathcal{Z}} - \lambda^2 d\right) = \lambda^2 \varepsilon(\lambda) - u_2 - u_2^2, \quad \partial_{u_2} \equiv \frac{d}{du_2}. \tag{10.17}$$

It is easy to find explicitly the first two terms of the expansion of $\tilde{\mathcal{Z}}$ in powers of λ^2 coming from equation (10.17), cf. (10.15),

$$\tilde{\mathcal{Z}}_0 = \sqrt{1+u_2}, \tag{10.18}$$

$$\tilde{\mathcal{Z}}_2(u_2) = \frac{u_2 + d\left(1 + u_2 - \sqrt{1+u_2}\right)}{2u_2(1+u_2)}. \tag{10.19}$$

Interestingly, from the polynomial form of the coefficient $c_{2k}^{(2n)}$ in d, see (10.12), one can find the structure of generating function,

$$\tilde{\mathcal{Z}}_{2k}(u_2) = \sum_{n=0}^{k} f_n^{(k)}(u_2) d^n, \tag{10.20}$$

where all $f_n^{(k)}(u_2)$, $n = 0, 1, \ldots k$ are real functions, as a polynomial in d of degree k.

The asymptotic behavior of the generating functions $\mathcal{Z}_{2k}(u)$, $k = 0, 1, 2, \ldots$ in expansion (10.15) at large u is related to the asymptotic behavior of the function y at large r in a quite interesting manner. It can be easily found that for fixed (effective) coupling constant $g(\lambda)$, the asymptotic expansion of y at large r, but rewritten in variable v, see (7.28), has the form

$$y = (2m\hbar^2)^{\frac{1}{4}} \left(\lambda v^2 + \frac{1}{2\lambda} + \frac{d+1}{2} v^{-1} - \frac{4\lambda^2\varepsilon + 1}{8\lambda^3} v^{-2} + \cdots \right), \tag{10.21}$$

where $v \to \infty$. Note that the first three terms of the expansion are independent on the energy ε, the first two terms do not depend on d, while the third term depends on $(d+1)$ as the factor. On the other hand, the first three terms in the expansion at large u of lowest generating function $(\frac{2m}{g^2})^{1/2} \mathcal{Z}_0(u)$ are

$$\left(\frac{2m}{g^2}\right)^{1/2} \mathcal{Z}_0(u) = \left(\frac{2m}{g^2}\right)^{1/2} \left(u^2 + \frac{1}{2} - \frac{1}{8} u^{-2} + \cdots \right), \quad u \to \infty, \tag{10.22}$$

see (1.39), and also they are ε- and d-independent. In order to compare the expansions (10.21) and (10.22), let us replace the classical coordinate u by the quantum coordinate v,

$$u = \lambda v.$$

(evidently, large v implies large u and vice versa as long as λ is fixed). Then the expansion (10.22) becomes

$$\left(\frac{2m}{g^2}\right)^{1/2} \mathcal{Z}_0(\lambda v) = (2m\hbar^2)^{\frac{1}{4}} \left(\lambda v^2 + \frac{1}{2\lambda} - \frac{1}{8\lambda^3} v^{-2} + \cdots \right), \tag{10.23}$$

where $v \to \infty$. It reproduces exactly the first two terms in (10.21) but fails to reproduce $O(v^{-1})$, this term is absent in the expansion. However, the next generating function $(\frac{2m}{g^2})^{1/2} \lambda^2 \mathcal{Z}_2(\lambda v)$ at large v reproduces the term $O(v^{-1})$ exactly in the original expansion (10.21),

$$\left(\frac{2m}{g^2}\right)^{1/2} \lambda^2 \mathcal{Z}_2(\lambda v) = (2m\hbar^2)^{\frac{1}{4}} \left(\frac{d+1}{2} v^{-1} - \frac{d}{2\lambda} v^{-2} + \cdots \right), \tag{10.24}$$

where $v \to \infty$. In turn, it fails to reproduce correctly the term $O(v^{-2})$. Thus, the expansion of the sum $\left(\frac{2m}{g^2}\right)^{1/2} (\mathcal{Z}_0(\lambda v) + \lambda^2 \mathcal{Z}_2(\lambda v))$ at large v reproduces exactly the first three, energy ε-independent terms in the expansion (10.21). These three terms are responsible for the square-integrability of the wavefunction at large v.

All higher generating functions $\mathcal{Z}_4(\lambda v)$, $\mathcal{Z}_6(\lambda v)\ldots$ contribute at large v to the same term $O(v^{-2})$ as follows

$$\left(\frac{2m}{g^2}\right)^{1/2}\lambda^{2n}\mathcal{Z}_{2n}(\lambda v) = (2m\hbar^2)^{\frac{1}{4}}\left(-\frac{\varepsilon_{2n-2}\lambda^{2n-3}}{2}v^{-2}+\cdots\right), \qquad (10.25)$$

at $v \to \infty$ for $n > 2$; here ε_{2n-2} is the energy PT correction of the order $(2n-2)$. As a consequence, no matter how many generating functions we consider in the expansion $(\frac{2m}{g^2})^{1/2}(\mathcal{Z}_0(\lambda v) + \lambda^2\mathcal{Z}_2(\lambda v) + \ldots)$, the term of order $O(v^{-2})$ of (10.21) can not be reproduced exactly.

10.1.2 *The Approximant and Variational Calculations*

The first two generating functions, $G_0(r)$ and $G_2(r)$, in the expansion (1.82) are given by

$$G_0(r;g) = \frac{1}{3g^2}\left(1+g^2r^2\right)^{3/2}, \qquad (10.26)$$

$$g^2G_2(r;g) = \frac{1}{4}\log[1+g^2r^2] + \frac{d}{2}\log\left[1+\sqrt{1+g^2r^2}\right], \qquad (10.27)$$

while the next two generating functions $G_{4,6}(r;g)$ are presented in Appendix I. The generating functions $G_0(r)$ and $G_2(r)$ serve as the building blocks for the construction of the Approximant. Following the general prescription of Chapter 8, see Eq. (2.2) we write the Approximant for the ground state eigenfunction in the exponential representation

$$\Psi^{(0,0)}_{approximation} = e^{-\Phi^{(0,0)}_{approximation}},$$

with

$$\Phi^{(0,0)}_{approximation}(r) = \frac{\tilde{a}_0 + \tilde{a}_2r^2 + \tilde{a}_4g^2r^4}{\sqrt{\tilde{b}_2^2 + \tilde{b}_4g^2r^2}} + \frac{1}{4}\log\left[\tilde{b}_2^2 + \tilde{b}_4g^2r^2\right]$$
$$+ \frac{d}{2}\log\left[\tilde{b}_2 + \sqrt{\tilde{b}_2^2 + \tilde{b}_4g^2r^2}\right], \qquad (10.28)$$

see [Del Valle and Turbiner (2020)]. Here $\tilde{a}_{0,2,4}, \tilde{b}_{2,4}$ are five free parameters. Without a loss of generality one can choose $\tilde{b}_2(\tilde{b}_4) = 1$, which is a normalization of the ratio in the second line of (10.28). It effectively reduces the number of free parameters to four. Two logarithmic terms, added in (10.28), generate the prefactors to the exponential function, in fact, those terms appear just as a certain *minimal* modification of the logarithmic terms, which occur in the second generating function $G_2(r;g)$ (10.27).

10.1.3 *The First Approximate Solution*

As a result, the Approximant of the ground state function for arbitrary $d = 1, 2, 3, \ldots$, which emerges from the Approximant (10.28), is given by

$$
{}_1\Psi^{(0,0)}_{approximation}(r) = \frac{1}{\left(\tilde{b}_2^2 + \tilde{b}_4 g^2 r^2\right)^{1/4} \left(\alpha \tilde{b}_2 + \sqrt{\tilde{b}_2^2 + \tilde{b}_4 g^2 r^2}\right)^{d/2}}
$$

$$
\times \exp\left(-\frac{\tilde{a}_0 + \tilde{a}_2 r^2 + \tilde{a}_4 g^2 r^4}{\sqrt{\tilde{b}_2^2 + \tilde{b}_4 g^2 r^2}} + \frac{\tilde{a}_0}{\tilde{b}_2}\right), \qquad (10.29)
$$

where $\alpha = 1$. This is the central formula of Section. It will be called the *first* approximate trial function for the ground state for quartic rAHO. We set $\tilde{b}_2 = 1$ for future convenience.

It must be emphasized that at $d = 1$ the exponent (10.28) in formula (10.29) coincides with the exponent found in [Turbiner (2005)], [Turbiner (2010)]. But it is slightly different in logarithmic terms, hence, in the form of pre-factors in (10.29), where $\alpha = 0$, with the same asymptotic behavior at $r \to \infty$. This is the consequence of the fact that in the time when [Turbiner (2005)], [Turbiner (2010)] were written, the rGB equation (1.41) was unknown as well as the expansion in generating functions (1.82), (10.15), see also (10.27). This difference leads to non-essential increase in accuracy in variational energy based on (10.29) with respect to ones used in [Turbiner (2005)], [Turbiner (2010)], while the local deviation of these functions does not make much difference for the variational energies.

Variational studies, carried out for different d and g, indicate the existence of the approximate relation between parameters,

$$
\tilde{b}_4 \approx 9\tilde{a}_4^2,
$$

which is fulfilled with very high accuracy. It is easy to check that by setting the relation to be exact,

$$
\tilde{b}_4 = 9\tilde{a}_4^2, \qquad (10.30)
$$

leads to the reproduction of the dominant term $\sim v^2$ in expansion (10.21), hence, the leading term in the asymptotic behavior of the phase at large distances $r \to \infty$. Note that by choosing

$$\tilde{a}_0 = \frac{1}{3g^2}, \quad \tilde{a}_2 = \frac{2}{3}, \quad \tilde{a}_4 = \frac{1}{3}, \tag{10.31}$$

the phase Approximant $\Phi^{(0,0)}_{approximation}$ reproduces exactly the first two terms (10.27) in the expansion in generating functions (1.82). Such a choice of parameters already leads to a highly accurate variational energies (equivalently, the expectation value of the radial Hamiltonian). However, all three parameters $\tilde{a}_0, \tilde{a}_2, \tilde{a}_4$ in (10.31) are far from being optimal from the viewpoint of the variational calculus. Making minimization of the energy with respect to these parameters one can see that they appear as smooth functions in g^2, simultaneously being slow-changing versus d for fixed g^2. Plots of the parameters $\tilde{a}_{0,2,4}$ *vs.* g^2 for $d = 1, 2, 3, 6$ are shown in Fig. 10.1, while $\tilde{a}_{0,2,4}$ *vs.* d for fixed g^2 are shown in Fig. 10.2. It is worth mentioning that at small $g \lesssim 0.1$ the parameters $\tilde{a}_{0,2,4}$ are d-independent.

Making analysis of the parameters $\tilde{a}_{0,2,4}$ *vs.* g^2 for different d one can see the appearance of another, d-independent approximate relation between the parameters,

$$\tilde{a}_2 \approx \frac{1 + 27\tilde{a}_4^2}{18\tilde{a}_4}.$$

Making this approximate relation exact

$$\tilde{a}_2 = \frac{1 + 27\tilde{a}_4^2}{18\tilde{a}_4}, \tag{10.32}$$

corresponds to the fact that the coefficient in front of r — subleading growing term at $r \to \infty$ in the trial phase (10.28) — is reproduced *exactly* in accordance to (10.21). Thus, it can be concluded that the trial phase (10.28), subject to constraints (10.32), (10.30), at large r reproduces exactly all three growing with r terms: r^3, r and $\log r$. Eventually, if we require to reproduce all those terms exactly, the Approximant (10.28) in its final form will contain two free parameters $\{\tilde{a}_0, \tilde{a}_4\}$ ONLY: parameters $\{\tilde{a}_2, \tilde{b}_4\}$ obey constraints (10.32), (10.30), respectively. This will occur as the *second* approximate trial function, see below.

As was indicated in Chapter 9, see also [Del Valle and Turbiner (2019, 2020)], the Approximant of the ground state function $\Psi^{(0,0)}_{approximation}$ can be served as a building block to construct the Approximants of the excited states. Now let us consider the excited state with quantum numbers (n_r, ℓ).

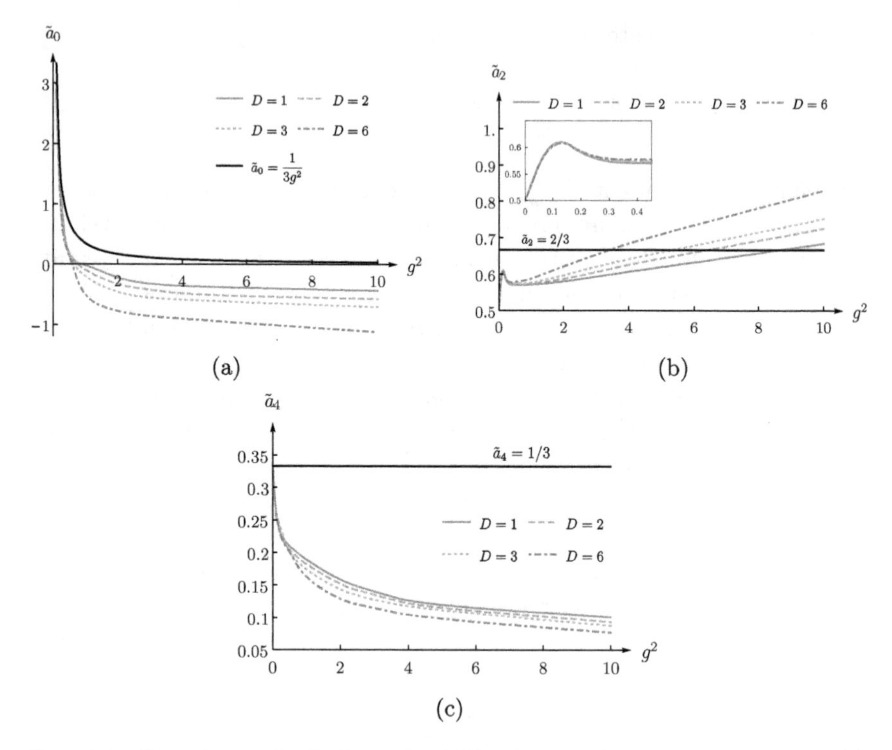

Fig. 10.1. Ground state $(0,0)$ of quartic rAHO: Variational parameters $\tilde{a}_0$ — plot (a), $\tilde{a}_2$ — plot (b) and $\tilde{a}_4$ — plot (c) *vs.* the coupling constant g for $D(\equiv d) = 1, 2, 3, 6$. Parameters (10.31), which allow to reproduce the first two terms G_0, G_2 (10.27) in expansion (10.21) (see text), shown by solid (black) line which is horizontal for $\tilde{a}_2$ — plot (b) and $\tilde{a}_4$ — plot (c).

By taking the mixed exponential representation (7.16),

$$\Psi = r^{\ell} P(r^2) e^{-\frac{1}{\hbar}\Phi(r)},$$

and the introducing $y(r) = \partial_r \Phi(r)$, see (7.18), one can show as a result of the analysis of the general radial Riccati equation (7.17) that the expansion (10.21) remains almost unchanged

$$y = (2m\hbar^2)^{\frac{1}{4}} \left(\lambda v^2 + \frac{1}{2\lambda} + \frac{d + 2n_r + \ell + 1}{2} v^{-1} + \cdots \right), \tag{10.33}$$

where v is given by (7.28), $v \to \infty$. In particular, the first three (shown explicitly) terms of this expansion remain independent on the energy ε. Integrating y over r we will arrive at the expansion of the phase $\Phi(r)$ at $r \to \infty$: the same three terms remain growing at $r \to \infty$, they can be

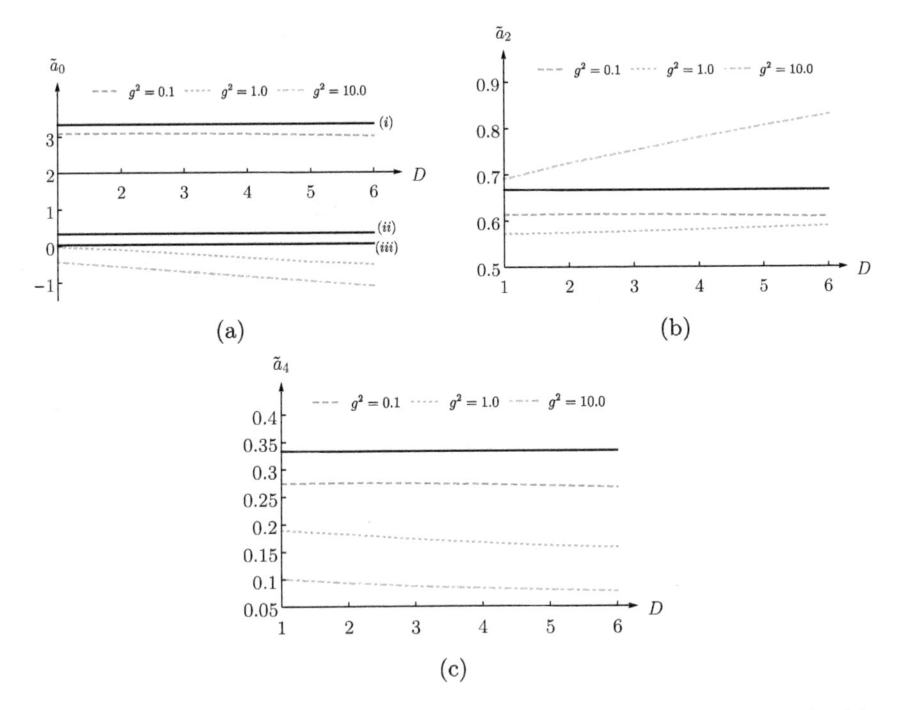

Fig. 10.2. Ground state $(0,0)$ of quartic rAHO: Variational parameters $\tilde{a}_0$ — plot (a), $\tilde{a}_2$ — plot (b), $\tilde{a}_4$ — plot (c) *vs.* $D(\equiv d)$ for fixed $g^2 = 0.1, 1, 10$. $D(\equiv d)$-independent parameters (10.31), which allow to reproduce the first two terms G_0, G_2 (10.27) in expansion (10.21) (see text), shown by solid (black) horizontal lines. In plot (a) the horizontal lines correspond to $\tilde{a}_0 = \frac{1}{3g^2}$ at $g^2 = 0.1(i), 1.(ii), 10.(iii)$.

found explicitly. The only change that occurs in those three terms is in the coefficient in front of the logarithmic term, which becomes

$$(1/4 + d/2 + 2n_r + \ell)\log r,$$

see the 3rd term in (10.33). Eventually, it leads to a modification of the coefficient in front of the second logarithmic term in G_2, see (10.27),

$$d \to d + 4n_r + 2\ell,$$

which is similar to the cubic rAHO, while G_0 remains unchanged,

$$g^2 G_2(r;g) = \frac{1}{4}\log[1 + g^2 r^2] + \frac{d + 4n_r + 2\ell}{2}\log\left[1 + \sqrt{1 + g^2 r^2}\right], \quad (10.34)$$

Final form of the phase approximant for the excited states reads

$$\Phi^{(n_r,\ell)}_{approximation}(r) = \frac{\tilde{a}_0 + \tilde{a}_2 r^2 + \tilde{a}_4 g^2 r^4}{\sqrt{\tilde{b}_2^2 + \tilde{b}_4 g^2 r^2}} + \frac{1}{4}\log\left[\tilde{b}_2^2 + \tilde{b}_4 g^2 r^2\right]$$

$$+ \frac{d + 4n_r + 2\ell}{2}\log\left[\tilde{b}_2 + \sqrt{\tilde{b}_2^2 + \tilde{b}_4 g^2 r^2}\right]. \qquad (10.35)$$

cf. (10.28). By taking this form into account the *first* Approximant for the excited states gets very compact form

$$_1\Psi^{(n_r,\ell)}_{approximation}(r) = \frac{r^\ell P^{(\ell)}_{n_r}(r^2)}{\left(1 + \tilde{b}_4 g^2 r^2\right)^{\frac{1}{4}}\left(1 + \sqrt{1 + \tilde{b}_4 g^2 r^2}\right)^{\frac{d}{2}+2n_r+\ell}}$$

$$\times \exp\left(-\frac{\tilde{a}_0 + \tilde{a}_2 r^2 + \tilde{a}_4 g^2 r^4}{\sqrt{1 + \tilde{b}_4 g^2 r^2}} + \tilde{a}_0\right), \qquad (10.36)$$

cf. (3.22) for cubic rAHO and (5.58) for one-dimensional quartic AHO, where the parameter $\tilde{b}_2$ in (10.28) is fixed equal to one without loosing generality, $\tilde{b}_2 = 1$. Here $P^{(\ell)}_{n_r}(r^2)$ is the polynomial of degree n_r with n_r real *positive* roots, where the coefficient in front of the leading degree is normalized to be equal to one. In a similar way as in the one-dimensional case, for fixed angular momentum ℓ, all n_r free parameters of $P^{(\ell)}_{n_r}(r^2)$ are found by imposing the orthogonality constraints,

$$(\Psi^{(n_r,\ell)}, \Psi^{(k_r,\ell)}) = 0, \quad k_r = 0, \ldots, (n_r - 1). \qquad (10.37)$$

Finally, in order to fix the remaining four free parameters $\tilde{a}_{0,2,4}, \tilde{b}_4$ in exponential we use the Approximant $\Psi^{(n_r,\ell)}_{approximation}$ as entry in variational calculations, where the condition (10.30) is imposed *a priori*. Eventual number of variational parameters is equal to three: $\tilde{a}_{0,2,4}$.

Formula (10.36) is one of the key formulas of the entire book. It provides a highly-accurate approximate solution for the spectra of the quantum radial quartic symmetric anharmonic oscillator in its full generality.

The variational energy calculations for four low-lying states with quantum numbers $(0,0)$, $(0,1)$, $(0,2)$, $(1,0)$ for different values of $d > 1$ and g^2 are presented in Tables 10.1–10.4. For some of these states, the variational energy $E_{var} = E_0^{(1)}$, the first correction E_2 to it, as well as its corrected value of variational energy $E_0^{(2)} = E_{var} + E_2$ are shown. Systematically,

Table 10.1. Ground state $(0,0)$ energy for the quartic rAHO potential $r^2+g^2r^4$ for $d=1,2,3,6$ and $g^2=0.1,1,10$. Variational energy $E_0^{(1)}$, the first correction E_2 (rounded to 3 s.d.) found with ${}_1\Psi^{(0,0)}_{approximation}$ (see text), the corrected energy $E_0^{(2)}=E_0^{(1)}+E_2$ shown. $E_0^{(2)}$ coincides with LMM results (see text and Appendix II. G) in all displayed 12 d.d.

	$d=1$			$d=2$		
g^2	$E_0^{(1)}$	$-E_2$	$E_0^{(2)}$	$E_0^{(1)}$	$-E_2$	$E_0^{(2)}$
0.1	1.065 285 509 544	3.00×10^{-14}	1.065 285 509 544	2.168 597 211 269	5.28×10^{-14}	2.168 597 211 269
1.0	1.392 351 641 563	3.37×10^{-11}	1.392 351 641 530	2.952 050 091 995	3.17×10^{-11}	2.952 050 091 962
10.0	2.449 174 072 588	4.69×10^{-10}	2.449 174 072 118	5.349 352 819 751	3.44×10^{-10}	5.349 352 819 462
	$d=3$			$d=6$		
g^2	$E_0^{(1)}$	$-E_2$	$E_0^{(2)}$	$E_0^{(1)}$	$-E_2$	$E_0^{(2)}$
0.1	3.306 872 013 152	2.20×10^{-13}	3.306 872 013 152	6.908 332 111 232	9.80×10^{-14}	6.908 332 111 232
1.0	4.648 812 704 237	2.69×10^{-11}	4.648 812 704 210	10.390 627 295 514	9.68×10^{-12}	10.390 627 295 504
10.0	8.599 003 455 030	2.22×10^{-10}	8.599 003 454 807	19.936 900 374 076	6.48×10^{-11}	19.936 900 374 011

Table 10.2. Energy of the 1st excited state in the potential $V = r^2 + g^2 r^4$ for different d and g^2 labeled by quantum numbers $(0,1)$. For $d = 1$ it corresponds to 1st negative parity state $n_r = 0$ at $p = 1$ defined in the whole line $(-\infty, \infty)$. Variational energy $E_0^{(1)}$, the first correction E_2 to it found with ${}_1\Psi^{(0,1)}_{approximation}$, the corrected energy $E_0^{(2)} = E_0^{(1)} + E_2$ shown. Displayed correction E_2 rounded to 3 s.d.; $E_0^{(2)}$ coincides with LMM results (see text and Appendix II. G) in all 12 displayed d.d.

	$d=1$			$d=2$		
g^2	$E_0^{(1)}$	$-E_2$	$E_0^{(2)}$	$E_0^{(1)}$	$-E_2$	$E_0^{(2)}$
0.1	3.306 872 013 236	8.33×10^{-11}	3.306 872 013 153	4.477 600 360 878	1.10×10^{-10}	4.477 600 360 768
1.0	4.648 812 707 206	2.99×10^{-9}	4.648 812 704 212	6.462 906 003 251	3.39×10^{-9}	6.462 905 999 864
10.0	8.599 003 467 556	1.27×10^{-8}	8.599 003 454 810	12.138 224 752 729	1.38×10^{-8}	12.138 224 738 901

	$d=3$			$d=6$		
g^2	$E_0^{(1)}$	$-E_2$	$E_0^{(2)}$	$E_0^{(1)}$	$-E_2$	$E_0^{(2)}$
0.1	5.678 682 663 377	1.33×10^{-10}	5.678 682 663 243	9.447 358 518 278	1.80×10^{-10}	9.447 358 518 099
1.0	8.380 342 533 658	3.56×10^{-9}	8.380 342 530 101	14.658 513 816 952	3.39×10^{-9}	14.658 513 813 563
10.0	15.927 096 988 667	1.40×10^{-8}	15.927 096 974 709	28.536 810 849 436	1.21×10^{-8}	28.536 810 837 360

Table 10.3. Energy of the 2nd excited state in the potential $V = r^2 + g^2 r^4$ for different d and g^2 labeled by quantum numbers $(0, 2)$ for $d > 1$, as for $d = 1$ it corresponds to 1st excitation $n_r = 1$ of positive parity $p = 0$: $(1, 0)$. Variational energy $E_0^{(1)}$ found with use of ${}_1\Psi^{(1,0)}_{approximation}$ for $d = 1$ and ${}_1\Psi^{(0,2)}_{approximation}$ for $d > 1$; the first correction E_2 and the corrected energy $E_0^{(2)} = E_0^{(1)} + E_2$ shown. Displayed correction E_2 rounded to 3 s.d. $E_0^{(2)}$ coincides with LMM results (see text and Appendix II. G) in all 12 displayed d.d.

	$d=1$			$d=2$		
g^2	$E_0^{(1)}$			$E_0^{(1)}$	$-E_2$	$E_0^{(2)}$
0.1	5.747 959 269 942			6.908 332 112 167	9.35×10^{-10}	6.908 332 111 232
1.0	8.655 049 995 062			10.390 627 321 799	2.63×10^{-8}	10.390 627 295 506
10	16.635 921 650 401			19.936 900 479 247	1.05×10^{-7}	19.936 900 374 040
	$d=3$			$d=6$		
g^2	$E_0^{(1)}$	$-E_2$	$E_0^{(2)}$	$E_0^{(1)}$	$-E_2$	$E_0^{(2)}$
0.1	8.165 006 438 494	1.00×10^{-9}	8.165 006 437 493	12.084 471 853 886	1.11×10^{-9}	12.084 471 852 776
1.0	12.485 556 075 670	2.47×10^{-8}	12.485 556 051 000	19.217 523 515 555	1.97×10^{-8}	19.217 523 495 879
10.0	24.145 857 689 623	9.48×10^{-8}	24.145 857 594 824	37.811 402 320 699	6.90×10^{-8}	37.811 402 251 702

Table 10.4. Energy E of the 3rd excited state $(1,0)$ for $d = 2, 3, 6$ — the first radial excitation — for the potential $V = r^2 + g^2 r^4$ at $g^2 = 0.1, 1.0, 10.0$ and its node r_0, calculated in LMM with 50,100,200 mesh points, respectively, see Appendix II. G. Variational energy $E_0^{(1)}$ and its node position $r_0^{(0)}$ correspond to underlined digits, both found with ${}_1\Psi^{(1,0)}_{approximation}$ at $d > 1$. First corrections E_2 and $r_0^{(1)}$ to variational results (not shown) contribute to the first non-underline figure (and next ones).

	$d=2$		$d=3$		$d=6$	
g^2	$E_0^{(1)}$	$r_0^{(0)}$	$E_0^{(1)}$	$r_0^{(0)}$	$E_0^{(1)}$	$r_0^{(0)}$
0.1	7.039 707 584	0.918 783 458	8.352 677 825	1.111 521 078	12.415 256 177	1.522 966 591
1.0	10.882 435 576	0.733 724 778	13.156 803 922	0.875 567 486	20.293 829 707	1.166 753 149
10.0	21.175 135 370	0.524 083 057	25.806 276 215	0.621 795 290	40.388 142 970	0.820 068 428

the variational energy $E_0^{(1)}$ is found with extremely high absolute accuracy: $10^{-8} - 10^{-14}$, which is found by calculating the correction E_2. The variational results are compared with numerical ones obtained via LMM, see Appendix II. G for technical details. LMM results are obtained taking 50, 100, 200 optimized mesh points for $g^2 = 0.1, 1., 10.0$, respectively, at different d. It allows us to reach, at least, 12 d.d. in the energies of $(0,0)$, $(0,1)$, $(0,2)$ states at $d = 1,2,3,6$, see Tables 10.1–10.3 denoted as $E_0^{(2)}$. Making analysis of numerical results suggests that, when E_2 is evaluated, the $E_0^{(2)}$ provides 12–13 correct d.d., at least. It implies that, once E_2 is taken into account, all digits of $E_0^{(2)}$ printed in Tables 10.1–10.4 are exact. These highly accurate results are confirmed independently by the calculating the second correction to the variational energy E_3 which is always $\lesssim 10^{-12}$ for all d and g^2 which were studied. It indicates a very fast rate of convergence in the Non-Linearization Procedure when trial function (10.36) is taken as the zero approximation. We must mention the hierarchy of eigenstates which holds for any fixed integer $d > 1$ and g^2:

$$(0,0) \to (0,1) \to (0,2) \to (1,0).$$

Interestingly (but not surprisingly), it coincides with the hierarchy of the first four eigenstates for the cubic rAHO established in Chapter 9.

There is a considerable number of calculations devoted to estimate the energy of the first low-lying states. The results, presented in Tables 10.1–10.4, are in complete agreement with [Turbiner (2005)], [Turbiner (2010)] for $d = 1$ and superior considerably of those obtained for $d > 1$ and different g^2, see e.g. [Taşeli and Eid (1996); Weniger (1996a)] and [Witwit (1992)].

Deviation of ${}_1\Psi^{(n_r,\ell)}_{approximation}$ from the exact (unknown) eigenfunction $\Psi_{(n_r,\ell)}$ can be estimated by using the Non-Linearization Procedure. In particular, it can be shown that for the ground state function this deviation is extremely small and bounded,

$$\left| \frac{\Psi_{(0,0)}(r) - {}_1\Psi^{(0,0)}_{approximation}(r)}{{}_1\Psi^{(0,0)}_{approximation}(r)} \right| \lesssim 10^{-6}, \tag{10.38}$$

in the range $r \in [0,\infty)$ at any dimension d and any g^2 that were considered. Therefore we can say that the first Approximant ${}_1\Psi^{(0,0)}_{approximation}$ is a locally accurate approximation of the exact wave function $\Psi_{(0,0)}$ once the optimal parameters are chosen. Similar situation occurs for the first Approximants for excited states at different d and g^2.

In all cases the first order correction y_1 to the logarithmic derivative of the ground state is a bounded function at different d and g^2 in $r \in [0, 15]$. For example, for $g^2 = 1$ the first correction y_1 has the upper bound

$$|y_1|_{max} \sim \begin{cases} 0.0106, & d = 1 \\ 0.0092, & d = 2 \\ 0.0086, & d = 3 \\ 0.0072, & d = 6 \end{cases} \qquad (10.39)$$

in above domain. It is the consequence of the fact that by construction the derivative of phase $\Phi^{(n_r,\ell)}_{approximation}(r)$ reproduces exactly the growing terms r^3 and $\log r$ at large r in expansion (10.21) and the coefficient in front of the growing term r differs very little from the exact one. "Boundness" of y_1 and its small value of the maximum in domain, which give a dominant contributions in all involved integrals, implies that we deal with smartly designed zero order-approximation of the exact $\Psi_{(n_r,\ell)}$ that leads, in framework of the Non-Linearization Procedure, to a fast convergent series for the energy and wave function. In Figs. 10.3–10.5 y_0 and y_1 *vs.* r are presented for $g^2 = 1$ in physics dimensions $d = 1, 2, 3$. It must be emphasized that all curves in these figures are slow-changing *vs.* d. Therefore, it is not a surprise that similar plots should appear for larger d as well as for other values of g. An analysis of these plots indicates that, in particular, $|y_1|$ is extremely small function in comparison with $|y_0|$ in the domain $0 \le r \lesssim 1.7$, thus, in domain which provides the essentially-dominant contribution in variational integrals. It is the consequence of minimization of the energy functional. It is the real reason why the energy correction E_2 is extremely small being of order $\sim 10^{-8}$, or sometimes even smaller, $\sim 10^{-10}$. Similar situation occurs for the phase (and its derivative) of the Approximants for the excited states.

The Approximant ${}_1\Psi^{(n_r,\ell)}_{approximation}$ (10.36) also allows us to get an accurate estimate of the position of the radial nodes of the exact wave function. For example, in the state (1, 0) where is a single positive node, $r_0 > 0$, the trial function (10.36) provides the zero order estimate of the radial node $r_0^{(0)}$ coming directly from the orthogonality constraint (10.37). Results are presented in Table 10.4. A comparison of these numerical results with those coming from the LMM indicates that the Approximant ${}_1\Psi^{(1,0)}_{approximation}$ defines the node with not less than 5 d.d. From Table 10.4 it can be noted that the radial node is an increasing function of d at fixed g^2, but decreasing with the increase of g^2 at fixed d.

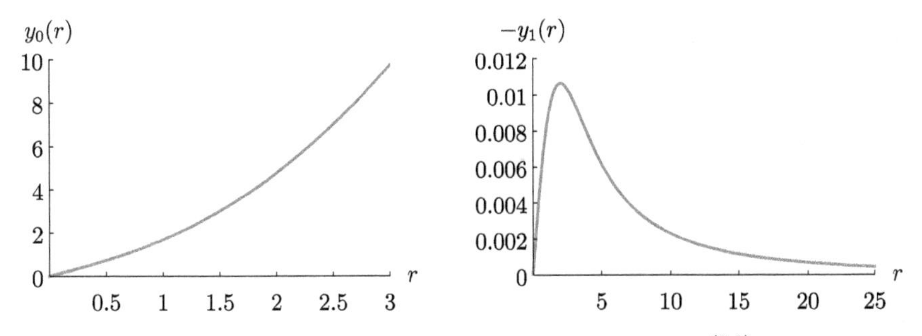

Fig. 10.3. Quartic rAHO at $d = 1$, ground state: function $y_0 = (\Phi^{(0,0)}_{approximation})'$ (on left) and its first correction y_1 (on right) *vs.* r for $g^2 = 1$.

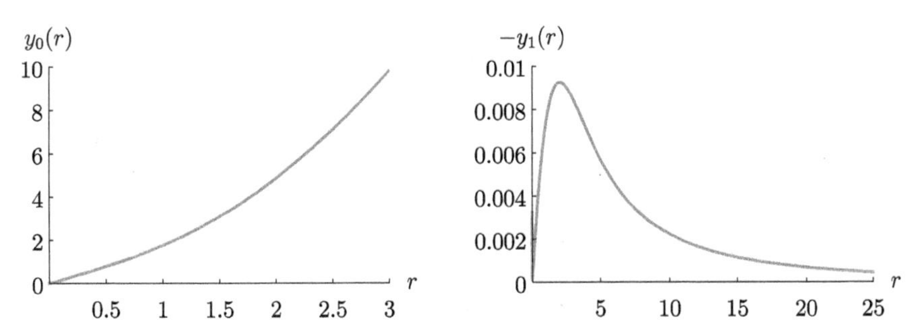

Fig. 10.4. Quartic rAHO at $d = 2$, ground state: function $y_0 = (\Phi^{(0,0)}_{approximation})'$ (on left) and its first correction y_1 (on right) *vs.* r for $g^2 = 1$.

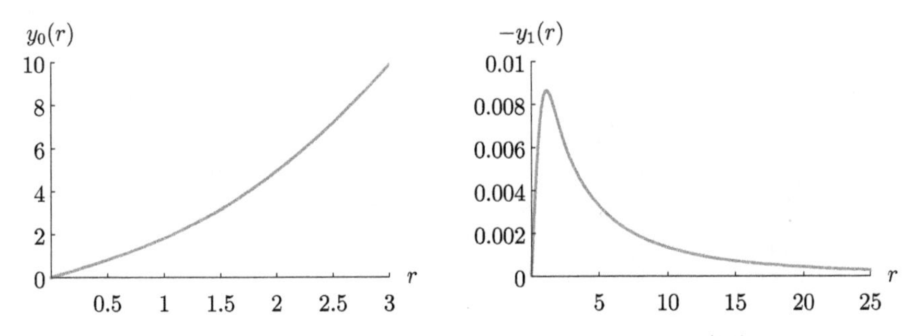

Fig. 10.5. Quartic rAHO at $d = 3$, ground state: function $y_0 = (\Phi^{(0,0)}_{approximation})'$ (on left) and its first correction y_1 (on right) *vs.* r for $g^2 = 1$.

10.1.4 *The Second Approximate Solution*

As was mentioned above the first three terms of the expansion of y (10.33) remain independent on the energy ε. Integrating y over r we will arrive at the expansion of the phase $\Phi(r)$ at $r \to \infty$: now the same three terms become growing at $r \to \infty$, all their coefficients can be found explicitly. Let us require that in the expansion of the approximate phase $\Phi^{(n_r,\ell)}_{approximation}(r)$ at $r \to \infty$ the same *three growing terms* are reproduced with the same coefficients as for the exact y. It leads to three constraints: two of them are on the parameters $\tilde{a}_{0,2,4}, \tilde{b}_{2,4}$ and the third one is on the degrees of the prefactors. By changing the normalization in the denominator in (10.28): $\tilde{b}_4 = 1$ and by renaming $\tilde{b}_2 \to B$, $\tilde{a}_0 \to A$ we arrive at the *second* approximate trial function,

$$
{}_2\Psi^{(n_r,\ell)}_{(approximation)} = \frac{r^\ell P_{n_r}(r^2)}{(B^2+g^2r^2)^{\frac{1}{4}}\left(B+\sqrt{B^2+g^2r^2}\right)^{2n_r+\ell+\frac{d}{2}}} \times \exp\left(-\frac{A+(B^2+3)r^2/6+g^2r^4/3}{\sqrt{B^2+g^2r^2}}+\frac{A}{B}\right), \qquad (10.40)
$$

with two free parameters A, B. Their optimal values are easily found in the variational calculus. They are smooth, slow-changing functions of d and g. It can be checked that y_0 remains almost unchanged, see for the ground state Figs. 10.3, 10.4, 10.5 (on left), while the first correction $y_1(r)$ changes dramatically: it becomes bounded function *everywhere*, $|y_1(r)| \leq$ const, cf. (1.57) as well as all higher y_n-corrections at $n > 1$. This guarantees convergence of the Non-Linearization Procedure in rigorous mathematical sense. Similar property holds for all excited states. However, this remarkable property comes with a certain cost: the accuracy of the variational energies deteriorates with respect to ones obtained with the first approximate trial function ${}_1\Psi^{(n_r,\ell)}_{(approximation)}(r)$ (10.36).

Variational energies obtained with function (10.40) for the first 4 eigenstates at $g = 0.1, 1, 10$ for $d = 1, 2, 3, 6$ are presented in Tables 10.5–10.5, where they are compared with the exact results from LMM, see Appendix G Part II. Accuracy of the variational energies remains on the level of *unprecedented* accuracy 10-11 s.d. (before rounding). Taking into account the first correction E_2 to the variational energy E_{var} allows us to get the extra 3-4 s.d. correctly in the corrected energy $E_0^{(2)} = E_{var} + E_2$.

Table 10.5. Ground state energy in LMM for the quartic AHO potential $V(r) = r^2 + g^2 r^4$ for $d = 1, 2, 3, 6$. Variational energy, found with the *second* approximate trial function ${}_2\Psi^{(0,0)}_{(approximation)}$, given by underlined digits.

g^2	$d = 1$	$d = 2$	$d = 3$	$d = 6$
0.1	1.065 285 509 54 4	2.168 597 211 269	3.306 872 013 153	6.908 332 111 232
1.0	1.392 351 641 530	2.952 050 091 963	4.648 812 704 212	10.390 627 295 504
10.0	2.449 174 072 118	5.349 352 819 406	8.599 003 454 808	19.936 900 374 011

Table 10.6. First excited state energy in LMM for the quartic AHO potential $V(r) = r^2 + g^2 r^4$ for $d = 1, 2, 3, 6$. Variational energy, found with the *second* approximate trial function ${}_2\Psi^{(0,1)}_{(approximation)}$, given by underlined digits.

g^2	$d = 1$	$d = 2$	$d = 3$	$d = 6$
0.1	3.306 872 013 153	4.477 600 360 768	5.678 682 663 243	9.447 358 518 099
1.0	4.648 812 704 212	6.462 905 999 864	8.380 342 530 102	14.658 513 813 566
10.0	8.599 003 454 808	12.138 224 738 900	15.927 096 974 707	28.536 810 837 358

Table 10.7. Second excited state energy in LMM for the quartic AHO potential $V(r) = r^2 + g^2 r^4$ for $d = 1, 2, 3, 6$. Variational energy, found with the *second* approximate trial function ${}_2\Psi^{(0,2)}_{(approximation)}$, given by underlined digits.

g^2	$d = 1$	$d = 2$	$d = 3$	$d = 6$
0.1	5.747 959 268 834	6.908 332 111 232	8.165 006 437 493	12.084 471 852 776
1.0	8.655 049 957 759	10.390 627 295 504	12.485 556 051 000	19.217 523 495 889
10.0	16.635 921 492 413	19.936 900 374 011	24.145 857 594 802	37.811 402 251 692

10.1.5 *The Strong Coupling Expansion*

In this Section assuming $2m = \hbar = 1$ we will focus on finding the first two terms of the strong coupling expansion of the ground state energy for the quartic AHO (10.1),

$$E \equiv g^{2/3}\tilde{\varepsilon} = g^{2/3}(\tilde{\varepsilon}_0 + \tilde{\varepsilon}_2 g^{-4/3} + \tilde{\varepsilon}_4 g^{-8/3} + \cdots). \qquad (10.41)$$

In contrast to the weak coupling expansion, see (10.5) at $\lambda = g$, the expansion (10.41) has a finite radius of convergence in $1/g$. This expansion corresponds to PT in powers of $\hat{\lambda}$ for the potential

$$V(w) = w^4 + \hat{\lambda}^2 w^2, \quad \hat{\lambda}^2 = g^{-4/3}, \qquad (10.42)$$

in the radial Schrödinger equation defined in $w \in [0, \infty)$.

Table 10.8. Third excited state energy in LMM for the quartic AHO potential $V(r) = r^2 + g^2\, r^4$ for $d = 1, 2, 3, 6$. Variational energy, found with the *second* approximate trial function ${}_2\Psi^{(1,0)}_{(approximation)}$, given by underlined digits.

g^2	$d = 2$	$d = 3$	$d = 6$
0.1	7.039 707 584 466	8.352 677 825 786	12.415 256 177 935
1.0	10.882 435 576 820	13.156 803 898 049	20.293 829 707 536
10.0	21.175 135 370 923	25.806 276 215 056	40.388 142 970 115

In order to calculate the first two terms $\tilde{\varepsilon}_0$ and $\tilde{\varepsilon}_2$ of the strong coupling expansion (10.41) we use the first Approximant (10.29). In Table 10.9 for different D the leading coefficient $\tilde{\varepsilon}_0$ and the second perturbative correction $\hat{\varepsilon}_2$ as well as $\tilde{\varepsilon}_0^{(2)} = \tilde{\varepsilon}_0^{(1)} + \hat{\varepsilon}_2$, calculated via the Non-Linearization Procedure, are presented. Numerical results for $\tilde{\varepsilon}_0$, based on the LMM and obtained with 12 d.d., indicate that $\tilde{\varepsilon}_0^{(2)}$ found in Non-Linearization procedure with the Approximant (10.29) reproduce not less that 10 d.d. This accuracy is verified independently by calculating the next correction $\hat{\varepsilon}_3$ which results in the order of

$$\hat{\varepsilon}_3 \sim 10^{-2}\hat{\varepsilon}_2 .$$

In turn, Table 10.10 contains the results of the first two approximations for the coefficient $\tilde{\varepsilon}_2$ in (10.41). It should be mentioned that our final results for the coefficient $\tilde{\varepsilon}_0$ reproduce and sometimes exceed the best results available in literature so far for $d = 1$, see e.g. [Turbiner and Ushveridze (1988); Fernández and Guardiola (1997); Weniger (1996b)].

10.1.6 *Quartic rAHO: Conclusions*

It is shown that the 2(3)-parametric Approximants (10.36), (10.40), taken as variational trial functions for the first four states $(0,0), (0,1), (0,2), (1,0)$ of the quartic radial d-dimensional rAHO with the potential (10.1) provide extremely high[1] relative accuracy in energy ranging from $\sim 10^{-14}$ to $\sim 10^{-8}$ for different coupling constants g and dimension d. Variational parameters depend on g and d in a smooth manner and their dependences can be easily interpolated.

[1] unprecedentedly high accuracy

Table 10.9. Ground state $(0,0)$ energy $\tilde{\varepsilon}_0$ for the power-like potential $W = r^4$ (see (10.42)) for $d = 1, 2, 3, 6$ found in PT based on the Approximant ${}_1\Psi^{(0,0)}_{(approximation)}$: $\tilde{\varepsilon}_0^{(1)}$ corresponds to the variational energy, $\hat{\varepsilon}_2$ is the second PT correction, $\tilde{\varepsilon}_0^{(2)} = \tilde{\varepsilon}_0^{(1)} + \hat{\varepsilon}_2$ is the corrected variational energy. 10 d.d. in $\tilde{\varepsilon}_0^{(2)}$ confirmed independently in LMM.

	$d = 1$			$d = 2$	
$\tilde{\varepsilon}_0^{(1)}$	$-\hat{\varepsilon}_2$	$\tilde{\varepsilon}_0^{(2)}$	$\tilde{\varepsilon}_0^{(1)}$	$-\hat{\varepsilon}_2$	$\tilde{\varepsilon}_0^{(2)}$
1.060 362 090 491	7.02×10^{-12}	1.060 362 090 484	2.344 829 072 753	9.27×10^{-12}	2.344 829 072 744
	$d = 3$			$d = 6$	
$\tilde{\varepsilon}_0^{(1)}$	$-\hat{\varepsilon}_2$	$\tilde{\varepsilon}_0^{(2)}$	$\tilde{\varepsilon}_0^{(1)}$	$-\hat{\varepsilon}_2$	$\tilde{\varepsilon}_0^{(2)}$
3.799 673 029 810	9.27×10^{-12}	3.799 673 029 801	8.928 082 199 890	4.07×10^{-11}	8.928082199850

Table 10.10. Subdominant coefficient $\tilde{\varepsilon}_2$ in the strong coupling expansion (10.41) for the ground state $(0, 0)$ energy for the quartic radial anharmonic potential (10.42) for different $d = 1, 2, 3, 6$. First order correction $\tilde{\varepsilon}_{2,1}$ in PT, see text, included. 10 d.d. in $\tilde{\varepsilon}_2^{(2)}$ confirmed independently in LMM.

	$D = 1$			$d = 2$	
$\tilde{\varepsilon}_2^{(1)}$	$\tilde{\varepsilon}_{2,1}$	$\tilde{\varepsilon}_2^{(2)}$	$\tilde{\varepsilon}_2^{(1)}$	$\tilde{\varepsilon}_{2,1}$	$\tilde{\varepsilon}_2^{(2)}$
0.362 022 648 388	3.96×10^{-10}	0.362 022 648 784	0.651 477 773 845	4.38×10^{-10}	0.651 477 774 283
	$D = 3$			$d = 6$	
$\tilde{\varepsilon}_2^{(1)}$	$\tilde{\varepsilon}_{2,1}$	$\tilde{\varepsilon}_2^{(2)}$	$\tilde{\varepsilon}_2^{(1)}$	$\tilde{\varepsilon}_{2,1}$	$\tilde{\varepsilon}_2^{(2)}$
0.901 605 894 682	2.03×10^{-9}	0.901 605 896 709	1.526 804 282 772	-3.06×10^{-8}	1.526 804 252 175

If variationally optimized Approximants are taken as zero approximation in Non-Linearization (iterative) Procedure, they lead to a fast convergent scheme with rate of convergence $\sim 10^{-3}$ (or higher). For the ground state it was calculated the relative deviation of the logarithmic derivative of variationally optimized Approximant from the exact one *vs* radial coordinate r for different g and D. It was always smaller than $\sim 10^{-6}$. It implies that the Approximants with interpolated parameters $\tilde{a}_{4,0}$ *vs.* g and d provide highly accurate uniform approximation of the eigenfunctions of the Quartic Radial Anharmonic Oscillator while the respective eigenvalues are given by ratio of two integrals with integrands proportional to the Approximants.

Chapter 11

Radial Sextic Anharmonic Oscillator

In this Chapter we consider radial anharmonic oscillator (rAHO) with sextic anharmonicity (1.22),

$$V(r) = r^2 + g^4 r^6. \tag{11.1}$$

The corresponding radial Schrödinger equation (without the centrifugal term) has the form

$$\left[-\frac{\hbar^2}{2m}\left(\partial_r^2 + \frac{2\ell + d - 1}{r}\partial_r\right) + r^2 + g^4 r^6\right] \Xi_{n_r,\ell}(r) = E_{n_r,\ell}\Xi_{n_r,\ell}(r), \tag{11.2}$$

see (1.9) at $p = 6$, where the radial wavefunction

$$\Psi_{n_r,\ell}(r) = r^{\ell}\Xi_{n_r,\ell}(r),$$

and n_r is the radial quantum number, ℓ is quantum angular momentum. For the logarithmic derivative of the radial function Ξ:

$$y = -\partial_r \log \Xi,$$

one can write the radial Riccati-Bloch (rRB) equation

$$\partial_v \mathcal{Y} - \mathcal{Y}\left(\mathcal{Y} - \frac{2\ell + d - 1}{v}\right) = \varepsilon\,(\lambda) - v^2 - \lambda^4 v^6, \partial_v \equiv \frac{d}{dv}, \tag{11.3}$$

cf. (1.30), where

$$v = \left(\frac{2m}{\hbar^2}\right)^{\frac{1}{4}} r, \quad y = (2m\hbar^2)^{\frac{1}{4}}\mathcal{Y}\,(v)\,, \quad E = \frac{\hbar}{(2m)^{\frac{1}{2}}}\varepsilon,$$

cf. (7.28), (1.29), or one can write the radial Generalized Bloch (rGB) equation

$$\lambda^2 \partial_u \mathcal{Z} - \mathcal{Z}\left(\mathcal{Z} - \frac{\lambda^2(2\ell + d - 1)}{u}\right) = \lambda^2 \varepsilon(\lambda) - u^2 - u^6, \partial_u \equiv \frac{d}{du}, \quad (11.4)$$

cf. (1.41), where

$$u = gr, \quad \mathcal{Z}(u) = \frac{g}{(2m)^{1/2}} y,$$

cf. (1.39), (1.40). In both equations

$$\lambda = \left(\frac{\hbar^2}{2m}\right)^{\frac{1}{4}} g,$$

plays the role of the *effective* coupling constant.

Study of quantum dynamics subject to the sextic potential (11.1) is the main topic of this Chapter. It will be presented in maximally self-contained form.

11.1 PT in the Weak Coupling Regime

In the Weak Coupling Regime the perturbative expansion for ε and $\mathcal{Y}(v)$, developed in rRB equation (11.3), are of the form

$$\varepsilon = \varepsilon_0 + \varepsilon_4 \lambda^4 + \varepsilon_8 \lambda^8 + \cdots, \quad \varepsilon_0 = d, \quad (11.5)$$

and

$$\mathcal{Y}(v) = \mathcal{Y}_0 + \mathcal{Y}_4 \lambda^4 + \mathcal{Y}_8 \lambda^8 + \cdots, \quad \mathcal{Y}_0 = v, \quad (11.6)$$

respectively. All coefficients in front of terms of the orders λ^{4n+1}, λ^{4n+2}, λ^{4n+3}, $n = 0, 1, 2, \ldots$ are equal to zero. Since the potential (11.1) is (formally) even, PT can be constructed by algebraic means. The first non-vanishing corrections for the ground state are

$$\varepsilon_4 = \frac{1}{8} d(d+2)(d+4), \quad \mathcal{Y}_4(v) = \frac{1}{2} v^5 + \frac{1}{4}(d+4)v^3 + \frac{1}{8}(d+d)(d+4)v, \quad (11.7)$$

while the next two corrections $\varepsilon_{8,12}$ and $\mathcal{Y}_{8,12}(v)$ are presented in Appendix II.J. It can be shown that the correction $\mathcal{Y}_{4n}(v)$ has the form of odd polynomial in v,

$$\mathcal{Y}_{4n}(v) = v \sum_{k=0}^{2n} c_{2k}^{(4n)} v^{2(2n-k)}, \tag{11.8}$$

with coefficients $c_{2k}^{(4n)}$ being polynomial in d of degree k,

$$c_{2k}^{(4n)} = P_k^{(4n)}(d), \qquad c_{4n}^{(4n)} = \frac{\varepsilon_{4n}}{d}. \tag{11.9}$$

The correction ε_{4n} has the factorization property

$$\varepsilon_{4n}(d) = d(d+2)(d+4)R_{2n-2}(d), \tag{11.10}$$

where $R_{2n-2}(d)$ is a polynomial in d of degree $(2n-2)$, cf. (4.12), in particular, $R_0 = \frac{1}{8}$.

Due to invariance $v \to -v$ of the original equation (11.3) it is convenient to simplify it by introducing a new unknown function and changing v-variable to its square,

$$\mathcal{Y} = v\tilde{\mathcal{Y}} \quad \text{and} \quad v_2 = v^2.$$

As a result, (11.3) is converted into

$$2v_2\partial_{v_2}\tilde{\mathcal{Y}} - \tilde{\mathcal{Y}}\left(v_2\tilde{\mathcal{Y}} - d\right) = \varepsilon\,(\lambda) - v_2 - \lambda^4 {v_2}^3, \quad \partial_{v_2} \equiv \frac{d}{dv_2}. \tag{11.11}$$

This is a convenient form of the rRB equation to carry out the PT consideration. It particular, the first correction (11.7) in expansion of $\tilde{\mathcal{Y}}$ becomes a second degree polynomial in v_2,

$$\tilde{\mathcal{Y}}_4(v_2) = \frac{1}{2}{v_2}^2 + \frac{d+4}{4}v_2 + \frac{(d+2)(d+4)}{8},$$

and, in general, $\tilde{\mathcal{Y}}_{4n}(v_2)$ is a polynomial in v_2 of degree $(2n)$. Corrections $\tilde{\mathcal{Y}}_{4,6}$ are presented in Appendix II.J.

From (11.9) one can see that all corrections ε_{4n} vanish at $d = 0, -2, -4$, hence, their formal sum results in $\varepsilon = 0, -2, -4$, respectively. In the case $d = 0$ the radial Schrödinger equation takes the form

$$-\frac{\hbar^2}{2m}\left(\frac{d^2\Psi(r)}{dr^2} - \frac{1}{r}\frac{d\Psi(r)}{dr}\right) + (r^2 + g^4 r^6)\Psi(r) = 0, \tag{11.12}$$

cf. (10.13). Its formal solution, cf. [Dolgov and Popov (1979)], is given in terms of the parabolic cylinder functions [Abramowitz and Stegun (1964)] (also known as the Weber functions), it reads

$$\Psi = C_1 D_{\nu_+}(\lambda v^2) + C_2 D_{\nu_-}(i\lambda v^2) , \quad \nu_\pm = -\frac{1}{2} \pm \frac{1}{4\lambda^2} , \tag{11.13}$$

if written in v and λ, see (7.28) and (1.31), respectively, cf. (10.14). It has the meaning of the zero mode of the radial Schrödinger operator at $d=0$. The function (11.13) cannot be made normalizable for any choice of the constants C_1 and C_2. Hence, the radial Schrödinger operator at $d=0$ for sextic rAHO potential (11.1) has no zero mode in the Hilbert space. It complements the similar statement made for quartic rAHO potential (10.1). One can guess that zero mode in the Hilbert space is absent for the radial Schrödinger operator with the anharmonicity r^{2m} at $d=0$.

Like it happened in the quartic rAHO the assumption that $E(d=0)=0$ is incorrect: non perturbative contribution in λ (or g) to energy should be present at $d=0$. Note even though at $d=-2$ and $d=-4$ all corrections ε_{4n} vanish and we formally have $\varepsilon = -2$ and $\varepsilon = -4$, respectively, no exact solutions have been found for the corresponding radial Schrödinger equation.

11.1.1 *Ground state: Generating Functions*

For the sextic rAHO using the rGB equation (11.4), the expansion of $\mathcal{Z}(u)$ in generating functions

$$\mathcal{Z}(u) = \mathcal{Z}_0(u) + \mathcal{Z}_2(u)\lambda^2 + \mathcal{Z}_4(u)\lambda^4 + \cdots , \tag{11.14}$$

can be constructed, where the reduced energy expansion is given by (11.5),

$$\varepsilon = \varepsilon_0 + \varepsilon_4\lambda^4 + \varepsilon_8\lambda^8 + \cdots , \quad \varepsilon_0 = d,$$

Interestingly, (11.14) has the same structure as the expansion for the quartic rAHO: all generating functions $\mathcal{Z}_{2k+1}(u)$, $k=1,2,\ldots$ of odd orders λ^{2k+1} are absent in expansion, cf. (4.15). This contrasts with the expansion of $\mathcal{Y}(v;\lambda)$ and $\varepsilon(\lambda)$ in which the powers λ^{4n}, $n=0,1,2,\ldots$ are present only. In fact, for any even rAHO potential, $V(r)=V(-r)$, $\mathcal{Z}(u)$ is written in terms of generating functions as an expansion in powers of λ^2. As the result there are two different families of generating functions,

$$\mathcal{Z}_{4k}(u) = u \sum_{n=k}^{\infty} c_{4k}^{(4n)} u^{4(n-k)}, \tag{11.15}$$

and

$$\mathcal{Z}_{4k+2}(u) = u \sum_{n=k+1}^{\infty} c_{4k+2}^{(4n)} u^{4(n-k)-2}, \tag{11.16}$$

those occur in correspondence to $\varepsilon_{4k+2} = 0$ and $\varepsilon_{4k} \neq 0$. Following (11.8) and (11.9) it is easy to see that for both families the generating function $\mathcal{Z}_{2p}(u)$ is a polynomial in d of degree p,

$$\mathcal{Z}_{2p}(u) = u \sum_{n=0}^{p} f_n^{(p)}(u^2) d^n, \tag{11.17}$$

where $f_n^{(p)}(u^2)$, $n = 0, 1, \ldots, k$ are some real functions. The first two terms in expansion (11.14) are

$$\mathcal{Z}_0(u) = u\sqrt{1+u^4}, \tag{11.18}$$

$$\mathcal{Z}_2(u) = \frac{2u^4 + d\left(1 + u^4 - \sqrt{1+u^4}\right)}{2u\left(1+u^4\right)}. \tag{11.19}$$

Due to invariance $u \to -u$ one can simplify the non-linear equation (11.4) by introducing

$$\mathcal{Z} = u\tilde{\mathcal{Z}}(u_2), \quad u_2 = u^2.$$

Finally, (11.4) is reduced to

$$2\lambda^2 u_2 \partial_{u_2} \tilde{\mathcal{Z}} - \tilde{\mathcal{Z}}\left(u_2 \tilde{\mathcal{Z}} - \lambda^2 d\right) = \lambda^2 \varepsilon(\lambda) - u_2 - u_2^3, \quad \partial_{u_2} \equiv \frac{d}{du_2}. \tag{11.20}$$

In this case the first two terms in expansion (11.14) are simplified,

$$\tilde{\mathcal{Z}}_0(u_2) = \sqrt{1+u_2^2}, \tag{11.21}$$

$$\tilde{\mathcal{Z}}_2(u_2) = \frac{2u_2^2 + d\left(1 + u_2^2 - \sqrt{1+u_2^2}\right)}{2u_2\left(1+u_2^2\right)}, \tag{11.22}$$

cf. (11.18), (11.19).

The asymptotic behavior at large u of $\mathcal{Z}_{2p}(u)$ is related with the expansion of y at large r. The expansion of y at large r and fixed (effective) coupling constant $g(\lambda)$ can be rewritten conveniently in variable v, see (7.28). Finally,

$$y = (2m\hbar^2)^{\frac{1}{4}} \left(\lambda^2 v^3 + \left(\frac{d}{2} + 1 + \frac{1}{2\lambda^2} \right) v^{-1} - \frac{\varepsilon}{2\lambda^2} v^{-3} + \cdots \right). \qquad (11.23)$$

The first two terms of this expansion are energy-independent, while, additionally, the first term is d-independent. Following an analogous procedure to one, which was used for the quartic rAHO, we can transform the expansion of $\mathcal{Z}_{2p}(u)$ at large u into an expansion at large v via the connection between the classical u and quantum v coordinates. The first two terms in (11.14) at large v are

$$\left(\frac{2m}{g^2} \right)^{1/2} \mathcal{Z}_0(\lambda v) = (2m\hbar^2)^{\frac{1}{4}} \left(\lambda^2 v^3 + \frac{1}{2\lambda^2} v^{-1} - \frac{1}{8\lambda^6} v^{-5} + \cdots \right), \qquad (11.24)$$

and

$$\left(\frac{2m}{g^2} \right)^{1/2} \lambda^2 \mathcal{Z}_2(\lambda v) = (2m\hbar^2)^{\frac{1}{4}} \left(\frac{d+2}{2} v^{-1} - \frac{d}{2\lambda^2} v^{-3} - \frac{1}{\lambda^4} v^{-5} + \cdots \right), \qquad (11.25)$$

respectively. We explicitly observe that the sum $(\frac{2m}{g^2})^{1/2}(\mathcal{Z}_0(\lambda v) + \lambda^2 \mathcal{Z}_2(\lambda v))$ at large v reproduces exactly the first two terms in the expansion (11.23). However, all higher generating functions, $\mathcal{Z}_4(\lambda v)$, $\mathcal{Z}_6(\lambda v), \ldots$ contribute to the same order $O(v^{-3})$ for large v,

$$\left(\frac{2m}{g^2} \right)^{1/2} \lambda^{2p} \mathcal{Z}_{2p}(\lambda v) = (2m\hbar^2)^{\frac{1}{4}} \left(-\frac{\varepsilon_{4p-4} \lambda^{4p-6}}{2} v^{-3} + \cdots \right), \quad v \to \infty, \qquad (11.26)$$

here ε_{4p-4} is the energy correction of order λ^{4p-4}. Therefore, it does not matter how many generating functions are considered in the expansion $\left(\frac{2m}{g^2}\right)^{1/2} (\mathcal{Z}_0(\lambda v) + \lambda^2 \mathcal{Z}_2(\lambda v) + \cdots)$ for large v, the ε-dependent coefficient in front of term of order $O(v^{-3})$ is never reproduced exactly. Similar situation had occurred in the quartic rAHO.

11.1.2 *The Approximant and Variational Calculations*

Let set $\hbar = 1$ and $m = 1/2$, which implies that $v = r$ and $\varepsilon = E$. Following formulas (11.18) and (11.19), the first two terms of expansion of the phase (1.17) can be calculated,

$$G_0(r;g) = \frac{r^2}{4g^2}\sqrt{1+g^4r^4} + \frac{1}{4g^2}\log\left[g^2r^2 + \sqrt{1+g^4r^4}\right], \qquad (11.27)$$

and

$$g^2G_2(r;g) = \frac{1}{4}\log\left[1+g^4r^4\right] + \frac{d}{4}\log\left[1+\sqrt{1+g^4r^4}\right], \qquad (11.28)$$

while the next generating functions $G_4(r;g)$ and $G_6(r;g)$ are presented Appendix J. Keeping the expressions for $G_0(r;g)$ and $G_2(r;g)$ in mind one can proceed to construct the Approximant $\Psi^{0,0}_{approximation}$. Following the general formula of Chapter 8, Eq. (2.2), we write the Approximant for the ground state eigenfunction in the exponential representation

$$\Psi^{(0,0)}_{approximation} = e^{-\Phi^{(0,0)}_{approximation}},$$

with

$$\begin{aligned}\Phi^{(0,0)}_{approximation}(r) = {} & \frac{\tilde{a}_0 + \tilde{a}_2r^2 + \tilde{a}_4g^2r^4 + \tilde{a}_6g^4r^6}{\sqrt{\tilde{b}_2^2 + \tilde{b}_4g^2r^2 + \tilde{b}_6g^4r^4}} \\ & + \frac{1}{4g^2}\log\left[\tilde{c}_2g^2r^2 + \sqrt{\tilde{b}_2^2 + \tilde{B}_4g^2r^2 + \tilde{b}_6g^4r^4}\right] \\ & + \frac{1}{4}\log\left[\tilde{b}_2^2 + \tilde{b}_4g^2r^2 + \tilde{b}_6g^4r^4\right] \\ & + \frac{d}{4}\log\left[\tilde{b}_2 + \sqrt{\tilde{b}_2^2 + \tilde{b}_4g^2r^2 + \tilde{b}_6g^4r^4}\right], \end{aligned} \qquad (11.29)$$

see [Del Valle and Turbiner (2020)], where in a straightforward way it was put

$$\tilde{B}_4 = \tilde{b}_4. \qquad (11.30)$$

Afterwards, the remaining $\tilde{a}_{0,2,4,6}, \tilde{b}_{2,4,6}, \tilde{c}_2$ are eight free parameters. The first log term in (11.29), see the second line, originates from $G_0(r;g)$ in (11.27), while the second and third log terms, see the third line, originate from $G_2(r;g)$ in (11.28). All three logarithmic terms in (11.29) generate the prefactors to the exponential function of the (reduced) phase. In fact, those terms appear just as a certain *minimal* modification of the logarithmic terms, which occur in the first and second generating functions $G_{0,2}(r;g)$ (11.27)–(11.28).

Without a loss of generality one can choose $\tilde{b}_2(\tilde{b}_6) = 1$, which is a normalization of the ratio in the first line of (11.29). It reduces the number of free parameters to seven.

11.1.3 *The First Approximate Solution*

As a result, the Approximant of the ground state function for arbitrary $d = 1, 2, 3, \ldots$, which emerges from the Approximant (11.29) with the imposed condition (11.30), is given by

$$
\begin{aligned}
&{}_1\Psi^{(0,0)}_{approximation} \\
&= \frac{1}{\left(\tilde{b}_2^2 + \tilde{b}_4 g^2 r^2 + \tilde{b}_6 g^4 r^4\right)^{1/4} \left(\tilde{b}_2 + \sqrt{\tilde{b}_2^2 + \tilde{b}_4 g^2 r^2 + \tilde{b}_6 g^4 r^4}\right)^{d/4}} \\
&\quad \times \frac{1}{\left(\tilde{c}_2 g^2 r^2 + \sqrt{\tilde{b}_2^2 + \tilde{b}_4 g^2 r^2 + \tilde{b}_6 g^4 r^4}\right)^{1/4g^2}} \\
&\quad \times \exp\left(-\frac{\tilde{a}_0 + \tilde{a}_2 r^2 + \tilde{a}_4 g^2 r^4 + \tilde{a}_6 g^4 r^6}{\sqrt{\tilde{b}_2^2 + \tilde{b}_4 g^2 r^2 + \tilde{b}_6 g^4 r^4}} + \frac{\tilde{a}_0}{\tilde{b}_2}\right). \qquad (11.31)
\end{aligned}
$$

This is the central formula for this Section. Effectively, it is seven-parametric function. Later it will be shown that if taken as the variational trial function, it leads to a highly accurate, uniform approximation of the exact ground state eigenfunction from one side and the extremely accurate variational energy of the ground state for any coupling constant from another side.

By making a preliminary variational study for different d and g^4 by using the trial function (11.31) with all seven free variational parameters one can see the appearance of the relation

$$\tilde{b}_6 = 16\tilde{a}_6^2, \qquad (11.32)$$

cf. (10.30) for quartic AHO, with very high accuracy: it guarantees that the approximate phase (11.29) reproduces exactly the only growing monomially, dominant term $\sim r^3$ in the expansion (11.23) of the exact phase. Hence, one can introduce the relation (11.32) into ${}_1\Psi^{(0,0)}_{approximation}$ as the constraint. It reduces the number of the free parameters to six. About those remaining six parameters one can see that their optimal values demonstrate smooth, slow-changing behavior *vs.* with respect to both the coupling constant g, see Fig. 11.1, and dimension d, see Fig. 11.2.

Further analysis of the preliminary 7-parametric variational study using the trial function (11.31) for different d and g^4 indicates the existence of additional relation between the variational parameters:

$$\tilde{b}_4 = 32\tilde{a}_6\tilde{a}_4, \tag{11.33}$$

cf. (10.32), which corresponds to absence of subdominant term $\sim r$ in expansion of the approximate phase (11.29) at $r \to \infty$, see (11.23). This relation for optimal variational parameters $\{\tilde{b}_4, \tilde{a}_6, \tilde{a}_4$ is satisfied with a certain, sufficiently high accuracy. Hence, this relation can be made exact and be inserted to ${}_1\Psi^{(0,0)}_{approximation}$ (11.29) as another constraint. It will be done later, see Section 11.1.4 for the second approximate trial function.

Finally, the Approximant (11.31) with constraint (11.32) inserted contains six free parameters, $\{\tilde{a}_0, \tilde{a}_2, \tilde{a}_4, \tilde{a}_6, \tilde{b}_4, \tilde{c}_2\}$. This 6-parametric Approximant will be called the *first* approximate trial function for the ground state for sextic rAHO. We set $\tilde{b}_2 = 1$ for future convenience.

It must be emphasized that by choosing the parameters

$$\tilde{a}_0 = 0, \quad \tilde{a}_2 = \frac{1}{4}, \quad \tilde{a}_4 = 0, \quad \tilde{b}_4 = 0, \quad \tilde{a}_6 = \frac{1}{4}, \quad \tilde{c}_2 = 1, \tag{11.34}$$

in ${}_1\Phi^{(0,0)}_{approximation}$ (11.29) allows us to reproduce exactly the first two terms in semiclassical expansion (11.14). Note that contrary to the quartic rAHO case (10.1), no single parameter in (11.34) has explicit dependence on the coupling constant g, see (4.31). However, this choice of parameters (11.34) is far from being optimal from the point of the variational calculations of the energy. Performing minimization of the expectation value of the radial Schrödinger operator (11.2) for different values of g^4 and d one can see that all six parameters $\{\tilde{a}_0, \tilde{a}_2, \tilde{a}_4, \tilde{a}_6, \tilde{b}_4, \tilde{c}_2\}$ are smooth, slow-changing functions *vs.* g^4 and d. Plots of the variational parameters for the ground state, as functions of g^4 for fixed d, are shown in Fig. 11.1. In turn, Fig. 11.2 presents the plots of the parameters as functions of d for fixed g^4. Similar behavior of parameters occurs for the excited states $(0,1), (0,2), (1,0)$, see below.

As was indicated in Chapter 9, see also [Del Valle and Turbiner (2019, 2020)], the Approximant of the ground state function $\Psi^{(0,0)}_{approximation}$ (11.31) can serve as a building block to construct the Approximants of the excited states. Now let us consider the excited state with quantum numbers (n_r, ℓ).

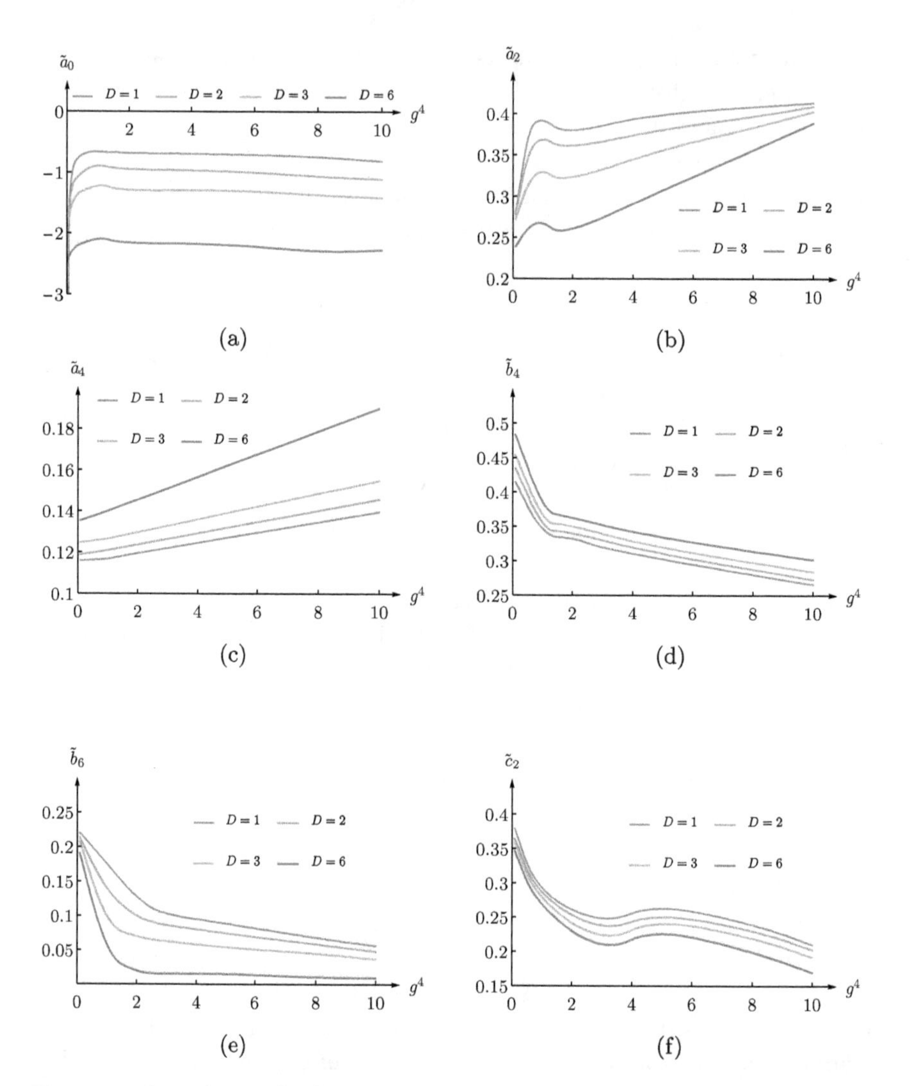

Fig. 11.1. Ground state (0, 0), the first trial function (11.38): Variational parameters $\tilde{a}_0(a)$, $\tilde{a}_2(b)$, $\tilde{a}_4(c)$, $\tilde{b}_4(d)$, $\tilde{b}_6(e)$, $\tilde{c}_2(f)$ *vs.* the coupling constant g^4 in domain $g^4 \in [0, 10]$ for $D(\equiv d) = 1, 2, 3, 6$

By taking the mixed exponential representation (7.16),

$$\Psi = r^{\ell} P(r^2) e^{-\frac{1}{\hbar}\Phi(r)},$$

and the introducing $y(r) = \partial_r \Phi(r)$, see (7.18), one can show as a result of the analysis of the general radial Riccati equation (7.17) that the expansion

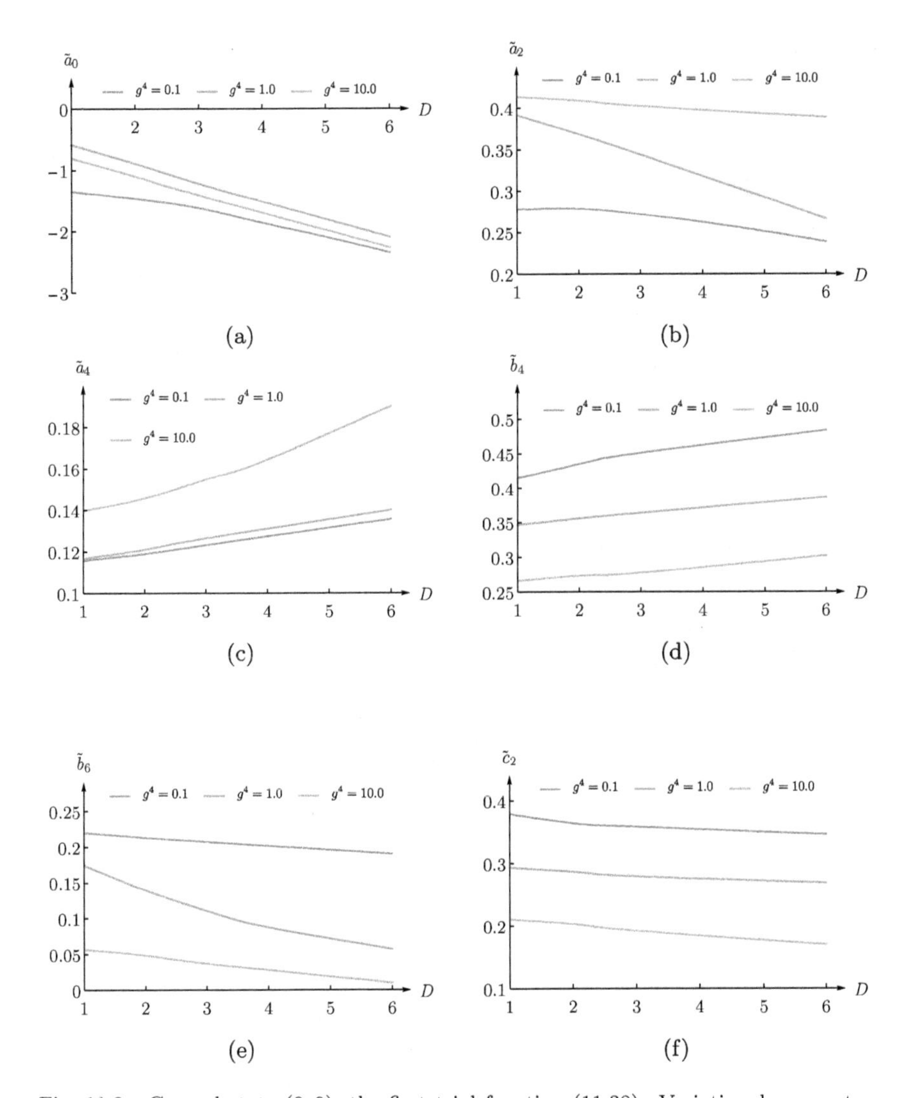

Fig. 11.2. Ground state $(0,0)$, the first trial function (11.38): Variational parameters $\tilde{a}_0(a)$, $\tilde{a}_2(b)$, $\tilde{a}_4(c)$, $\tilde{b}_4(d)$, $\tilde{b}_6(e)$, $\tilde{c}_2(f)$ *vs.* dimension $D(\equiv d)$ for $g^4 = 0.1, 1.0, 10.0$

(11.23) remains almost unchanged

$$y = (2m\hbar^2)^{\frac{1}{4}} \left(\lambda^2 v^3 + \left(\frac{d + 4n_r + 2\ell}{2} + 1 + \frac{1}{2\lambda^2} \right) v^{-1} - \frac{\varepsilon}{2\lambda^2} v^{-3} + \cdots \right), \tag{11.35}$$

where v is given by (7.28), $v \to \infty$. In particular, the first two (shown explicitly) terms of this expansion remain independent on the energy ε. Integrating y over r we will arrive at the expansion of the phase $\Phi(r)$ at $r \to \infty$: the same two terms remain growing at $r \to \infty$, they can be found explicitly. The only change that occurs in those two terms is the change of the coefficient in front of the logarithmic term, which becomes

$$(1 + d/2 + 2n_r + \ell) \log r,$$

see the 2nd term in (11.35). Eventually, it leads to a modification of the coefficient in front of the second logarithmic term in G_2, see (11.28),

$$d \to d + 4n_r + 2\ell,$$

which is similar to the cubic and quartic rAHO,[1]

$$g^2 G_2(r; g) = \frac{1}{4} \log\left[1 + g^4 r^4\right] + \frac{d + 4n_r + 2\ell}{4} \log\left[1 + \sqrt{1 + g^4 r^4}\right], \quad (11.36)$$

while G_0 remains unchanged.

Final form of the phase approximant for the excited states reads

$$\begin{aligned} \Phi^{(n_r,\ell)}_{approximation}(r) = {} & \frac{\tilde{a}_0 + \tilde{a}_2 r^2 + \tilde{a}_4 g^2 r^4 + \tilde{a}_6 g^4 r^6}{\sqrt{\tilde{b}_2^2 + \tilde{b}_4 g^2 r^2 + \tilde{b}_6 g^4 r^4}} + \frac{1}{4g^2} \\ & \times \log\left[\tilde{c}_2 g^2 r^2 + \sqrt{\tilde{b}_2^2 + \tilde{B}_4 g^2 r^2 + \tilde{b}_6 g^4 r^4}\right] \\ & + \frac{1}{4} \log\left[\tilde{b}_2^2 + \tilde{b}_4 g^2 r^2 + \tilde{b}_6 g^4 r^4\right] + \frac{d + 4n_r + 2\ell}{4} \\ & \times \log\left[\tilde{b}_2 + \sqrt{\tilde{b}_2^2 + \tilde{b}_4 g^2 r^2 + \tilde{b}_6 g^4 r^4}\right], \end{aligned} \quad (11.37)$$

cf. (11.29). Based on this formula the *first* Approximant for the *excited* states gets a very compact form

[1]It can be shown that this modification holds for arbitrary rAHO.

$$
{}_1\Psi^{(n_r,\ell)}_{approximation}
$$

$$
= r^\ell P_{n_r}(r^2) \times \frac{1}{\left(\tilde{b}_2^2 + \tilde{b}_4 g^2 r^2 + \tilde{b}_6 g^4 r^4\right)^{\frac{1}{4}} \left(\tilde{b}_2 + \sqrt{\tilde{b}_2^2 + \tilde{b}_4 g^2 r^2 + \tilde{b}_6 g^4 r^4}\right)^{\frac{d+4n_r+2\ell}{4}}}
$$

$$
\times \frac{1}{\left(\tilde{c}_2 g^2 r^2 + \sqrt{\tilde{b}_2^2 + \tilde{b}_4 g^2 r^2 + \tilde{b}_6 g^4 r^4}\right)^{1/4g^2}}
$$

$$
\times \exp\left(-\frac{\tilde{a}_0 + \tilde{a}_2 r^2 + \tilde{a}_4 g^2 r^4 + \tilde{a}_6 g^4 r^6}{\sqrt{\tilde{b}_2^2 + \tilde{b}_4 g^2 r^2 + \tilde{b}_6 g^4 r^4}} + \frac{\tilde{a}_0}{\tilde{b}_2}\right), \tag{11.38}
$$

where the condition $\tilde{B}_4 = \tilde{b}_4$ (11.30) is already taken into account, the conditions (11.32) and $\tilde{b}_2 = 1$ will be imposed in the concrete calculations; cf. (3.22) for cubic rAHO, (5.58) for one-dimensional quartic AHO, (4.36) for quartic rAHO, (6.50) for one-dimensional sextic AHO. Here $P^{(\ell)}_{n_r}(r^2)$ is the polynomial of degree n_r with all real *positive* roots in r^2, where the coefficient in front of the leading degree $(r^2)^{n_r}$ is normalized to be equal to one. For fixed angular momentum ℓ, all n_r free parameters of $P^{(\ell)}_{n_r}(r^2)$ (which are the coefficients of the polynomial, for instance) are found by imposing the orthogonality constraints,

$$
(\Psi^{(n_r,\ell)}, \Psi^{(k_r,\ell)}) = 0, \quad k_r = 0, \ldots, (n_r - 1). \tag{11.39}
$$

Finally, in order to fix the remaining free parameters we impose the constraint (11.32) and use the Approximant $\Psi^{(n_r,\ell)}_{approximation}$ as entry in variational calculation. Final number of variational parameters is equal to six: $\tilde{a}_{0,2,4,6}, \tilde{b}_4, \tilde{c}_2$. Like in the quartic rAHO case it is easily demonstrated that lowest eigenenergies $E_{(n_r,\ell)}$ support the following hierarchy

$$
E_{(n_r,0)} < E_{(n_r,1)} < E_{(n_r+1,0)},
$$

for any coupling constant $g^4 > 0$ and any integer d.

In Tables 11.1–11.4 there are presented the variational energies for the first four low-lying states with quantum numbers $(0,0), (1,0), (0,1)$ and $(0,2)$ for different values of $d > 1$ and g^4. As for $d = 1$ the states $(0,0), (1,0), (0,1)$ are studied for the one-dimensional operator defined in Chapter I. 6. The variational energy $E_0^{(1)}$, the first correction E_2 to it, and the corrected value $E_0^{(2)}$ are presented for some of above states. The variational energy $E_0^{(1)}$ provides unprecedented absolute accuracy of order

$10^{-9}-10^{-13}$ for six-parametric trial function: it is confirmed by calculating the order of the correction E_2. Using the LMM, see Appendix G, one can evaluate independently the energy on these states. The same number of optimized mesh points that was used for the quartic rAHO is used for the sextic case, i.e. 50, 100 and 200 for $g^4 = 0.1, 1.0, 10.0$, respectively. As for energy the LMM allows us to reach not less than 12 d.d. which all coincide with $E_0^{(2)}$. In other words, all digits for $E_0^{(2)}$ printed in Tables 11.1–11.4 are exact. The calculation of the next correction to variational energy, E_3, confirms it: systematically $E_3 \lesssim 10^{-12}$ for any d and g^4 considered. This indicates an extremely fast rate of convergence of the energy expansion when first trial function (11.31) for the ground state or (11.38) for the excited state are taken as zero approximation in the Non-Linearization procedure. Note that the hierarchy of eigenstates for fixed integer $d > 1$ and positive g^4 is the same as the cubic and quartic rAHO potentials: $(0,0)$, $(0,1)$, $(0,2)$, $(1,0)$. Probably, the same hierarchy should hold for any two-term rAHO potential.

In contrast with the quartic rAHO case, the available literature on the energies of the low-lying states for the sextic rAHO is very limited. Most of calculations were carried for the one-two dimensional case. In general, our results reproduce all known numerical ones that can be found for $d = 1, 2, 3$, see e.g. [Weniger (1996a,b); Taşeli and Eid (1996); Meißner and Steinborn (1997)] and [Witwit (1992)]. At $d = 6$ the calculations presented here are carried out seemingly for the first time.

The relative deviation of the ground state trial function $\Psi^{(0,0)}_{approximation}$ from the exact (unknown) ground state eigenfunction $\Psi_{(0,0)}$, if estimated within the Non-Linearization procedure, it is bounded and very small,

$$\left| \frac{\Psi_{(0,0)}(r) - \Psi^{(0,0)}_{approximation}(r)}{\Psi^{(0,0)}_{approximation}(r)} \right| \lesssim 10^{-6}, \tag{11.40}$$

in the whole range of $r \in (0, \infty]$ at any integer d and at any coupling constant g^4 which is considered. Thus, the above-described first Approximant (11.31) leads to a locally accurate approximation of the exact ground state wave function once optimal parameters are chosen. A similar situation occurs for excited states for different d and g^4.

For all d and g^4, the correction $|y_1|$ to the logarithmic derivative of the ground state is very small function in comparison with $|y_0|$ in the domain $0 \lesssim r \lesssim 1.7$ in which the dominant contribution to the integrals - required by the variational method - occurs. In all cases the first order correction

Table 11.1. Ground state $(0,0)$ energy for the sextic rAHO potential $V = r^2 + g^4 r^6$ for $d = 1, 2, 3, 6$ and $g^4 = 0.1, 1.0, 10.0$. Variational energy $E_0^{(1)}$, the first correction E_2 (rounded to 3 s.d.) found with ${}_1\Psi^{(0,0)}_{approximation}$, the corrected energy $E_0^{(2)} = E_0^{(1)} + E_2$ shown. $E_0^{(2)}$ coincides with LMM results, see App.F in 12 d.d., thus, in all printed digits.

	$d=1$			$d=2$		
g^4	$E_0^{(1)}$	$-E_2$	$E_0^{(2)}$	$E_0^{(1)}$	$-E_2$	$E_0^{(2)}$
0.1	1.109 087 078 465	1.20×10^{-13}	1.109 087 078 465	2.307 218 600 932	7.04×10^{-13}	2.307 218 600 931
1.0	1.435 624 619 003	3.22×10^{-13}	1.435 624 619 003	3.121 935 474 246	9.81×10^{-13}	3.121 935 474 246
10.0	2.205 723 269 598	3.22×10^{-12}	2.205 723 269 595	4.936 774 524 584	1.72×10^{-12}	4.936 774 524 582
	$d=3$			$d=6$		
g^4	$E_0^{(1)}$	$-E_2$	$E_0^{(2)}$	$E_0^{(1)}$	$-E_2$	$E_0^{(2)}$
0.1	3.596 036 921 222	1.76×10^{-12}	3.596 036 921 220	7.987 905 269 800	7.11×10^{-13}	7.987 905 269 799
1.0	5.033 395 937 721	5.21×10^{-12}	5.033 395 937 720	11.937 202 695 862	9.62×10^{-13}	11.937 202 695 862
10.0	8.114 843 118 826	7.60×10^{-12}	8.114 843 118 819	19.880 256 604 739	3.12×10^{-12}	19.880 256 604 736

Table 11.2. The 1st excited state energy for the sextic rAHO $V = r^2 + g^4 r^6$ for different d and g^4 labeled by quantum numbers $(0, 1)$. For $d = 1$ it corresponds to 1st negative parity state $n_r = 0$ and $p = 1$ on the whole line. Variational energy $E_0^{(1)}$, the first correction E_2 found with use of ${}_1\Psi^{(0,1)}_{approximation}$; the corrected energy $E_0^{(2)} = E_0^{(1)} + E_2$ shown. Displayed correction E_2 rounded to 3 s.d. $E_0^{(2)}$ coincides with LMM results, see App.F, in all 12 displayed d.d.

	$d=1$			$d=2$		
g^4	$E_0^{(1)}$	$-E_2$	$E_0^{(2)}$	$E_0^{(1)}$	$-E_2$	$E_0^{(2)}$
0.1	3.596 036 921 295	7.50×10^{-11}	3.596 036 921 220	4.974 197 493 807	9.01×10^{-11}	4.974 197 493 717
1.0	5.033 395 937 795	7.52×10^{-11}	5.033 395 937 720	7.149 928 601 496	5.84×10^{-11}	7.149 928 601 438
10.0	8.114 843 118 966	1.48×10^{-10}	8.114 843 118 818	11.688 236 0345 77	1.81×10^{-10}	11.688 236 034 396
	$d=3$			$d=6$		
g^4	$E_0^{(1)}$	$-E_2$	$E_0^{(2)}$	$E_0^{(1)}$	$-E_2$	$E_0^{(2)}$
0.1	6.439 143 322 388	6.64×10^{-11}	6.439 143 322 321	11.324 899 788 818	8.15×10^{-10}	11.324 899 788 004
1.0	9.455 535 276 950	1.09×10^{-10}	9.455 535 276 841	17.387 207 808 723	1.26×10^{-9}	17.387 207 807 460
10.0	15.619 579 279 334	5.05×10^{-10}	15.619 579 278 830	29.302 506 554 618	1.22×10^{-9}	29.302 506 553 402

Table 11.3. The second excited state energy for the sextic AHO $V = r^2 + g^4 r^6$ for different d and g^4 labeled by quantum numbers $(0, 2)$ for $d > 1$, as for $d = 1$ it corresponds to 1st excitation $n_r = 1$ of positive parity $p = 0$: $(1, 0)$. Variational energy $E_0^{(1)}$ found with $\Psi^{(1,0)}_{approximation}$ for $d = 1$ and ${}_1\Psi^{(0,2)}_{approximation}$ for $d > 1$, the first correction E_2 to it and the corrected energy $E_0^{(2)} = E_0^{(1)} + E_2$ shown. Displayed correction E_2 rounded to 3 s.d. E_0^2 coincides with LMM results, App.F, in all 12 displayed d.d.

	$d = 1$	$d = 2$		
g^4	$E_0^{(1)}$	$E_0^{(1)}$	$-E_2$	$E_0^{(2)}$
0.1	6.644 391 710 782	7.987 905 270 111	3.12×10^{-10}	7.987 905 269 799
1.0	9.966 622 004 356	11.937 202 696 127	2.66×10^{-10}	11.937 202 695 862
10.0	16.641 218 168 076	19.880 256 605 756	1.02×10^{-9}	19.880 256 604 742

	$d = 3$			$d = 6$		
g^4	$E_0^{(1)}$	$-E_2$	$E_0^{(2)}$	$E_0^{(1)}$	$-E_2$	$E_0^{(2)}$
0.1	9.617 462 285 440	1.50×10^{-10}	9.617 462 285 290	14.962 630 328 506	1.60×10^{-10}	14.962 630 328 346
1.0	14.584 132 948 883	3.15×10^{-9}	14.584 132 945 729	23.431 551 835 405	2.19×10^{-9}	23.431 551 833 215
10.0	24.447 468 037 325	5.42×10^{-9}	24.447 468 031 906	39.815 551 142 800	7.05×10^{-9}	39.815 551 135 750

Table 11.4. Energy E of the 3rd excited state $(1,0)$ - the first radial excitation — for sextic rAHO potential $V = r^2 + g^4 r^6$ for $d = 2, 3, 6$ at $g^4 = 0.1, 1.0, 10.0$ and its radial node r_0 calculated in LMM, see App.F, with 9 correct d.d.. Variational energy $E_0^{(1)}$ and its radial node $r_0^{(0)}$ given by underlined digits, both found with ${}_1\Psi^{(1,0)}_{approximation}$.

	$d = 2$		$d = 3$		$d = 6$	
g^4	E	r_0	E	r_0	E	r_0
0.1	$\underline{8.402\,580\,46}2$	$\underline{0.837\,31}0\,052$	$\underline{10.237\,873\,7}21$	$\underline{0.995\,78}7\,872$	$\underline{16.154\,260\,6}10$	$\underline{1.308\,54}3\,484$
1.0	$\underline{12.914\,938\,79}3$	$\underline{0.671\,82}1\,606$	$\underline{15.989\,440\,78}7$	$\underline{0.790\,36}4\,964$	$\underline{25.938\,441\,0}37$	$\underline{1.019\,16}6\,200$
10.0	$\underline{21.792\,578\,2}51$	$\underline{0.515\,91}4\,526$	$\underline{27.155\,08}5\,604$	$\underline{0.604\,32}2\,682$	$\underline{44.521\,781}\,513$	$\underline{0.773\,860}\,964$

$(-y_1)$ to the logarithmic derivative of the ground state function y is positive and bounded function in above-mentioned domain in r for all studied d and g^4. For example, for $g^4 = 1$ and $d = 1, 2, 3, 6$ the first correction y_1 has the upper bound

$$|y_1|_{max} \sim \begin{cases} 0.0078, & d = 1 \\ 0.0065, & d = 2 \\ 0.0048, & d = 3 \\ 0.0031, & d = 6 \end{cases} \tag{11.41}$$

cf. (10.39). It is the consequence of the fact that by construction in the derivative $y_{trial} = (\Phi_{trial})'$ the first two dominant terms at large v in expansion (11.23) are reproduced exactly: $\sim v^3$ and $\sim 1/v$. Moreover, the minimization of the variational energy leads to the approximate fulfillment of the condition (11.33), hence, the coefficient in front of $\sim v$ — this term is absent in the expansion (11.23) — is sufficiently small. Correspondingly, $|y_1|$ grows at large v with a small rate. This deficiency will be fixed in the second approximate trial function, see below.

"Boundness" of $y_1(r)$ at the domain $r \lesssim 2$, which gives essential contribution to the variational integrals, and its small value of the maximum, and slow growth at large r implies that we deal with a smartly designed zero order approximation $\Psi^{(0,0)}_{approximation}$. It leads, in framework of the Non-Linearization Procedure, to a fast convergent iteration procedure for the energy and wave function. In Figs. 11.3–11.5, y_0 and y_1 *vs.* r are presented for $g^4 = 1$ in physics dimensions $d = 1, 2, 3$[2]. We emphasize that all curves in these figures are slow-changing *vs.* d. Therefore, it is not a surprise that similar plots should appear for $d = 6$ (not shown) as well as for other values of d as well as $g^4 > 0$. An analysis of these plots indicates that $(-y_1)^2$ is extremely small function in comparison with y_0 in the domain $0 \leq r \lesssim 1.7$, thus, in domain which provides the dominant contribution to the variational integrals. It is the real reason why the energy correction E_2 is extremely small being of order $\sim 10^{-8}$, or sometimes even smaller, $\sim 10^{-10}$. Similar situation occurs for the phase (and its derivative) of the Approximants for the excited states.

---

[2]In these plots (left side ones) in the correction $|y_1|$ the growing terms $\sim r$ at large r are subtracted. In reality, the plots correspond to the second approximate trial function, see Section 11.1.4.

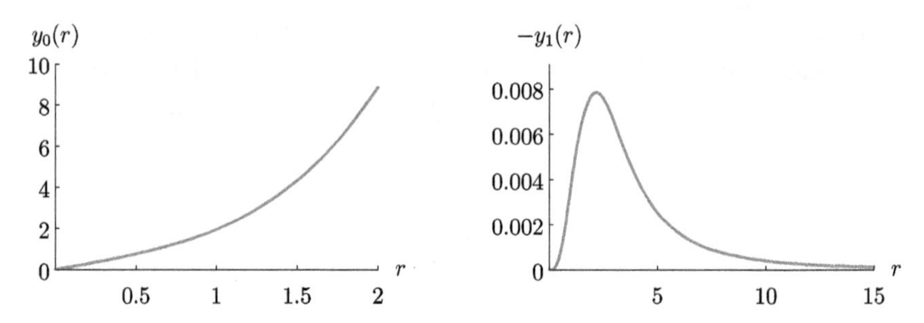

Fig. 11.3. Sextic rAHO at $d = 1$, ground state: function $y_0 = (\Phi^{(0,0)}_{approximation})'$ (on the left, for both approximations (11.38) and (11.42) the plots coincide) and its first correction y_1 (on the right, for the approximation (11.42)) *vs.* r for $g^4 = 1$.

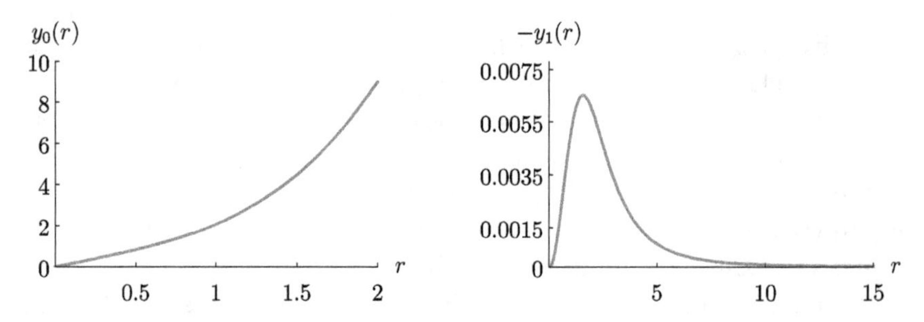

Fig. 11.4. Sextic rAHO at $d = 2$, ground state: function $y_0 = (\Phi^{(0,0)}_{approximation})'$ (on the left, for both approximations (11.38) and (11.42) the plots coincide) and its first correction y_1 (on the right, for the approximation (11.42)) *vs.* r for $g^4 = 1$.

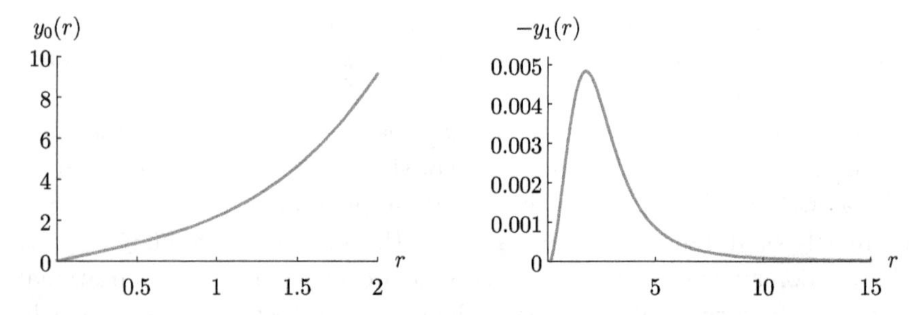

Fig. 11.5. Sextic rAHO at $d = 3$, ground state: function $y_0 = (\Phi^{(0,0)}_{approximation})'$ (on the left, for both approximations (11.38) and (11.42) the plots coincide) and its first correction y_1 (on the right, for the approximation (11.42)) *vs.* r for $g^4 = 1$.

In a similar way as it was discussed for the quartic rAHO, the Approximant for the first radial excited state ${}_1\Psi^{(1,0)}_{approximation}$ provides an estimate of the position of the radial node by imposing the orthogonality constraint to the Approximant of the ground state. For sextic AHO potential the zero order approximation $r_0^{(0)}$ gives not less than 5 d.d. of accuracy. The first order correction $r_0^{(1)}$ gives a contribution to the 6 d.d. Variational results are presented in Table 11.4, they compared with ones obtained in LMM, see App.F, with 50 optimized mesh points for $g^4 = 0.1$, with 100 optimized mesh points for $g^4 = 1$, with 200 optimized mesh points for $g^4 = 10$. It can be noted that the radial node r_0 grows with an increase of d at fixed g^4, but decreases with the increase of g^4 at fixed d.

11.1.4 *The Second Approximate Trial Function*

As was mentioned above the first three terms of the expansion of y (11.35),

$$y = (2m\hbar^2)^{\frac{1}{4}} \times \left(\lambda^2 v^3 + 0 \cdot v + \left(\frac{d + 4n_r + 2\ell}{2} + 1 + \frac{1}{2\lambda^2}\right) v^{-1} - \frac{\varepsilon}{2\lambda^2} v^{-3} + \cdots\right),$$

where the term $(0 \cdot v)$ was artificially added[3], remain independent on the energy ε. Integrating y over r we will arrive at the expansion of the phase $\Phi(r)$ at $r \to \infty$: now the same three terms become growing at $r \to \infty$, all their coefficients can be found explicitly. Let us require that in the expansion of the approximate phase $\Phi^{(n_r,\ell)}_{approximation}(r)$ at $r \to \infty$ the same *three growing terms* are reproduced with the same coefficients as in the exact y. It leads to three constraints: two of them are on the parameters $\tilde{a}_{4,6}, \tilde{b}_{4,6}$, see (11.32) and (11.33), and the third one is on the sum of degrees of the prefactors. By changing the normalization in the denominator in (11.29): $\tilde{b}_6 = 1$, which leads to $\tilde{a}_6 = 1/4$, see (11.32), and by renaming $\tilde{b}_4 \to C$, $\tilde{a}_0 \to A$, $\tilde{a}_2 \to B$ and $\tilde{b}_2 \to F$. Resulting from the constraint (11.33) the parameter $\tilde{a}_4 = 1/6$. Finally, we arrive at the *second* approximate trial function, see below. However, one more fact should be taken into account. The preliminary variational study, when the constraints (11.32) and (11.33) are imposed and degrees of the prefactors are chosen accordingly, leads to relaxing the condition (11.30): $\tilde{B}_4 = \tilde{b}_4$, it is compatible with

$$\tilde{B}_4 = 0.$$

As the result we arrive at the final, 4-parametric expression for the *second* approximate trial function

[3]It would occur, if the original sextic rAHO potential (11.1) would contain r^4 term

Table 11.5. Ground state energy found in LMM, see App.F, for the sextic rAHO potential $V(r) = r^2 + g^4 r^6$ for $d = 1, 2, 3, 6$. All printed digits are correct. Variational energy found with ${}_2\Psi^{(0,0)}_{approximation}$ given by underlined digits.

g^4	$d=1$	$d=2$	$d=3$	$d=6$
0.1	1.109 087 078 466	2.307 218 600 932	3.596 036 921 220	7.987 905 269 800
1.0	1.435 624 619 003	3.121 935 474 246	5.033 395 937 720	11.937 202 695 862
10.0	2.205 723 269 596	4.936 774 524 583	8.114 843 118 820	19.880 256 604 734

$$
{}_2\Psi^{(n_r,\ell)}_{approximation} = r^{\ell} P_{n_r}(r^2) \times \frac{1}{(F^2 + Cr^2 + g^4 r^4)^{\frac{1}{4}} (F + \sqrt{F^2 + Cr^2 + g^4 r^4})^{\frac{d}{4} + n_r + \frac{\ell}{2}}}
\times \frac{\exp\left(-\dfrac{A + Br^2 + r^4/6 + g^4 r^6/4}{\sqrt{F^2 + Cr^2 + g^4 r^4}} + \dfrac{A}{F}\right)}{\left(g^2 r^2 + \sqrt{F^2 + g^4 r^4}\right)^{\frac{1}{4g^2}}}, \tag{11.42}
$$

with free parameters A, B, C, F. Their optimal values are easily found in the variational calculus. They are smooth, slow-changing functions of d and g. It can be checked that y_0 remains almost unchanged, see for the ground state Figs. 11.3, 11.4, 11.5 (on left), while the first correction $y_1(r)$ (on right) changes dramatically: it becomes bounded function *everywhere*, $|y_1(r)| \leq$ const, cf. (1.57) as well as all higher y_n-corrections at $n > 1$. This guarantees convergence of the Non-Linearization Procedure in rigorous mathematical sense. Similar property holds for all excited states. However, this remarkable property comes with a certain cost: the accuracy of the variational energies deteriorates with respect to ones obtained with the first approximate trial function ${}_1\Psi^{(n_r,\ell)}_{(approximation)}(r)$ (11.38).

Variational energies obtained with function (11.42) for the first 4 eigenstates at $g = 0.1, 1, 10$ for $d = 1, 2, 3, 6$ are presented in Tables 11.5–11.8, where they are compared with the exact results from LMM, see Appendix G, Part II. Accuracy of the variational energies remains on the level of *unprecedented* accuracy 10-11 s.d. (before rounding). Taking into account the first correction E_2 to the variational energy E_{var} allows us to get the extra 3-4 s.d. correctly in the corrected energy $E_0^{(2)} = E_{var} + E_2$.

Table 11.6. First excited state energy in LMM, see App.F, for the sextic rAHO potential $V(r) = r^2 + g^4 r^6$ for $d = 1, 2, 3, 6$. All printed digits are correct. Variational energy found with ${}_2\Psi^{(0,1)}_{approximation}$ given by underlined digits.

g^4	$d = 1$	$d = 2$	$d = 3$	$d = 6$
0.1	3.596 036 921 220	4.974 197 493 717	6.439 143 322 322	11.324 899 788 004
1.0	5.033 395 937 720	7.149 928 601 438	9.455 535 276 841	17.387 207 807 460
10.0	8.114 843 118 820	11.688 236 034 396	15.619 579 278 830	29.302 506 553 402

Table 11.7. Second excited state energy in LMM, see App.F, for the sextic rAHO potential $V(r) = r^2 + g^4 r^6$ for $d = 1, 2, 3, 6$. All printed digits are correct. Variational energy found with ${}_2\Psi^{(0,2)}_{approximation}$ given by underlined digits.

g^4	$d = 1$	$d = 2$	$d = 3$	$d = 6$
0.1	6.644 391 708 657	7.987 905 269 800	9.617 462 285 290	14.962 630 328 347
1.0	9.966 621 999 718	11.937 202 695 862	14.584 132 945 728	23.431 551 833 213
10.0	16.641 218 108 251	19.880 256 604 734	24.447 468 031 898	39.815 551 135 727

Table 11.8. Third excited state energy in LMM, see App.F, for the sextic rAHO potential $V(r) = r^2 + g^4 r^6$ for $d = 2, 3, 6$. All printed digits are correct. Variational energy found with ${}_2\Psi^{(1,0)}_{approximation}$ given by underlined digits. Radial node not shown.

g^4	$d = 2$	$d = 3$	$d = 6$
0.1	8.402 580 462 131	10.237 873 721 423	16.154 260 610 103
1.0	12.914 938 793 085	15.989 440 787 826	25.938 441 037 662
10.0	21.792 578 251 378	27.155 085 604 631	44.521 781 513 351

11.2 The Strong Coupling Expansion

In this section we calculate the first two terms in the strong coupling expansion for the energies of a few low-lying states.

For the sextic rAHO potential (11.1) the strong coupling (convergent) expansion has the following form

$$\varepsilon = g(\tilde{\varepsilon}_0 + \tilde{\varepsilon}_6 g^{-3} + \tilde{\varepsilon}_{12} g^{-6} + \cdots), \tag{11.43}$$

as long as $2m = \hbar = 1$. Evidently, this expansion corresponds to PT in power of $\hat{\lambda} \equiv g^{-3}$ for the potential

$$V(w) = w^6 + \hat{\lambda} w^2, \tag{11.44}$$

Table 11.9. Ground state (0,0) energy $\tilde{\varepsilon}_0$ for the potential $W = w^6$ (see (11.44)) for $d = 1, 2, 3, 6$ found in PT, based on the Approximant ${}_1\Psi^{(0,0)}_{approximation}$ as the zero-approximation: $\tilde{\varepsilon}_0^{(1)}$ corresponds to the variational energy, $\hat{\varepsilon}_2$ is the second PT correction, $\tilde{\varepsilon}_0^{(2)} = \tilde{\varepsilon}_0^{(1)} + \hat{\varepsilon}_2$ is the corrected variational energy. 10 d.d. in $\tilde{\varepsilon}_0^{(2)}$ confirmed independently in LMM, see App.F.

$d = 1$			$d = 2$		
$\tilde{\varepsilon}_0^{(1)}$	$-\hat{\varepsilon}_2$	$\tilde{\varepsilon}_0^{(2)}$	$\tilde{\varepsilon}_0^{(1)}$	$-\hat{\varepsilon}_2$	$\tilde{\varepsilon}_0^{(2)}$
1.144 802 453 800	3.21×10^{-12}	1.144 802 453 797	2.609 388 463 259	5.72×10^{-12}	2.609 388 463 253
$d = 3$			$d = 6$		
$\tilde{\varepsilon}_0^{(1)}$	$-\hat{\varepsilon}_2$	$\tilde{\varepsilon}_0^{(2)}$	$\tilde{\varepsilon}_0^{(1)}$	$-\hat{\varepsilon}_2$	$\tilde{\varepsilon}_0^{(2)}$
4.338 598 711 518	4.73×10^{-12}	4.338 598 711 513	10.821 985 609 895	7.21×10^{-12}	10.821 985 609 888

Table 11.10. Subdominant (subleading) coefficient $\tilde{\varepsilon}_6$ in the strong coupling expansion (11.43) of the ground state $(0,0)$ energy for the sextic rAHO potential (11.1) for $d = 1, 2, 3, 6$. First order correction $\tilde{\varepsilon}_{6,1}$ in PT, see text, included. 10 d.d. in $\tilde{\varepsilon}_6$ confirmed independently on LMM, see App.F.

$d=1$			$d=2$		
$\tilde{\varepsilon}_6^{(1)}$	$-\tilde{\varepsilon}_{6,1}$	$\tilde{\varepsilon}_6^{(2)}$	$\tilde{\varepsilon}_6^{(1)}$	$-\tilde{\varepsilon}_{6,1}$	$\tilde{\varepsilon}_6^{(2)}$
0.307 920 304 114	3.83×10^{-10}	0.307 920 303 731	0.534 591 069 789	2.85×10^{-10}	0.534 591 069 504
$d=3$			$d=6$		
$\tilde{\varepsilon}_6^{(1)}$	$-\tilde{\varepsilon}_{6,1}$	$\tilde{\varepsilon}_6^{(2)}$	$\tilde{\varepsilon}_6^{(1)}$	$-\tilde{\varepsilon}_{6,1}$	$\tilde{\varepsilon}_6^{(2)}$
0.718 220 134 970	1.55×10^{-9}	0.718 220 133 425	1.137 762 108 070	2.68×10^{-10}	1.137 762 107 802

in the radial Schrödinger equation defined in $w \in [0, \infty)$.

In order to calculate $\tilde{\varepsilon}_0$ and $\tilde{\varepsilon}_6$ one can use one of the the Approximants ${}_{1,2}\Psi^{(n_r,\ell)}_{(approximation)}(r)$ (11.38), (11.42) for variational estimates and then develop the perturbation procedure already described in Chapter II. 9, Section 3.4 and Chapter II. 4, Section 3.4. In Table 11.9 we present for different d the variational estimate for $\tilde{\varepsilon}_0$ (denoted by $\tilde{\varepsilon}_0^{(1)}$), the correction $\hat{\varepsilon}_2$ and the corrected value $\tilde{\varepsilon}_0^{(2)}$ calculated via the Non-Linearization Procedure. Using the LMM, App.F it is verified that $\tilde{\varepsilon}_0^{(2)}$ provides not less than 12 exact d.d. at every d considered. Hence, the first 10 d.d. in variational energy $\varepsilon_0^{(1)}$ printed in Table 11.9 are exact. In this case, the $\tilde{\varepsilon}_3 \sim 10^{-2}\varepsilon_2$, it indicates a high rate of convergence in PT. Finally, in Table 11.10 we present the first two approximations of the coefficient in front of subdominant term $\tilde{\varepsilon}_6$, see (11.43). Comparing the results for $\tilde{\varepsilon}_0$ and $\tilde{\varepsilon}_6$ with those available in literature, see e.g. [Taşeli and Eid (1996)] and [Meißner and Steinborn (1997)], one can see that they are easily reproduced, furthermore, they can be made superior by calculating a few higher order corrections.

11.3 Sextic rAHO: Conclusions

It is shown in this Chapter that the 6(4)-parametric Approximants (11.38), (11.42) taken as variational trial functions for the first four eigenstates $(0,0), (0,1), (0,2), (1,0)$ of the sextic d-dimensional rAHO with the radial potential (11.1) provide extremely high relative accuracy in energy spectra ranging from $\sim 10^{-14}$ to $\sim 10^{-8}$ for different coupling constants g and dimension d. These variational results are unprecedented for so few parametric trial functions. Variational parameters depend on g and d in a smooth manner and can be easily interpolated. If variationally optimized Approximants are taken as the zero approximation in the Non-Linearization (iteration) procedure, they lead to fast convergent procedure with rate of convergence $\sim 10^{-4}$. For the ground state it was calculated explicitly the relative deviation of the logarithmic derivative of variationally optimized Approximant from the exact one *vs.* radial coordinate r for different g and d. It was always smaller than $\sim 10^{-6}$ at any $r \in [0, r_{max})$ with $r_{max} \approx 2-5$ for the first trial function (11.38) and $r_{max} = \infty$ for the second trial function (11.42). It implies that the first trial function for the ground state with easily interpolated parameters $\{\tilde{a}_0, \tilde{a}_2, \tilde{a}_4, \tilde{a}_6, \tilde{b}_4, \tilde{c}_2\}$ *vs.* g and d as well as the second trial function for the ground state with easily interpolated parameters A, B, C, F provide highly accurate approximation of the ground state eigenfunctions of the Sextic Radial Anharmonic Oscillator while the respective eigenvalues are given by ratio of two, numerically evaluated integrals. In the straightforward manner an interested reader can extend these results to the excited states.

The results presented in this Chapter can be easily generalized to the so-called double sextic rAHO

$$V = r^2 + ag^2r^4 + g^4r^6, \tag{11.45}$$

where a is a parameter; at $a = 0$ the potential is converted to (1.22). Both trial functions (11.38) and (11.42) remain appropriate: the only modification is related with changing the constraint (11.33).

Conclusions

Let us take a general anharmonic oscillator potential bounded from below and confining, but not necessarily a polynomial,

$$\boxed{\begin{aligned} V(r;g;\{a\}) &= \frac{1}{g^2}\,\hat{V}(g\,r) = \frac{1}{g^2}\sum_{k=2} a_k\, g^k\, r^k \\ &= a_2 r^2 + a_3 g r^3 + a_4 g^2 r^4 + \cdots + a_p g^{p-2} r^p + \cdots\,, \end{aligned}}$$

where $a_2 > 0, a_3, a_4, \ldots$ are real parameters. If $r \in (-\infty, \infty)$, this potential corresponds to the one-dimensional Schrödinger equation, while if $r \in [0, \infty)$, this is the potential of the radial Schrödinger equation or, saying differently, of the spherically symmetric d-dimensional Schrödinger equation with r as the hyperradius. The main result, presented in this book, is the approximate, locally-accurate, uniform in r expression for the eigenfunctions of the following form

$$\boxed{\begin{aligned} \Psi^{(n_r,\ell)}_{approximation}(r) &= r^{\ell}\, P^{(\ell)}_{n_r}(r)[\text{prefactor}_{(n_r,\ell)}\,(r,g;\{\tilde{b}\})] \\ &\quad\times \exp\left(-\frac{A + \tilde{a}_1 r + \hat{V}(r,g;\{\tilde{a}\})}{\sqrt{\frac{1}{r^2}\,\hat{V}(r,g;\ \{\tilde{b}\})}} + \frac{A}{\tilde{b}_0^2}\right), \end{aligned}}$$

where (n_r, ℓ) are the radial and angular quantum numbers for the case $d > 1$ whilst for $d = 1$ the angular quantum number $\ell = 0$ and $n_r \equiv K$ is the (principal) quantum number.

Here $\hat{V}(r, g;\ \{\tilde{a}\})$, $\hat{V}(r, g;\ \{\tilde{b}\})$ are two potentials, both of them were defined via $V(r; g; \{a\})$ with the coefficients $a_2, a_3, \ldots, a_p, \ldots$ (denoted here as $\{a\}$), but with different sets of coefficients $\{\tilde{a}\}$ or $\{\tilde{b}\}$, respectively, instead. In general, if the potential is a polynomial of degree $2p$, the parameters $\{\tilde{a}\}, \{\tilde{b}\}$ and $\tilde{a}_1, A$ are $2p$ free (variational) parameters subject to $[p/2]$ constraints: for the phase $\Phi(r)$ (1.17) and its (reduced) logarithmic derivative $\mathcal{Z}$ (3.2) all $[p/2]$ growing terms at large distances r in (3.8) do not depend on the energy and can be reproduced exactly. If it is like that the approximate eigenfunction does not deviate from the exact one at large distances *exponentially*. This leads to constraints on the coefficients $\{\tilde{a}\}, \{\tilde{b}\}$. An important additional constraint occurs for $d > 1$: it is related to the absence of the linear term in the expansion of wavefunction at small distances, it constraints the parameter $\tilde{a}_1$ in $\Psi^{(n_r,\ell)}_{approximation}(r)$. Remaining parameters can be found in the variational calculus. It is a highly-non-trivial fact that the exponential contains the ratio of a polynomial to the square root of another polynomial as a result of matching the small and large distances expansions. This fact seems to be of a general nature: it appears in *all* studied problems, both the one-dimensional and the multi-dimensional (even without the radial symmetry of potentials).

The prefactor $[\text{prefactor}_{(n_r,\ell)}(r, g; \{\tilde{b}\})]$ in (1.1) depends on the concrete potential under consideration, it is usually defined by the subdominant term in the semiclassical expansion or, saying differently, by the second correction $\mathcal{Z}_2$ in (1.71).

The compact formula $\Psi^{(n_r,\ell)}_{approximation}(r)$ was successfully applied to several anharmonic oscillators in different dimensions d and with different coupling constants g: the cubic rAHO, quartic AHO and quartic rAHO, sextic AHO and sextic rAHO. For all these cases the approximate eigenfunction provides a number of correct significant figures at any point in coordinate space and many more correct significant figures in the eigenvalue (energy). Needless to say that many other anharmonic oscillator potentials as well as the potentials of different nature can be studied by using $\Psi^{(n_r,\ell)}_{approximation}(r)$. Formula $\Psi^{(0,0)}_{approximation}(r)$ for the ground state can be easily generalized to the radially perturbed Coulomb problem (the funnel-type potentials) and to the problem of the Hydrogen atom in a constant uniform magnetic or electric field.

Appendices of Part I

Appendix A

Classical Quartic Anharmonic Oscillator

The classical counterpart of the one-dimensional quantum quartic anharmonic oscillator is described by the Newton equation

$$\ddot{x}(t) + 2x(t) + 4g^2 x(t)^3 = 0, \quad \dot{x}(t) = \frac{dx(t)}{dt}. \tag{A.1}$$

Here $x(t)$ denotes the displacement with respect to the equilibrium position ($x = 0$) as function of time t. Equation (A.1) is a particular case of the well-studied Duffing equation [Duffing (1918)] without dissipation and external force. It is straightforward to prove that (A.1) is equivalent to the equation

$$\frac{1}{2}\dot{x}(t)^2 + x(t)^2 + g^2 x(t)^4 = E = constant, \tag{A.2}$$

which explicitly exhibits the conservation of energy ($E \geq 0$) of the system. A classical standard analysis of the quartic anharmonic potential shows that all trajectories in the phase space are periodic, see Fig. A.1.

The general solution of (A.1) may be written in terms of the Jacobi elliptic cosine,

$$x(t) = \mathcal{C}_1 \,\mathrm{cn}(\omega\, t + \mathcal{C}_2, \mu), \tag{A.3}$$

taking ω and the elliptic modulus μ as follows

$$\omega = \sqrt{2\,(1 + 2\,\mathcal{C}_1^2 g^2)}, \quad \mu = \frac{\mathcal{C}_1^2 g^2}{1 + 2\,\mathcal{C}_1^2 g^2}. \tag{A.4}$$

The value of the constant $\mathcal{C}_{1,2}$ is fixed through initial conditions. Needless to say that at $g^2 = 0$ the elliptic modulus vanishes and (A.3) describes harmonic oscillations of natural frequency $\omega = \sqrt{2}$. The energy associated

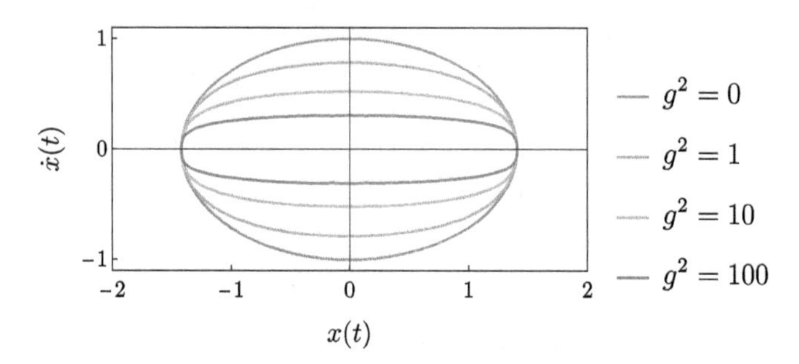

Fig. A.1. Trajectories in phase space for the classical quartic anharmonic oscillator with energy $E = 1$ and coupling constants $g^2 = 1, 10, 100$. The harmonic trajectory ($g^2 = 0$) bounds all anharmonic ones ($g^2 > 0$).

to the solution (A.3) only depends on $\mathcal{C}_1$ and the coupling constant g^2,

$$E = \mathcal{C}_1^2(1 + \mathcal{C}_1^2 g^2). \tag{A.5}$$

The classical anharmonic oscillator is the simplest symmetric model (under the transformation $x \to -x$) that describes non-linear oscillations. In fact, many properties that appear for this case can also be present in a more general one. That explains why, despite the existence of the solution (A.3), approximate methods are frequently applied to study the quartic anharmonic oscillator, see e.g. [Landau and Lifshitz (1976)] and [Krylov and Bogoliubov (1950)]. For example, using perturbation theory, the multiple-scale analysis and the so-called secular terms emerge, see [Bender and Orszag (1978)]. To the best of our knowledge, the explicit solutions are still unknown when higher order non-linearities are considered.

Appendix B

Computational Realization of PT on the Riccati-Bloch Equation

B.1 Ground State of the One-Dimensional Quartic Anharmonic Oscillator

For a given n — the number of correction — the exact solution of equation (5.21) can be easily constructed and written in MATHEMATICA-12 code. It involves the calculations of all corrections of the lower order than n.

The following code computes the first corrections up to the order (**max**)$\equiv n$ defined by the user, where the same mathematical notations as in the Chapter 5 are used. To simplify the presentation we choose the 3rd correction (**max**=3):

```
max = 3;
Ŷ[0] = 1;
Ŷ[n_] := Sum[a_k^n v2^k, {k, 0, n}];
Q[1] = v2^2;
Q[n_] := -v2 Sum[Ŷ[n - k] Ŷ[k], {k, 1, n-1}]
Eq[n_] := 2 v2 D[Ŷ[n], v2] + (2 p + 1) Ŷ[n] - 2 v2 Ŷ[0] Ŷ[n] == ϵn - Q[n]
Do[sol = SolveAlways[Eq[n], v2]; Set @@@ sol[[1]], {n, 1, max}]
Do[Print["ϵ"n, " = ", Factor[ϵn], "\n"], {n, 1, max}]
Do[Print["Ŷ"n, " = ", Collect[Ŷ[n], v2, Factor], "\n"], {n, 1, max}]
```

In this case, we request making the calculation of the first 3 corrections ε_i and $\hat{Y}_i(v_2), i = 1, 2, 3$. One can immediately note that the first 6 lines of the code are literally the translation of the equations (5.20), (5.21) and (5.22) into the MATHEMATICA's language. The command **SolveAlways** solves the recurrence relations between coefficients $a_k^{(n)}$'s, while **Set** assigns the obtained solutions to the corresponding variables. With the last two lines, we print on the screen the results for the first 3 corrections for ε_i and $\hat{Y}_i(v_2), i = 1, 2, 3$:

$$\epsilon_1 = \frac{1}{4} \times (1 + 2\,p) \times (3 + 2\,p)$$

$$\epsilon_2 = -\frac{1}{16} \times (1 + 2\,p) \times (3 + 2\,p) \times (7 + 4\,p)$$

$$\epsilon_3 = \frac{1}{64} \times (1 + 2\,p) \times (3 + 2\,p) \times \left(111 + 118\,p + 32\,p^2\right)$$

$$\hat{Y}_1 = \frac{1}{4} \times (3 + 2\,p) + \frac{v_2}{2}$$

$$\hat{Y}_2 = -\frac{1}{16} \times (3 + 2\,p) \times (7 + 4\,p) + \frac{1}{16} \times (-11 - 6\,p)\, v_2 - \frac{v_2^2}{8}$$

$$\hat{Y}_3 = \frac{1}{64} \times (3 + 2\,p) \times \left(111 + 118\,p + 32\,p^2\right) + \frac{1}{16} \times \left(45 + 46\,p + 12\,p^2\right) v_2 + \frac{1}{32} \times (21 + 10\,p)\, v_2^2 + \frac{v_2^3}{16}$$

By changing the value of the external parameter **max** one can compute any desired number of corrections. In a standard today's laptop (of the year 2022), the calculation of the first 100 corrections takes less than 1 minute of CPU time, 200 takes ~3 minutes, and 300 requires ~ 8 minutes. The only limitation of the procedure is related with the amount of the memory needed to store the resulting coefficients which appear in the form of rational numbers. It can become significant for calculation of the higher-order corrections where rational numbers of the enormous size occur.

B.2 Ground State of the One-Dimensional Sextic Anharmonic Oscillator

As for the one-dimensional sextic AHO, see Chapter 6, two minor modifications of the above code are required: (i) changing the explicit form of the correction $\hat{\mathcal{Y}}_i(v_2) \equiv \hat{Y}_i(v_2)$, now it is a polynomial of degree $2i$, and (ii) replacing the definition of $Q_1(v_2)$ to v_2^3, see line 3 of the code, accordingly. Taking into account these modifications, the following MATHEMATICA-12 code calculates the first 3 corrections for ε_i and $\hat{Y}_i(v_2), i = 1, 2, 3$:

```
max = 3;
Ŷ[0] = 1;
Ŷ[n_] := Sum[a[n][k] v2^k, {k, 0, 2 n}];
Q[1] = v2^3;
Q[n_] := -v2 Sum[Ŷ[n - k] Ŷ[k], {k, 1, n-1}]
Eq[n_] := 2 v2 D[Ŷ[n], v2] + (2 p + 1) Ŷ[n] - 2 v2 Ŷ[0] Ŷ[n] == ϵn - Q[n]
Do[sol = SolveAlways[Eq[n], v2]; Set @@@ sol[[1]], {n, 1, max}]
Do[Print["ϵ"n, " = ", Factor[ϵn], "\n"], {n, 1, max}]
Do[Print["Ŷ"n, " = ", Collect[Ŷ[n], v2, Factor], "\n"], {n, 1, max}]
```

where the output of the calculations, printed on screen, is shown here:

$$\epsilon_1 = \frac{1}{8} \times (1+2p) \times (3+2p) \times (5+2p)$$

$$\epsilon_2 = -\frac{1}{128} \times (1+2p) \times (3+2p) \times (5+2p) \times \left(233+180p+36p^2\right)$$

$$\epsilon_3 = \frac{(1+2p) \times (3+2p) \times (5+2p) \times \left(82\,645+109\,872\,p+57\,288\,p^2+13\,824\,p^3+1296\,p^4\right)}{1024}$$

$$\hat{Y}_1 = \frac{1}{8} \times (3+2p) \times (5+2p) + \frac{1}{4} \times (5+2p)\, v_2 + \frac{v_2^2}{2}$$

$$\hat{Y}_2 = -\frac{1}{128} \times (3+2p) \times (5+2p) \times \left(233+180p+36p^2\right) - \frac{1}{32} \times (5+2p) \times \left(109+82p+16p^2\right) v_2 + \frac{1}{8} \times \left(-47-33p-6p^2\right) v_2^2 + \frac{1}{16} \times (-19-6p)\, v_2^3 - \frac{v_2^4}{8}$$

$$\hat{Y}_3 = \frac{(3+2p) \times (5+2p) \times \left(82\,645+109\,872\,p+57\,288\,p^2+13\,824\,p^3+1296\,p^4\right)}{1024} + \frac{1}{256} \times (5+2p) \times \left(39\,575+51\,722\,p+26\,468\,p^2+6264\,p^3+576\,p^4\right) v_2 + \frac{1}{256} \times \left(72\,385+91\,068\,p+44\,608\,p^2+10\,064\,p^3+880\,p^4\right) v_2^2 + \frac{1}{32} \times \left(2165+1924\,p+596\,p^2+64\,p^3\right) v_2^3 + \frac{5}{64} \times \left(141+80p+12p^2\right) v_2^4 + \frac{1}{32} \times (37+10p)\, v_2^5 + \frac{v_2^6}{16}$$

In a similar way as for quartic AHO by increasing the value of **max** one can calculate any desired number of corrections. Limitation due to the size of memory usage for this case becomes more severe.

The code can be easily modified for calculations of the perturbation theory corrections for the ground state of any two-term AHO.

Appendix C

One-dimensional Quartic AHO; The First 6 Eigenstates: Interpolating Parameters, Nodes, Energies

In this Appendix for the function (5.58) the interpolating parameters A and B *versus* the coupling constant $g^2 \geq 0$ are presented in explicit form for the first six lowest eigenstates as the result of the fit of their numerical data with relative accuracy 10^{-4}, see Figs. 5.4 and 5.5 as the illustration. Fit of node positions for the excited states $(1,0),(1,1),(2,0),(2,1)$, found also with relative accuracy 10^{-4}, are presented as well for $g^2 \geq 0$.

C.1 Interpolations for Parameters A and B in (5.58)

For the sake of convenience let us introduce a new parameter

$$\tilde{\lambda} = (0.008 + g^2)^{1/3}, \tilde{\lambda}(0) = \frac{1}{5},$$

see (5.61) instead of the coupling constant g^2.

C.1.1 *Ground State* (0, 0)

$$A = \frac{-0.0171 + 0.4205\,\tilde{\lambda} - 0.1990\,\tilde{\lambda}^2 + 1.039\,\tilde{\lambda}^3 - 0.0567\,\tilde{\lambda}^4 - 1.797\,\tilde{\lambda}^5}{g^2\,\tilde{\lambda}}$$

$$B = \frac{0.3716 + 5.476\,\tilde{\lambda} + 2.231\,\tilde{\lambda}^2 + 33.51\,\tilde{\lambda}^3}{1 + 0.9981\,\tilde{\lambda} + 15.61\,\tilde{\lambda}^2}$$

Relative Deviation in energy is 10^{-9} or less

(C.1)

C.1.2 *First Excited State* (0, 1)

$$A = \frac{0.0005 + 0.2662\,\tilde{\lambda} + 0.1410\,\tilde{\lambda}^2 + 1.031\,\tilde{\lambda}^3 + 0.0612\,\tilde{\lambda}^4 - 3.399\,\tilde{\lambda}^5}{g^2\,\tilde{\lambda}}$$

$$B = \frac{-15.85 - 135.4\,\tilde{\lambda} + 70.20\,\tilde{\lambda}^2 + 803.0\,\tilde{\lambda}^3}{1 - 56.35\,\tilde{\lambda} + 338.3\,\tilde{\lambda}^2}$$

Relative Deviation in energy 10^{-9} or less

(C.2)

C.1.3 *Second Excited State* (1, 0)

$$A = \frac{-0.0179 + 0.4196\,\tilde{\lambda} - 0.2791\,\tilde{\lambda}^2 + 1.696\,\tilde{\lambda}^3 - 0.1108\,\tilde{\lambda}^4 - 4.918\,\tilde{\lambda}^5}{g^2\,\tilde{\lambda}}$$

$$B = \frac{-0.3739 + 13.30\,\tilde{\lambda} + 2.168\,\tilde{\lambda}^2 + 152.9\,\tilde{\lambda}^3}{1 + 0.999\,\tilde{\lambda} + 59.92\,\tilde{\lambda}^2}$$

Relative Deviation in energy 10^{-9} or less

(C.3)

C.1.4 *Third Excited State* (1, 1)

$$A = \frac{-0.0349 + 0.5604\,\tilde{\lambda} - 0.6622\,\tilde{\lambda}^2 + 2.313\,\tilde{\lambda}^3 - 0.2707\,\tilde{\lambda}^4 - 6.500\,\tilde{\lambda}^5}{g^2\,\tilde{\lambda}}$$

$$B = \frac{-1.182 + 13.64\,\tilde{\lambda} - 25.79\,\tilde{\lambda}^2 + 221.3\,\tilde{\lambda}^3}{1 - 9.942\,\tilde{\lambda} + 81.83\,\tilde{\lambda}^2}$$

Relative Deviation in energy 10^{-9} or less

(C.4)

C.1.5 *Fourth Excited State* (2, 0)

$$A = \frac{-0.0289 + 0.4963\,\tilde{\lambda} - 0.4867\,\tilde{\lambda}^2 + 2.346\,\tilde{\lambda}^3 - 0.2114\,\tilde{\lambda}^4 - 8.175\,\tilde{\lambda}^5}{g^2\,\tilde{\lambda}}$$

$$B = \frac{-1.655 + 17.83\,\tilde{\lambda} - 33.12\,\tilde{\lambda}^2 + 329.9\,\tilde{\lambda}^3}{1 - 12.09\,\tilde{\lambda} + 116.0\,\tilde{\lambda}^2}$$

Relative Deviation in energy 10^{-8} or less

(C.5)

C.1.6 *Fifth Excited State* (2, 1)

$$A = \frac{-0.0299 + 0.4933\,\tilde{\lambda} - 0.4715\,\tilde{\lambda}^2 + 2.534\,\tilde{\lambda}^3 - 0.2145\,\tilde{\lambda}^4 - 9.915\,\tilde{\lambda}^5}{g^2\,\tilde{\lambda}}$$

$$B = \frac{-3.234 + 28.25\,\tilde{\lambda} - 65.84\,\tilde{\lambda}^2 + 613.1\,\tilde{\lambda}^3}{1 - 22.87\,\tilde{\lambda} + 206.5\,\tilde{\lambda}^2}$$

Relative Deviation in energy 10^{-8} or less

(C.6)

C.2 Numerical Results (Energies)

C.2.1 *Energies of the First Six States (Tables)*

Energies of the first six low-lying states of the one-dimensional quartic AHO potential $V(x) = x^2 + g^2x^4$ are presented. Variational energy E_{var} and the first correction E_2 (rounded to 3 significant digits) in Non-Linearization procedure is shown. E_{var} can be calculated using fitted parameters A, B coming from interpolation of variational parameters (C.1)–(C.6) as entry ones. However, for preparation of Tables (see below) the variational energies found with the optimal parameters A, B, obtained in the variational calculations as a result of minimization, are employed. The energy E_{mesh} represents the results coming from the Lagrange Mesh Method, see Appendix F: all their printed digits are correct.

Tables:

	(0, 0)			(0, 1)		
g^2	E_{var}	$-E_2$	E_{mesh}	E_{var}	$-E_2$	E_{mesh}
0.1	1.065 285 509 557	3.36×10^{-12}	1.065 285 509 544	3.306 872 013 179	2.67×10^{-11}	3.306 872 013 153
1.0	1.392 351 641 982	4.52×10^{-10}	1.392 351 641 530	4.648 812 705 790	1.58×10^{-9}	4.648 812 704 212
10.0	2.449 174 076 214	4.10×10^{-9}	2.449 174 072 118	8.599 003 459 642	4.83×10^{-9}	8.599 003 454 808
100.0	4.999 417 553 553	8.42×10^{-9}	4.999 417 545 138	17.830 192 727 276	1.13×10^{-8}	17.830 192 715 950
	(1, 0)			(1, 1)		
g^2	E_{var}	$-E_2$	E_{mesh}	E_{var}	$-E_2$	E_{mesh}
0.1	5.747 959 268 861	2.79×10^{-11}	5.747 959 268 833	8.352 677 826 590	8.04×10^{-10}	8.352 677 825 786
1.0	8.655 049 968 779	1.10×10^{-8}	8.655 049 957 760	13.156 803 978 761	8.07×10^{-8}	13.156 803 898 050
10.0	16.635 921 505 008	1.26×10^{-8}	16.635 921 492 414	25.806 276 257 744	4.27×10^{-8}	25.806 276 215 055
100.0	34.873 984 279 901	1.79×10^{-8}	34.873 984 262 030	54.385 291 600 246	2.86×10^{-8}	54.385 291 571 642
	(2, 0)			(2, 1)		
g^2	E_{var}	$-E_2$	E_{mesh}	E_{var}	$-E_2$	E_{mesh}
0.1	11.098 595 626 554	3.92×10^{-9}	11.098 595 622 633	13.969 926 205 512	7.77×10^{-9}	13.969 926 197 743
1.0	18.057 557 528 638	9.23×10^{-8}	18.057 557 436 302	23.297 441 519 457	6.82×10^{-8}	23.297 441 451 228
10.0	35.885 171 276 032	5.38×10^{-8}	35.885 171 222 252	46.729 080 947 321	4.65×10^{-8}	46.729 080 900 818
100.0	75.877 004 068 470	3.98×10^{-8}	75.877 004 028 663	99.032 837 370 491	5.51×10^{-8}	99.032 837 315 402

C.2.2 *Energies of the First Six States (Fits)*

Fits of the energies of six low-lying states from Tables obtained with relative accuracy $\sim 10^{-5}$ (or better) for $\tilde{\lambda} \in [0, \infty)$:

Ground State

$$E_{0,0} = \frac{0.0785 + 0.2581\tilde{\lambda} + 0.6650\tilde{\lambda}^2 + 0.5436\tilde{\lambda}^3 + 1.0604\tilde{\lambda}^4}{0.0775 + 0.2837\,\tilde{\lambda} + 0.5129\,\tilde{\lambda}^2 + \tilde{\lambda}^3} \tag{C.7}$$

First Excited State

$$E_{0,1} = \frac{0.0972 + 0.4093\,\tilde{\lambda} + 1.5214\,\tilde{\lambda}^2 + 1.2993\,\tilde{\lambda}^3 + 3.7997\,\tilde{\lambda}^4}{0.0313 + 0.1623\,\tilde{\lambda} + 0.3421\,\tilde{\lambda}^2 + \tilde{\lambda}^3} \tag{C.8}$$

Second Excited State

$$E_{1,0} = \frac{0.0785 + 0.3138\,\tilde{\lambda} + 1.9360\,\tilde{\lambda}^2 + 1.4159\,\tilde{\lambda}^3 + 7.4557\,\tilde{\lambda}^4}{0.0143 + 0.0922\,\tilde{\lambda} + 0.1900\,\tilde{\lambda}^2 + \tilde{\lambda}^3} \tag{C.9}$$

Third Excited State

$$E_{1,1} = \frac{0.0396 + 0.1614\,\tilde{\lambda} + 2.1828\,\tilde{\lambda}^2 + 1.1098\,\tilde{\lambda}^3 + 11.6448\,\tilde{\lambda}^4}{0.0041 + 0.0535\,\tilde{\lambda} + 0.0953\,\tilde{\lambda}^2 + \tilde{\lambda}^3} \tag{C.10}$$

Fourth Excited State

$$E_{2,0} = \frac{0.0220 + 0.0712\,\tilde{\lambda} + 2.4383\,\tilde{\lambda}^2 + 0.8869\,\tilde{\lambda}^3 + 16.2618\,\tilde{\lambda}^4}{0.0010 + 0.0366\,\tilde{\lambda} + 0.0545\,\tilde{\lambda}^2 + \tilde{\lambda}^3} \tag{C.11}$$

Fifth Excited State

$$E_{2,1} = \frac{0.0075 - 0.0180\,\tilde{\lambda} + 2.6445\,\tilde{\lambda}^2 + 0.5449\,\tilde{\lambda}^3 + 21.2381\,\tilde{\lambda}^4}{-0.0006 + 0.0254\,\tilde{\lambda} + 0.0257\,\tilde{\lambda}^2 + \tilde{\lambda}^3} \tag{C.12}$$

Appendix D

One-Dimensional Sextic AHO; The First 6 Eigenstates: Interpolating Parameters, Nodes, Energies

In this Appendix for the function (6.50) the interpolation parameters A, B, C, D *versus* the coupling constant $g^4 \geq 0$ are presented in explicit form for the first six lowest eigenstates as the result of the fit of their numerical data with relative accuracy 10^{-4}, see Figs. 6.2 as illustration. Fits of node positions for the states $(1,0), (1,1), (2,0), (2,1)$, see Fig. 6.3, which were found also with relative accuracy 10^{-4}, are presented as well for $g^4 \geq 0$.

D.1 Numerical Results (Energies)

D.1.1 *Energies of the First Six States (Tables)*

Energies of the first six low-lying states of the one dimensional sextic AHO potential $V(x) = x^2 + g^4 x^6$. Variational energy E_{var} and the first correction E_2 (rounded to 3 significant digits) in Non-Linearization procedure are shown.

E_{var} can be calculated using fitted parameters A, B, C, D coming from interpolation of variational parameters (D.1), (D.3), (D.5), (D.8), (D.11), (D.15) as entry ones. These fits reproduce the optimal variational parameters in such a way that the energies (the expectation values of the Hamiltonian) are obtained with relative accuracy $\sim 10^{-6}$ (or, sometimes, better). However, for preparation of Tables (see below) the variational energies found with the optimal parameters A, B, C, D, obtained in the variational calculations as a result of minimization, are employed. The energy E_{mesh} represents the results coming from the Lagrange Mesh Method, see Appendix F: all their printed digits are correct.

Tables:

	(0, 0)			(0, 1)		
g^4	E_{var}	$-E_2$	E_{mesh}	E_{var}	$-E_2$	E_{mesh}
0.1	1.109 087 078 519	5.37×10^{-11}	1.109 087 078 466	3.596 036 921 410	1.90×10^{-10}	3.596 036 921 220
1	1.435 624 619 121	1.18×10^{-10}	1.435 624 619 003	5.033 395 938 000	2.79×10^{-10}	5.033 395 937 720
10	2.205 723 269 688	9.22×10^{-11}	2.205 723 269 596	8.114 843 119 041	2.22×10^{-10}	8.114 843 118 820
100	3.716 974 729 262	5.43×10^{-11}	3.716 974 729 209	13.946 206 623 107	2.12×10^{-10}	13.946 206 622 895
	(1, 0)			(1, 1)		
g^4	E_{var}	$-E_2$	E_{mesh}	E_{var}	$-E_2$	E_{mesh}
0.1	6.644 391 709 017	3.60×10^{-10}	6.644 391 708 657	10.237 873 721 866	4.43×10^{-10}	10.237 873 721 424
1	9.966 622 000 185	4.67×10^{-10}	9.966 621 999 718	15.989 440 788 522	6.97×10^{-10}	15.989 440 787 826
10	16.641 218 108 750	4.99×10^{-10}	16.641 218 108 251	27.155 085 605 486	8.54×10^{-10}	27.155 085 604 631
100	28.977 293 818 164	5.30×10^{-10}	28.977 293 817 634	47.564 984 583 713	2.14×10^{-9}	47.564 984 581 583
	(2, 0)			(2, 1)		
g^4	E_{var}	$-E_2$	E_{mesh}	E_{var}	$-E_2$	E_{mesh}
0.1	14.307 040 046 522	4.01×10^{-10}	14.307 040 046 121	18.801 758 335 242	1.88×10^{-9}	18.801 758 333 358
1	22.910 180 432 340	1.61×10^{-9}	22.910 180 430 729	30.622 590 572 188	1.65×10^{-9}	30.622 590 570 533
10	39.289 330 661 067	3.70×10^{-9}	39.289 330 657 370	52.849 512 680 783	2.52×10^{-9}	52.849 512 678 258
100	69.046 576 530 485	4.45×10^{-9}	69.046 576 526 035	93.073 891 698 518	3.14×10^{-9}	93.073 891 695 378

D.2 Interpolations for Parameters A,B,C,D and Energies E

For the sake of convenience we introduce a new parameter

$$\tilde{\lambda} = (0.3^4 + g^4)^{1/4},\ \tilde{\lambda}(0) = \frac{3}{10},$$

see (6.51), instead of the coupling constant g^4.

D.2.1 *Ground State Energy* $(0, 0)$

D.2.1.1 *Interpolations*

$$A_{fit}(\tilde{\lambda}) = \frac{-0.138932 - 0.0252327\tilde{\lambda} + 2.25711\tilde{\lambda}^2 + 4.85799\tilde{\lambda}^3 + 1.49841\tilde{\lambda}^4 - 8.06441\tilde{\lambda}^5 + 3.28001\tilde{\lambda}^6}{0.879601 - 1.70247\tilde{\lambda} + 3.44368\tilde{\lambda}^2 - 0.123163\tilde{\lambda}^3 - 2.42948\tilde{\lambda}^4 + \tilde{\lambda}^5},$$

$$B_{fit}(\tilde{\lambda}) = \frac{-0.0221282 + 4.2595\tilde{\lambda} - 4.28958\tilde{\lambda}^2 + 1.98455\tilde{\lambda}^3 - 1.40278\tilde{\lambda}^4 + 0.633034\tilde{\lambda}^5}{2.09973 - 0.272598\tilde{\lambda} - 2.04854\tilde{\lambda}^2 + \tilde{\lambda}^3},$$

$$C_{fit}(\tilde{\lambda}) = \frac{-0.994641 + 3.17288\tilde{\lambda} - 0.36222\tilde{\lambda}^2 + 0.450852\tilde{\lambda}^3 + 5.24763\tilde{\lambda}^4 + 6.83944\tilde{\lambda}^5 + \mathcal{P}_C}{3.80802 - 5.58396\tilde{\lambda} + 7.01453\tilde{\lambda}^2 + 0.505917\tilde{\lambda}^3 - 3.34007\tilde{\lambda}^4 + \tilde{\lambda}^5},$$

$$D_{fit}(\tilde{\lambda}) = \frac{0.604186 + 11.6987\tilde{\lambda} - 1.04516\tilde{\lambda}^2 - 2.33556\tilde{\lambda}^3 - 4.89785\tilde{\lambda}^4 + 3.02721\tilde{\lambda}^5}{2.29503 + 1.78158\tilde{\lambda} - 1.55166\tilde{\lambda}^2 - 1.5818\tilde{\lambda}^3 + \tilde{\lambda}^4}, \tag{D.1}$$

where

$$\mathcal{P}_{C_{fit}} = 9.56132\tilde{\lambda}^6 - 14.4263\tilde{\lambda}^7 + 4.01581\tilde{\lambda}^8.$$

D.2.1.2 *Fit for Ground State Energy*

$$E^{(0,0)}_{fit} = \frac{0.0721397 + 0.254139\tilde{\lambda} + 0.807002\tilde{\lambda}^2 + 1.25703\tilde{\lambda}^3 + 2.17277\tilde{\lambda}^4 + 1.60413\tilde{\lambda}^5 + 1.14482\tilde{\lambda}^6}{0.0730737 + 0.269887\tilde{\lambda} + 0.728111\tilde{\lambda}^2 + 1.62709\tilde{\lambda}^3 + 1.4015\tilde{\lambda}^4 + \tilde{\lambda}^5} \tag{D.2}$$

Expansions at $g^4 = 0$:

$$E^{(0,0)}_{fit} = 1 + 1.8742\, g^4 - 26.87 g^8 + 1078.00\, g^{12} - 68231.8\, g^{16} + \cdots$$

$$E^{(0,0)}_{PT} = 1 + 1.875\, g^4 - 27.30\, g^8 + 1210.62\, g^{12} - 102000\, g^{16} + \cdots$$

D.2.2 *First Excited State Energy* $(0, 1)$

D.2.2.1 *Interpolations*

$$A_{fit}(\tilde{\lambda}) = \frac{\begin{array}{c}0.776413 - 7.4512\tilde{\lambda} + 22.9476\tilde{\lambda}^2 - 32.3546\tilde{\lambda}^3 + 39.6379\tilde{\lambda}^4 \\ -23.5164\tilde{\lambda}^5 + 4.54653\tilde{\lambda}^6\end{array}}{-0.540049 + 2.99726\tilde{\lambda} - 6.04521\tilde{\lambda}^2 + 8.64586\tilde{\lambda}^3 - 5.20622\tilde{\lambda}^4 + \tilde{\lambda}^5},$$

$$B_{fit}(\tilde{\lambda}) = \frac{\begin{array}{c}0.0266796 - 0.153017\tilde{\lambda} + 0.0552796\tilde{\lambda}^2 + 0.474244\tilde{\lambda}^3 + 0.228534\tilde{\lambda}^4 \\ +1.35813\tilde{\lambda}^5\end{array}}{0.0452865 - 0.314626\tilde{\lambda} + 0.333194\tilde{\lambda}^2 + \tilde{\lambda}^3},$$

$$C_{fit}(\tilde{\lambda}) = \frac{\begin{array}{c}-0.311151 + 0.949529\tilde{\lambda} + 0.186288\tilde{\lambda}^2 - 0.423312\tilde{\lambda}^3 - 0.0231932\tilde{\lambda}^4 \\ +1.76996\tilde{\lambda}^5 + \mathcal{P}_C\end{array}}{0.267694 + 1.30085\tilde{\lambda} - 0.93576\tilde{\lambda}^2 - 0.653347\tilde{\lambda}^3 + 10.7508\tilde{\lambda}^4 + \tilde{\lambda}^5},$$

$$D_{fit}(\tilde{\lambda}) = \frac{\begin{array}{c}-4.82622 - 0.636268\tilde{\lambda} + 69.0222\tilde{\lambda}^2 + 21.1224\tilde{\lambda}^3 - 21.4241\tilde{\lambda}^4 \\ +3.15522\tilde{\lambda}^5\end{array}}{-5.60167 + 20.8253\tilde{\lambda} + 8.11418\tilde{\lambda}^2 - 7.02189\tilde{\lambda}^3 + \tilde{\lambda}^4}, \quad \text{(D.3)}$$

where

$$\mathcal{P}_{C_{fit}} = 6.67928\tilde{\lambda}^6 + 30.8223\tilde{\lambda}^7 + 3.58005\tilde{\lambda}^8.$$

D.2.2.2 *Fit for First Excited State Energy*

$$E^{(0,1)}_{fit} = \frac{\begin{array}{c}-0.0308515 + 0.17334\tilde{\lambda} + 1.48256\tilde{\lambda}^2 + 1.47828\tilde{\lambda}^3 + 7.7528\tilde{\lambda}^4 \\ +2.74419\tilde{\lambda}^5 + 4.33860\tilde{\lambda}^6\end{array}}{\begin{array}{c}-0.0126759 + 0.100159\tilde{\lambda} + 0.22504\tilde{\lambda}^2 + 1.62465\tilde{\lambda}^3 \\ +0.632118\tilde{\lambda}^4 + \tilde{\lambda}^5\end{array}} \quad \text{(D.4)}$$

Expansions at $g^4 = 0$:

$$E^{(0,1)}_{fit} = 3 + 13.091\, g^4 - 350.50\, g^8 + 21428.1\, g^{12} - 1.7844 \times 10^6\, g^{16} + \cdots$$

$$E^{(0,1)}_{PT} = 3 + 13.125\, g^4 - 368.32\, g^8 + 27165.2\, g^{12} - 3.3848 \times 10^6\, g^{16} + \cdots$$

D.2.3 *Second Excited State Energy* (1, 0)

D.2.3.1 *Interpolations*

$$A_{fit}(\tilde{\lambda}) = \frac{\begin{array}{c}-6.32074 + 23.5502\tilde{\lambda} - 0.803697\tilde{\lambda}^2 + 2.99098\tilde{\lambda}^3 + 0.707753\tilde{\lambda}^4 \\ -6.17378\tilde{\lambda}^5 + 7.00325\tilde{\lambda}^6\end{array}}{-0.499705 + 5.98347\tilde{\lambda} - 4.13868\tilde{\lambda}^2 + 1.64311\tilde{\lambda}^3 - 1.03692\tilde{\lambda}^4 + \tilde{\lambda}^5},$$

$$B_{fit}(\tilde{\lambda}) = \frac{\begin{array}{c}-0.0196037 + 2.05232\tilde{\lambda} - 0.539267\tilde{\lambda}^2 + 1.85027\tilde{\lambda}^3 + 2.11129\tilde{\lambda}^4 \\ +2.31187\tilde{\lambda}^5\end{array}}{1.29048 - 0.752778\tilde{\lambda} + 1.54449\tilde{\lambda}^2 + \tilde{\lambda}^3},$$

$$C_{fit}(\tilde{\lambda}) = \frac{\begin{array}{c}-0.433703 + 1.09102\tilde{\lambda} + 0.865348\tilde{\lambda}^2 + 0.781654\tilde{\lambda}^3 + 0.65987\tilde{\lambda}^4 \\ +0.118817\tilde{\lambda}^5 - \mathcal{P}_C\end{array}}{1.21218 - 0.132722\tilde{\lambda} - 0.489634\tilde{\lambda}^2 + 0.12007\tilde{\lambda}^3 - 0.697951\tilde{\lambda}^4 + \tilde{\lambda}^5},$$

$$D_{fit}(\tilde{\lambda}) = \frac{\begin{array}{c}0.445703 + 4.53216\tilde{\lambda} - 15.222\tilde{\lambda}^2 + 23.0325\tilde{\lambda}^3 - 15.7647\tilde{\lambda}^4 \\ +3.83933\tilde{\lambda}^5\end{array}}{1.43632 - 4.25345\tilde{\lambda} + 6.17397\tilde{\lambda}^2 - 4.1474\tilde{\lambda}^3 + \tilde{\lambda}^4}, \tag{D.5}$$

where

$$\mathcal{P}_{C_{fit}} = 1.04071\tilde{\lambda}^6 - 1.77882\tilde{\lambda}^7 + 3.13756\tilde{\lambda}^8.$$

D.2.3.2 *Fit for Node*

$$x_{node} = \frac{\begin{array}{c}-0.0052284089 - 0.0446675\tilde{\lambda}^{1/2} + 0.25983184\tilde{\lambda} \\ -0.55789749\tilde{\lambda}^{3/2} + 0.52149606\tilde{\lambda}^2\end{array}}{\begin{array}{c}-0.0076223396 - 0.0047567385\tilde{\lambda}^{1/2} - 0.062028445\tilde{\lambda} \\ +0.49126191\tilde{\lambda}^{3/2} - 1.0689497\tilde{\lambda}^2 + \tilde{\lambda}^{5/2}\end{array}} \tag{D.6}$$

D.2.3.3 *Fit for Second Excited State Energy*

$$E_{fit}^{(1,0)} = \frac{\begin{array}{c}-0.0207776 + 0.0716303\tilde{\lambda} + 0.107858\tilde{\lambda}^2 + 1.19363\tilde{\lambda}^3 + 2.42536\tilde{\lambda}^4 \\ +3.97098\tilde{\lambda}^5 \\ +9.07308\tilde{\lambda}^6\end{array}}{\begin{array}{c}-0.00322749 + 0.0024088\tilde{\lambda} + 0.0855064\tilde{\lambda}^2 + 0.168375\tilde{\lambda}^3 \\ +0.437573\tilde{\lambda}^4 + \tilde{\lambda}^5\end{array}} \tag{D.7}$$

Expansions at $g^4 = 0$:

$$E_{fit}^{(1,0)} = 5 + 46.568\, g^4 - 2103.71\, g^8 + 188719\, g^{12} - 2.11 \times 10^7\, g^{16} + \cdots$$

$$E_{PT}^{(1,0)} = 5 + 46.875\, g^4 - 2305.43\, g^8 + 270441\, g^{12} - 4.97 \times 10^7\, g^{16} + \cdots$$

D.2.4 *Third Excited State Energy* (1, 1)

D.2.4.1 *Interpolations*

$$
\begin{aligned}
A_{fit}(\tilde{\lambda}) &= \frac{\begin{array}{c}-8.94052 + 24.6352\tilde{\lambda} + 21.3912\tilde{\lambda}^2 + 13.7528\tilde{\lambda}^3 - 1.78042\tilde{\lambda}^4 \\ -15.8004\tilde{\lambda}^5 + 9.69478\tilde{\lambda}^6\end{array}}{-2.05626 + 11.4159\tilde{\lambda} - 6.526\tilde{\lambda}^2 + 2.87359\tilde{\lambda}^3 - 2.03726\tilde{\lambda}^4 + \tilde{\lambda}^5}, \\
B_{fit}(\tilde{\lambda}) &= \frac{\begin{array}{c}0.152403 + 1.75993\tilde{\lambda} + 2.40653\tilde{\lambda}^2 - 3.27552\tilde{\lambda}^3 - 3.47089\tilde{\lambda}^4 \\ +3.08333\tilde{\lambda}^5\end{array}}{2.27339 - 2.26003\tilde{\lambda} - 0.811606\tilde{\lambda}^2 + \tilde{\lambda}^3}, \\
C_{fit}(\tilde{\lambda}) &= \frac{\begin{array}{c}-1.11212 + 1.54133\tilde{\lambda} + 5.28863\tilde{\lambda}^2 + 9.13219\tilde{\lambda}^3 + 8.42601\tilde{\lambda}^4 \\ +4.78943\tilde{\lambda}^5 - \mathcal{P}_C\end{array}}{6.34055 + 1.14496\tilde{\lambda} + 0.423302\tilde{\lambda}^2 + 0.766876\tilde{\lambda}^3 - 1.77109\tilde{\lambda}^4 + \tilde{\lambda}^5}, \\
D_{fit}(\tilde{\lambda}) &= \frac{\begin{array}{c}-0.720626 - 0.349496\tilde{\lambda} + 9.25175\tilde{\lambda}^2 + 1.74039\tilde{\lambda}^3 \\ -9.8468\tilde{\lambda}^4 + 4.60313\tilde{\lambda}^5\end{array}}{-0.960843 + 3.57817\tilde{\lambda} - 0.711459\tilde{\lambda}^2 - 1.85894\tilde{\lambda}^3 + \tilde{\lambda}^4},
\end{aligned} \tag{D.8}
$$

where

$$
\mathcal{P}_{C_{fit}} = 0.481369\tilde{\lambda}^6 - 5.18183\tilde{\lambda}^7 + 3.24693\tilde{\lambda}^8.
$$

D.2.4.2 *Fit for Node*

$$
x_{node} = \frac{\begin{array}{c}-0.078519178 + 0.29478467\tilde{\lambda}^{1/2} - 0.38341644\tilde{\lambda} - 0.17585344\tilde{\lambda}^{3/2} \\ +0.81343335\tilde{\lambda}^2\end{array}}{\begin{array}{c}-0.015527401 - 0.066075186\tilde{\lambda}^{1/2} + 0.3654928\tilde{\lambda} - 0.47377282\tilde{\lambda}^{3/2} \\ -0.21544986\tilde{\lambda}^2 + \tilde{\lambda}^{5/2}\end{array}} \tag{D.9}
$$

D.2.4.3 *Fit for Third Excited State Energy*

$$
E^{(1,1)}_{fit} = \frac{\begin{array}{c}0.00444154 - 0.106686\tilde{\lambda} + 0.0677536\tilde{\lambda}^2 - 0.256703\tilde{\lambda}^3 + 2.78581\tilde{\lambda}^4 \\ +7.32196\tilde{\lambda}^5 + 14.93516\lambda^6\end{array}}{-0.000823052 - 0.00280831\tilde{\lambda} - 0.050806\tilde{\lambda}^2 + 0.114815\tilde{\lambda}^3 + 0.490378\tilde{\lambda}^4 + \tilde{\lambda}^5} \tag{D.10}
$$

Expansions at $g^4 = 0$:

$$
E^{(1,1)}_{fit} = 7 + 118.121\, g^4 - 9191.37\, g^8 + 1.46438 \times 10^6\, g^{12} - 2.98356 \times 10^8\, g^{16} + \cdots
$$

$$
E^{(1,1)}_{PT} = 7 + 118.125\, g^4 - 9358.95\, g^8 + 1.65528 \times 10^6\, g^{12} - 4.35485 \times 10^8\, g^{16} + \cdots
$$

D.2.5 *Fourth Excited State Energy* (2, 0)

D.2.5.1 *Interpolations*

$$A_{fit}(\lambda) = \frac{-1.1621 + 2.91078\tilde{\lambda} + 3.22204\tilde{\lambda}^2 + 2.59733\tilde{\lambda}^3 + 2.40959\tilde{\lambda}^4 + 1.39212\tilde{\lambda}^5 + 11.1933\tilde{\lambda}^6}{0.250543 - 0.99043\tilde{\lambda} + 3.14416\tilde{\lambda}^2 - 1.67364\tilde{\lambda}^3 + 0.26843\tilde{\lambda}^4 + \tilde{\lambda}^5},$$

$$B_{fit}(\lambda) = \frac{0.155905 + 0.730759\tilde{\lambda} + 5.15154\tilde{\lambda}^2 - 1.63934\tilde{\lambda}^3 - 6.70232\tilde{\lambda}^4 + 3.68394\tilde{\lambda}^5}{1.75358 - 0.585583\tilde{\lambda} - 1.82103\tilde{\lambda}^2 + \tilde{\lambda}^3},$$

$$C_{fit}(\lambda) = \frac{-1.5300 + 2.71284\tilde{\lambda} + 2.60589\tilde{\lambda}^2 + 0.663905\lambda^3 - 2.05659\tilde{\lambda}^4 + 3.2263\tilde{\lambda}^5 - \mathcal{P}_C}{1.92757 - 0.124914\tilde{\lambda} - 0.864681\tilde{\lambda}^2 - 0.743069\tilde{\lambda}^3 - 0.711006\tilde{\lambda}^4 + \tilde{\lambda}^5},$$

$$D_{fit}(\lambda) = \frac{0.453449 + 6.99453\tilde{\lambda} + 1.26319\tilde{\lambda}^2 - 4.38674\tilde{\lambda}^3 - 6.15877\tilde{\lambda}^4 + 4.60464\tilde{\lambda}^5}{1.76093 - 0.152394\tilde{\lambda} - 0.555061\tilde{\lambda}^2 - 1.47439\tilde{\lambda}^3 + \tilde{\lambda}^4}, \quad \text{(D.11)}$$

where

$$\mathcal{P}_{C_{fit}} = 5.91774\tilde{\lambda}^6 - 1.28952\tilde{\lambda}^7 + 3.11765\tilde{\lambda}^8.$$

D.2.5.2 *Fit for Nodes*

$$x_{node_{1,2}} = \frac{-0.063519455 + 0.26366392\tilde{\lambda}^{1/2} - 0.43432486\tilde{\lambda} + 0.1576067\tilde{\lambda}^{3/2} + 0.3370921\tilde{\lambda}^2}{-0.023958765 - 0.13600149\tilde{\lambda}^{1/2} + 0.77131778\tilde{\lambda} - 1.2857382\tilde{\lambda}^{3/2} + 0.46725241\tilde{\lambda}^2 + \tilde{\lambda}^{5/2}} \quad \text{(D.12)}$$

$$= \frac{-1.7572129 + 8.5369108\tilde{\lambda}^{1/2} - 17.406099\tilde{\lambda} + 14.319408\tilde{\lambda}^{3/2} + 1.0135369\tilde{\lambda}^2}{-0.14935374 - 1.364222\tilde{\lambda}^{1/2} + 8.2529269\tilde{\lambda} - 17.126994\tilde{\lambda}^{3/2} + 14.122656\tilde{\lambda}^2 + \tilde{\lambda}^{5/2}} \quad \text{(D.13)}$$

D.2.5.3 *Fit for Fourth Excited State Energy*

$$E^{(2,0)}_{fit} = \frac{0.01788667 - 0.1122653\tilde{\lambda} - 0.04332162\tilde{\lambda}^2 - 0.755475\tilde{\lambda}^3 + 2.101246\tilde{\lambda}^4 + 8.713058\tilde{\lambda}^5 + 21.71468\tilde{\lambda}^6}{0.001143266 - 0.004765047\tilde{\lambda} - 0.05486334\tilde{\lambda}^2 + 0.04036842\tilde{\lambda}^3 + 0.4014973\tilde{\lambda}^4 + \tilde{\lambda}^5} \quad \text{(D.14)}$$

Expansions at $g^4 = 0$:

$$E_{fit}^{(2,0)} = 9 + 242.072\, g^4 - 28066.8\, g^8 + 6.22229 \times 10^6\, g^{12} - 1.67742 \times 10^9\, g^{16} + \cdots$$

$$E_{PT}^{(2,0)} = 9 + 241.875\, g^4 - 28733.6\, g^8 + 7.29099 \times 10^6\, g^{12} - 2.65704 \times 10^9\, g^{16} + \cdots$$

D.2.6 *Fifth Excited State Energy* (2, 1)

D.2.6.1 *Interpolations*

$$A_{fit}(\tilde{\lambda}) = \frac{\begin{array}{c}-0.46657 + 0.864642\tilde{\lambda} + 1.15364\tilde{\lambda}^2 + 1.23493\tilde{\lambda}^3 + 1.24505\tilde{\lambda}^4 \\ +1.32474\tilde{\lambda}^5 + 14.5619\tilde{\lambda}^6\end{array}}{\begin{array}{c}-0.00665345 + 0.170138\tilde{\lambda} - 0.0210755\tilde{\lambda}^2 + 0.236557\tilde{\lambda}^3 \\ +0.0137864\tilde{\lambda}^4 + \tilde{\lambda}^5\end{array}},$$

$$B_{fit}(\tilde{\lambda}) = \frac{\begin{array}{c}-0.433234 + 2.28167\tilde{\lambda} + 3.21087\tilde{\lambda}^2 + 3.31109\tilde{\lambda}^3 + 3.5273\tilde{\lambda}^4 \\ +4.50982\tilde{\lambda}^5\end{array}}{1.2167 + 0.359486\tilde{\lambda} + 0.868032\tilde{\lambda}^2 + \tilde{\lambda}^3},$$

$$C_{fit}(\tilde{\lambda}) = \frac{\begin{array}{c}-0.944526 + 1.11349\tilde{\lambda} + 2.21114\tilde{\lambda}^2 + 2.72861\tilde{\lambda}^3 + 2.89208\tilde{\lambda}^4 \\ +2.86531\tilde{\lambda}^5 + \mathcal{P}_C\end{array}}{\begin{array}{c}1.97051 + 0.765764\tilde{\lambda} + 0.601285\tilde{\lambda}^2 + 0.860067\tilde{\lambda}^3 \\ +0.961337\tilde{\lambda}^4 + \tilde{\lambda}^5\end{array}},$$

$$D_{fit}(\tilde{\lambda}) = \frac{\begin{array}{c}0.563821 + 2.50238\tilde{\lambda} + 2.84348\tilde{\lambda}^2 + 2.80648\tilde{\lambda}^3 + 3.59646\tilde{\lambda}^4 \\ +4.92802\tilde{\lambda}^5\end{array}}{1.00435 - 0.35311\tilde{\lambda} + 1.22132\tilde{\lambda}^2 + 0.459138\tilde{\lambda}^3 + \tilde{\lambda}^4}, \tag{D.15}$$

where

$$\mathcal{P}_{C_{fit}} = 2.81886\tilde{\lambda}^6 + 2.95361\tilde{\lambda}^7 + 3.31926\tilde{\lambda}^8.$$

Fit for Nodes

$$x_{node_1} = \frac{\begin{array}{c}0.036520925 - 0.24915157\tilde{\lambda}^{1/2} + 0.70745156\tilde{\lambda} \\ -1.0008121\tilde{\lambda}^{3/2} + 0.58040962\tilde{\lambda}^2\end{array}}{\begin{array}{c}0.0046164832 + 0.043258509\tilde{\lambda}^{1/2} - 0.40450562\tilde{\lambda} \\ +1.2105151\tilde{\lambda}^{3/2} - 1.7231219\tilde{\lambda}^2 + \tilde{\lambda}^{5/2}\end{array}} \tag{D.16}$$

$$x_{node_2} = \frac{\begin{array}{c}0.071402072 - 0.49383773\tilde{\lambda}^{1/2} + 1.4140645\tilde{\lambda} - 2.0076375\tilde{\lambda}^{3/2} \\ +1.1662567\tilde{\lambda}^2\end{array}}{\begin{array}{c}0.00351383 + 0.046292989\tilde{\lambda}^{1/2} - 0.40466624\tilde{\lambda} \\ +1.2061251\tilde{\lambda}^{3/2} - 1.7205262\tilde{\lambda}^2 + \tilde{\lambda}^{5/2}\end{array}} \tag{D.17}$$

D.2.6.2 *Fit for Fifth Excited State Energy*

$$E^{(2,1)}_{fit} = \frac{\begin{array}{c}0.00402556 + 0.05040186\tilde{\lambda} - 0.3733187\tilde{\lambda}^2 + 0.3756896\tilde{\lambda}^3 \\ -1.804176\tilde{\lambda}^4 + 0.1025348\tilde{\lambda}^5 + 29.29965\tilde{\lambda}^6\end{array}}{\begin{array}{c}0.001245224 - 0.005715938\tilde{\lambda} + 0.01270724\tilde{\lambda}^2 - 0.1069739\tilde{\lambda}^3 \\ +0.003502232\tilde{\lambda}^4 + \tilde{\lambda}^5\end{array}} \quad \text{(D.18)}$$

Expansions at $g^4 = 0$:

$$E^{(2,1)}_{fit} = 11 + 432.187g^4 - 70000.0g^8 + 2.12604 \times 10^7 g^{12} - 7.85325 \times 10^9 g^{16} + \cdots$$

$$E^{(2,1)}_{PT} = 11 + 433.125g^4 - 72792.1g^8 + 2.53780 \times 10^7 g^{12} - 1.23982 \times 10^{10} g^{16} + \cdots$$

Appendix E

Numerical Evaluation of nth Correction ε_n and $\mathcal{Y}_n(v)$: Ground State

As discussed in Chapter 2, when considering the ground state, the *Non-Linearization Procedure* provides closed expressions for corrections ε_n and $\mathcal{Y}_n(v)$. Furthermore, in this scheme it is only required the zero-order approximation $\mathcal{Y}_0(v)$ to calculate corrections iteratively. This feature makes the Non-Linearization Procedure a powerful tool to develop perturbation series efficiently.

In this Appendix, we present a MATHEMATICA code which calculates, in principle,[1] an arbitrary finite number of corrections ε_n and $\mathcal{Y}_n(v)$ for a given input $\mathcal{Y}_0(v)$. As an illustration, we take the quartic AHO to show how the method is implemented. In what follows, we set that $\hbar = 1$ and $m = 1$. Note that with this choice $x = v$ according to (2.1).

E.1 Generalities

Let us consider the ground state of the quartic anharmonic potential

$$V(x) = x^2 + g^2\, x^4 .$$

Based on the Riccati-Bloch equation, we have seen that constructing perturbation theory only requires $\mathcal{Y}_0(x)$. If we choose

$$\mathcal{Y}_0(x) = x, \tag{E.1}$$

[1]The accumulation of numerical errors in each iteration may limit the largest order that we can calculate reliably.

the Non-Linearization Procedure becomes algebraic and corrections can be calculated up to a very high order using the code presented in Appendix B. However, the choice (E.1) leads to a divergent perturbation theory in powers of g^2 for both ε and $\mathcal{Y}(x)$. A natural question emerges: is it possible to choose a different zero-order approximation $\mathcal{Y}_0(x)$ other than (E.1) that leads to a convergent series? The answer to this question is: yes, we can. In fact, the Approximant (4.2)

$$\Psi^{(0,0)}_{(approximation)} = \frac{\exp\left(-\frac{A+(B^2+3)\,x^2/6+g^2\,x^4/3}{\sqrt{B^2+g^2\,x^2}}+\frac{A}{B}\right)}{(B^2+g^2\,x^2)^{\frac{1}{4}}\left(B+\sqrt{B^2+g^2\,x^2}\right)^{\frac{1}{2}}},$$

provides a suitable formula for $\mathcal{Y}_0(x)$ via

$$\mathcal{Y}_0(x) = -\frac{\partial_x \Psi^{(0,0)}_{(approximation)}}{\Psi^{(0,0)}_{(approximation)}}. \tag{E.2}$$

Owing to the constraints imposed at large x, condition (2.31) is automatically fulfilled. Therefore, it is guaranteed that perturbation theory will be convergent. As we will numerically demonstrate below, this choice leads to an extremely fast convergent perturbation theory.

To apply the Non-Linearization Procedure, we identify

$$V_0(x) = \frac{\partial_x^2 \Psi^{(0,0)}_{(approximation)}}{\Psi^{(0,0)}_{(approximation)}}, \tag{E.3}$$

as the unperturbed potential, and the difference

$$V_1(x) = (x^2 + g^2 x^4) - V_0(x) \tag{E.4}$$

as perturbation.[2] Note that perturbation theory is now developed in terms of a formal parameter λ_f which is inserted in front of $V_1(x)$: $\lambda_f V_1(x)$[3] different from g^2.

As an example, we calculate the corrections in Non-Linearization Procedure when $g^2 = 1$. According to the fits presented in Appendix C,

[2]For the next terms in expansion (2.20), we vanish $V_n(x)$=0, $n = 2, 3, \ldots$.
[3]In equation (2.20), $\lambda_f = 1$.

parameters A and B take the values

$$A = -0.6276, \quad B = 2.3684. \tag{E.5}$$

Using the Approximant, the expectation value of the Hamiltonian leads to

$$\varepsilon_0 + \varepsilon_1 = 1.392\,351\,649\,.... \tag{E.6}$$

In the framework of the Non-Linearization Procedure, the variational energy (E.6) coincides with the sum of the first two corrections $\varepsilon_0 + \varepsilon_1$, see [Turbiner (1984)] for details. If we consider in the perturbation series the partial sums of the first N corrections,

$$\sum_{n=0}^{N} \varepsilon_n, \quad \sum_{n=0}^{N} \mathcal{Y}_n(x),$$

then, thanks to convergence, they will provide an increasingly more accurate estimates of ε and $\mathcal{Y}(x)$ as N increases.

E.2 Code

The MATHEMATICA-12 code for the realization of the Non-Linearization Procedure is presented below. To simplify the presentation, we have split the code into two blocks.

(I) Block One:

```
(* Block I*)
max = 10; g = 1;

ψ0[x_] := 1/((B^2 + g^2 x^2)^(1/4) (B + Sqrt[B^2 + g^2 x^2])^(1/2)) Exp[-(A + (B^2+3) x^2/6 + g^2 x^4/3)/Sqrt[B^2 + g^2 x^2] + A/B];

A = Rationalize[-0.6276, 10^-20];
B = Rationalize[2.3684, 10^-20];
y0[x_] := -ψ0'[x]/ψ0[x]
norm2 = NIntegrate[ψ0[x]^2, {x, 0, ∞}, WorkingPrecision → 200];
V0[x_] := Simplify[ψ0''[x]/ψ0[x]]
V1[x_] := x^2 + g^2 x^4 - V0[x]
ε1 = 1/norm2 NIntegrate[V1[x] ψ0[x]^2, {x, 0, ∞}, WorkingPrecision → 200];
V[x_, n_] := V0[x] KroneckerDelta[n, 0] + V1[x] KroneckerDelta[n, 1]
Q[x_, n_] := -Sum[yk[x] yn-k[x], {k, 1, n-1}]
eq[n_] := D[yn[x], x] - 2 y0[x] yn[x] - (εn - V[x, n] - Q[x, n])
```

The first block (shown above) contains all definitions needed to realize the Non-Linearization Procedure. In the first line, we choose the maximum order of corrections **max** that we want to calculate (we have chosen **max=10**). Then we include the expressions that define the wave function for $g^2 = 1$. In the following lines, one can note that formulas (2.20), (2.25), and (2.26) were basically translated to MATHEMATICA language. At the end of this block, we define **eq(n)** in the same way it was defined in Appendix B. This function generates the equation that defines the nth correction, which is solved in the second block. Note that, thanks to the symmetry of the potential, it is enough to solve the equation in the semi-line $x > 0$.

(II) Block Two:

```
(* Block II *)
Quiet[Do[
  sol = NDSolve[
    {eq[i] == 0, y_i[20] == 0}, y_i, {x, 0, 20},
    WorkingPrecision → 100, InterpolationOrder → 20,
    MaxSteps → 100 000, Method → "StiffnessSwitching"];
  Set @@@ sol[[1]];
  ε_{i+1} = 1/norm2 NIntegrate[Q[x, i + 1] ψ_0[x]^2, {x, 0, 10},
     WorkingPrecision → 100,
     Method → {"GaussKronrodRule", "SymbolicProcessing" → 0},
     MaxRecursion → 100],
  {i, 1, max}]]
Do[Print["ε"_i, "=", N[ε_i, 40]], {i, 1, max}]
```

In the second block, we solve iteratively the equations defined in the first one. In contrast with perturbation theory developed in powers of g^2, where corrections can be calculated in exact form, here we need to solve the *perturbative* equations (2.25) numerically[4]. Therefore, it is evident that numerical errors could appear. In order to reduce them, we have inserted a number of options (**InterpolationOrder, MaxSteps,**

[4]In order to reduce numerical errors, it is convenient to solve numerically the equation (2.25) as the first step instead of using its solution in the form of integrals (2.27).

Method,WorkingPrecision) inside the command **NSolve** and **NIntegrate**. The interested reader is referred to the MATHEMATICA-12 documentation to check the scope of each option. The main goal of this block is the translation of (2.29) into MATHEMATICA language.

E.2.1 *Output*

The output of the program is shown below. Although it only presents on-screen the value of the first ten ε-corrections, the program stores in RAM all corrections $\mathcal{Y}_n(x)$, $n = 1, \ldots, 10$. They are useful when investigating the local accuracy of the Approximant (4.2), since they can be easily transformed into corrections to the phase and, ultimately, to the wave function. It is important to point out that the calculation of the first ten corrections takes about two hours of CPU time in today's standard desktop computer. If we are not interested in extremely high accuracy, the value for **WorkingPrecision** can be reduced throughout the two blocks of code **(I)** and **(II)**. In this way, CPU time can be drastically decreased to a few minutes.

ϵ_1=1.392351649204829943101599920695978863765

ϵ_2=-7.674075492260325345685456184602782368223 × 10^{-9}

ϵ_3=-4.625341178962971270707697626877770386246 × 10^{-13}

ϵ_4=-6.105387166974798019350487807085918798740 × 10^{-17}

ϵ_5=-1.199600347479270210272381225615198345036 × 10^{-20}

ϵ_6=-2.727376774537019111379827915566538510945 × 10^{-24}

ϵ_7=-6.737935314272449665689056054410354986482 × 10^{-28}

ϵ_8=-1.767437999496565804016181893785382988528 × 10^{-31}

ϵ_9=-4.850431855308680612584079419162433635829 × 10^{-35}

ϵ_{10}=-1.378339685945383032761420123964956284897 × 10^{-38}

The corrections[5] shown above present certain systematic. First of all, except for ε_1, all of them are negative. Furthermore, they satisfy

$$\left|\frac{\varepsilon_{n+1}}{\varepsilon_n}\right| \sim 10^{-3}, \qquad n = 1, ..., 9, \tag{E.7}$$

which clearly indicates a fast rate of convergence. The result of the partial sums of energy corrections of different orders is presented in Table E.1. It

[5]Without loss of generality, we define $\varepsilon_0 = 0$.

Table E.1. Quartic AHO, the Ground State $(0,0)$: First 8 partial sums of energy corrections. Digits in bold remain unchanged when the next order corrections are taken into account. The value of ε_{exact} was calculated in LMM, see F.7.

Approximation	Value
$\varepsilon_0+\varepsilon_1$	**1.392 351 649** 204 829 943 101 599 920 695 978
$\varepsilon_0+\varepsilon_1+\varepsilon_2$	**1.392 351 641 530** 754 450 841 274 575 010 522
$\varepsilon_0+\cdots+\varepsilon_3$	**1.392 351 641 530 291** 916 723 378 277 883 451
$\varepsilon_0+\cdots+\varepsilon_4$	**1.392 351 641 530 291 855 669** 506 608 135 471
$\varepsilon_0+\cdots+\varepsilon_5$	**1.392 351 641 530 291 855 657 5**10 604 660 679
$\varepsilon_0+\cdots+\varepsilon_6$	**1.392 351 641 530 291 855 657 507 877** 283 904
$\varepsilon_0+\cdots+\varepsilon_7$	**1.392 351 641 530 291 855 657 507 876 6**10 110
$\varepsilon_0+\cdots+\varepsilon_8$	**1.392 351 641 530 291 855 657 507 876 609 934**
ε_{exact}	**1.392 351 641 530 291 855 657 507 876 609 934**

is interesting to note that these sums can allow us to reach extremely high accuracy. To compare our results, we have inserted the energy ε_{exact}, which was calculated using the LMM with 200 points. The corresponding code for the LMM calculation can be found in Appendix F. It is worth mentioning that ε_{exact} calculated via the LMM indicates that ε_9 and ε_{10} are not reliable. In this case, the accumulation of numerical error generates wrong values for them. It can be fixed using a larger value in **WorkingPrecision** in the code. Certainly, this will lead to the increase of CPU time.

To conclude this Appendix, we show the plots of the first eight corrections $\mathcal{Y}_n(x)$, $n=0,\ldots,7$, see Fig. E.1. From those plots, one can note that

$$|\mathcal{Y}_{n+1}(x)| \ll |\mathcal{Y}_n(x)|, \quad n=1,\ldots,7. \tag{E.8}$$

More precisely,

$$\left|\frac{\max\{\mathcal{Y}_{n+1}(x)\}}{\max\{\mathcal{Y}_n(x)\}}\right| \simeq 10^{-4}, \quad n=1,\ldots,7, \tag{E.9}$$

which clearly indicates a fast convergence of perturbation theory developed for $\mathcal{Y}(x)$ with (E.2) as input for $\mathcal{Y}_0(x)$.

Needless to say, the code shown above can be easily modified to construct convergent perturbation theory for any one-dimensional system. An extension to the d-dimensional radial case is also straightforward.

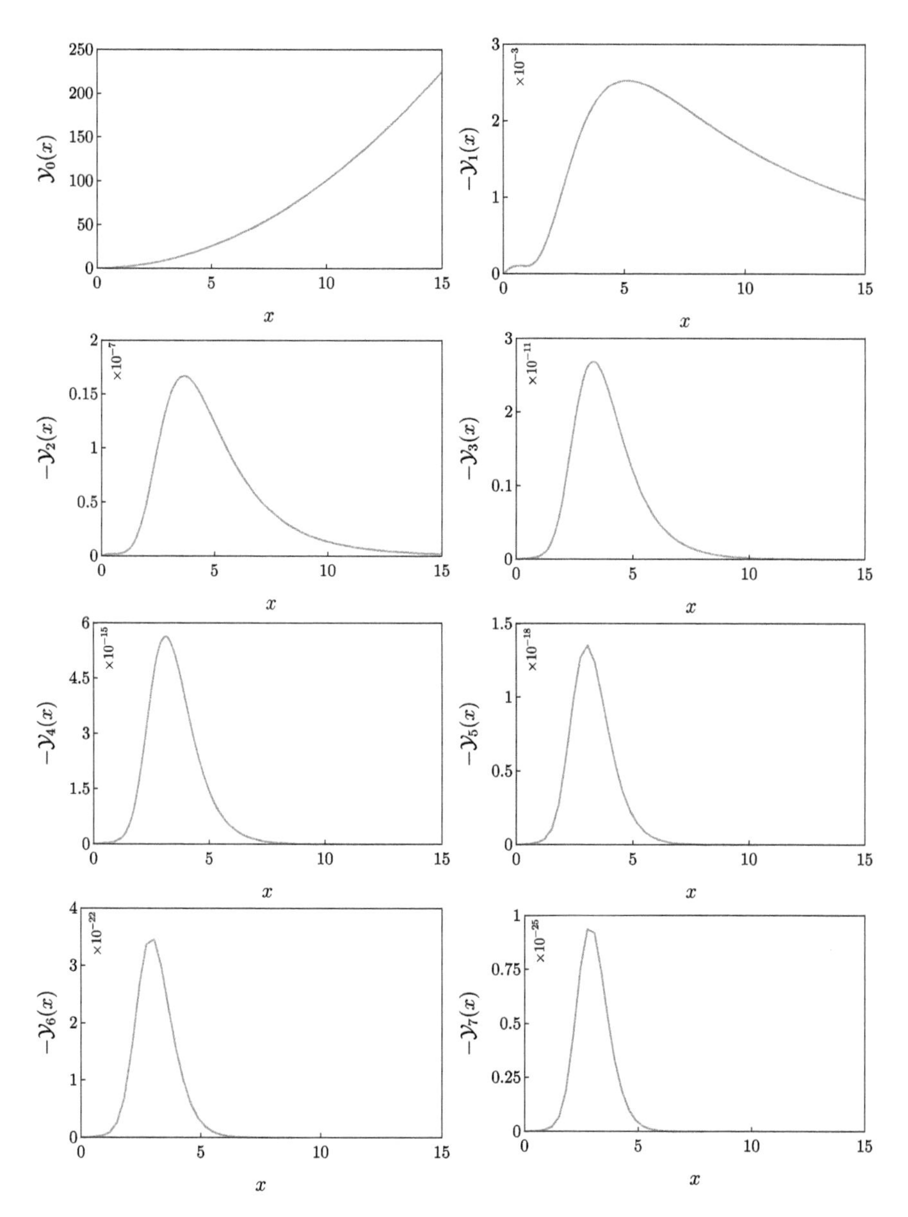

Fig. E.1. Quartic AHO. Plots of corrections $\mathcal{Y}_n(x)$ at $n = 0, ..., 7$. Except for the plot of $\mathcal{Y}_0$(x), the order of the correction is printed in the upper left corner.

Appendix F

The Lagrange Mesh Method

The Lagrange Mesh Method, see for review [Baye (2015)], was developed as an accurate, efficient and simple numerical method to solve the time-independent one-dimensional Schrödinger equation, in particular. It can be regarded as an approximate variational method. Three main ingredients are the basic building blocks for this method: (i) the Gauss quadrature, (ii) the Lagrange functions and (iii) the secular equation.

F.1 Gauss Quadrature

Consider one-dimensional integral over a smooth function $f(x)$ on the domain $[a,b]$,[1] namely,

$$\int_a^b f(x)\,w(x)\,dx, \tag{F.1}$$

where $w(x) \geq 0$ is a given weight function. With a few exceptions, see e.g.[Abramowitz and Stegun (1964)], this integral can not be computed analytically. Of course, numerical integration methods circumvent this inconvenience. Among numerous methods of numerical integration, the Gauss quadrature stands alone due to its simplicity, efficiency and ease to reach high accuracy. The basic idea of this method is to approximate the integral (F.1) by a certain finite sum:

$$\int_a^b f(x)\,w(x)\,dx \approx \sum_{k=1}^{N} w_k\, f(x_k), \tag{F.2}$$

[1]This domain can be finite (interval), semi-infinite (half-line) or infinite (entire line).

Table F.1. Meshes for classical orthogonal polynomials.

Domain	Weight Function	Mesh
$[-1,1]$	$(1-x)^{\alpha}(1+x)^{\beta}$	Chebyshev,[a] Legendre,[b] Jacobi[c]
$[0,\infty)$	$x^{\nu}\exp(-x)$	Laguerre[d]
$(-\infty,\infty)$	$\exp(-x^2)$	Hermite

[a] $\alpha=\pm1, \beta=0$, [b] $\alpha=\beta=0$, [c] $\alpha,\beta>-1$
[d] $\nu>-1$

where the *mesh points* $\{x_k\}_{k=1}^{N}$ are distributed (in general, non-uniformly) over the interval $[a,b]$, while their *weights* $-w_k-$ are some numbers that can be found by linear algebra means (see below). The expression in the right hand side of (F.2) is the so-called the *Gauss quadrature approximation.*

In order to choose x_k and w_k we impose the condition that the Gauss quadrature approximation at given N should be *exact* when the function $f(x)$ is a polynomial of the degree $(2N-1)$. Therefore, x_k should be chosen appropriately, following the taste of worker, and w_k may be obtained by solving linear algebraic equations. However, there are other, more specific ways to define x_k and w_k. One of such ways is related to choosing the zeroes of the classical orthogonal polynomials as the mesh points.

It can be shown if the N mesh points $\{x_k\}_{k=1}^{N}$ are chosen as the zeroes of the Nth-degree orthogonal polynomial $P_N(x)$ the function $w(x)$ is the weight function (or measure) in the corresponding space.[2] Thus, the weights w_k are given in terms of x_k, $P_{N-1}(x_k)$, and $P'_N(x_k)$, see [Baye (2015)]. This is the key observation, which is the basis of the Lagrange Mesh Method. As we will see later, the known explicit form of the weights is not relevant in the Lagrange Mesh Method, hence, we omit to present it constructively.

Once we choose an interval $[a,b]$ and a non-negative weight function $w(x)\geq 0$, the mesh is unambiguously defined. For example, if $w(x)=1$ and the interval is $[-1,1]$, the mesh points are the N zeroes of the Legendre polynomial of order N in variable x. The name of the mesh is given according to the polynomials involved. Therefore, in the above example those zeroes constitute the *Legendre mesh.* In Table F.1, we have summarized the different types of meshes for classical orthogonal polynomials.

[2] For a given weight function $w(x)>0$, one can always construct a set of orthogonal polynomials using the Gram-Schmidt procedure.

It is worth noting that the accuracy of the Gauss quadrature approximation (F.2) depends on how well $f(x)$ is approximated by a polynomial of degree $(2N-1)$ inside $[a,b]$. As a consequence, the Gauss quadrature approximation may fail if the original function $f(x)$ has singularities or discontinuities in $[a,b]$.

In practice, the appearance of the weight function $w(x)$ in the integrand of (F.1) can be omitted by defining a new function

$$g(x) = f(x)\, w(x).$$

In this way, (F.1) takes the form

$$\int_a^b g(x)\, dx \approx \sum_{k=1}^{N} \lambda_k g(x_k), \tag{F.3}$$

where now

$$\lambda_k = \frac{w_k}{w(x_k)}, \quad k = 1, 2, \ldots, N \tag{F.4}$$

plays the role of weights.

F.2 Lagrange Functions

The second ingredient of the Lagrange Mesh Method is a special type of infinitely differentiable real functions, which are called the *Lagrange Functions*. To define those functions, first we need to choose a particular mesh $\{x_k\}_{k=1}^{N}$ with its corresponding weights λ_k. By definition, a Lagrange function should satisfy two requirements:

(I). The Lagrange condition:

$$f_i(x_j) = \lambda_i^{-1/2} \delta_{ij}, \quad i, j = 1, 2, \ldots, N \tag{F.5}$$

(II). In the Gauss quadrature approximation, the integral of the product of any two Lagrange functions is exact:

$$\int_a^b f_i(x) f_j(x)\, dx \equiv \sum_{k=1}^{N} \lambda_k f_i(x_k) f_j(x_k). \tag{F.6}$$

As a result from properties (**I**) and (**II**), the Lagrange functions are orthonormal in or out the Gauss approximation, since

$$\begin{aligned}\int_a^b f_i(x)f_j(x)\,dx &= \sum_{k=1}^{N}\lambda_k f_i(x_k)f_j(x_k)\\ &= \sum_{k=1}^{n}\lambda_k\left(\lambda_i^{-1/2}\delta_{ik}\right)\left(\lambda_j^{-1/2}\delta_{jk}\right)\\ &= \delta_{ij}.\end{aligned} \tag{F.7}$$

There are well-known methods to construct the Lagrange functions. They are presented explicitly in [Baye (2015)]. Those and other explicit formulas associated with the Lagrange functions will be presented later in this Appendix D.

F.3 The Secular Equation and Lagrange Mesh Equations

The third ingredient of the Lagrange Mesh Method is the so-called secular equation in quantum mechanics. This equation is useful to find in approximate form the spectrum of a given the time-independent Schrödinger equation,

$$H\psi = E\,\psi, \tag{F.8}$$

where ψ is the wave function, E the energy, and H the Hamiltonian. In particular, we focus our discussion on one-dimensional Hamiltonians of the form

$$H = -\frac{\hbar^2}{2m}\partial_x^2 + V, \tag{F.9}$$

where m is the mass of the particle, V the potential, and $\hbar$ the Planck constant. We assume that H is defined in domain $[a,b]$.

Let us suppose that the wave function can be approximated as follows,

$$\psi(x) = \sum_{k=1}^{N} c_k\,\phi_k(x) \tag{F.10}$$

where $\{\phi_k\}_{k=1}^{N}$ is a set of orthogonal functions which are chosen in such a way that boundary conditions are fulfilled accordingly, while c_k are coefficients to be determined. In fact, using the variational principle, one can easily show that coefficients c_k are determined by the equation

$$\sum_{j=1}^{N} \left(\frac{\hbar^2}{2m} T_{ij} + V_{ij} \right) c_j = E\, c_i \tag{F.11}$$

where

$$T_{ij} = -\int_a^b \phi_i(x)\, \partial_x^2 \phi_j(x)\, dx, \quad V_{ij} = -\int_a^b \phi_i(x) V(x) \phi_j(x)\, dx. \tag{F.12}$$

Equation (F.11) is the so-called *secular equation*, see [Landau and Lifshitz (1977)] for details. In fact, (F.11) establishes that, in order to find c_k, we have to diagonalize the matrix representation of the Hamiltonian constructed with the functions $\{\phi_k\}_{k=1}^N$. Therefore, T_{ij} and V_{ij} play the role of kinetic and potential matrix elements, respectively. In the Lagrange Mesh Method, those matrix elements are calculated taking $\phi_k(x)$ as Lagrange functions $f_k(x)$ and using the Gauss quadrature to compute the involved integrals. Thus, under these considerations (F.11) now reads

$$\sum_{j=1}^{N} \left(\frac{\hbar^2}{2m} T_{ij}^{(G)} + V_{ij}^{(G)} \right) c_j = E\, c_i, \tag{F.13}$$

where the super-index (G) indicates that we are using the Gauss quadrature when integrating. In this sense, the Lagrange Mesh Method is an approximate variational method. The set of equations presented in (F.13) are called the *Lagrange equations.*

As a consequence of using Lagrange functions and the Gauss quadrature approximation, the potential elements are diagonal,

$$V_{ij}^{(G)} = V(x_i)\, \delta_{ij}. \tag{F.14}$$

Hence, its computation is straightforward, the potential is evaluated in the mesh points. On the other hand, the kinetic matrix elements $T_{ij}^{(G)}$ acquire simple expressions written in terms of N and the mesh points $\{x_k\}_{k=1}^N$. Explicit formulas for Hermite and Laguerre meshes are presented below.

F.4 Optimization of the Mesh: Rescaling of Mesh Points

It was observed empirically that the higher accuracy in the energies obtained in the LMM can be reached if the mesh points lie inside the compact domain, where the wave function is not too small. Using a global rescaling of the mesh points position in the form

$$x_k \to h\, x_k, \quad k = 1, 2, \ldots, N, \tag{F.15}$$

see [Baye (2015)], where h is called the scale factor (or the scaling parameter) one can move mesh points to this region. Since this region is not well defined, it is the state-of-art to choose it. If h is chosen properly, it can increase *considerably* the performance of the LMM and reduce CPU time. Usually, it requires a careful investigation of the dependences on h.

The usage of h modifies the Lagrange equations, which now reads

$$\sum_{j=1}^{N} \left(\frac{\hbar^2}{2mh^2}\, T_{ij}^{(G)} + V(hx_i)\, \delta_{ij} \right) c_j = E\, c_i. \tag{F.16}$$

Sometimes, h can be considered as a variational parameter. However, the Gauss quadrature approximation used in (F.16) may lead to a fake minimum in the energy with respect to h. Further details can be found in [Baye (2015)].

F.5 Hermite and Laguerre Meshes

F.5.1 *One-dimensional Case*

For the one-dimensional case, if the domain is $(-\infty, \infty)$, it is natural to consider the Hermite mesh by taking as the mesh points the zeroes $\{x_i\}$ of the Hermite polynomials. In this case the Lagrange functions read

$$f_i(x) = \frac{(-1)^{N-i}}{(2\, h_N)^{1/2}}\, \frac{H_N(x)}{x - x_i}\, e^{-x^2/2}, \quad i = 1, 2, \ldots, N, \tag{F.17}$$

where

$$h_N = 2^N\, N! \sqrt{\pi}. \tag{F.18}$$

These functions certainly satisfy properties (**I**) and (**II**), see (F.5) and (F.6). Note that the explicit form of the weights λ_k can be extracted form (F.17). The kinetic matrix elements T_{ij} are given by

$$T_{i\neq j}^{(G)} = (-1)^{i-j}\, \frac{2}{(x_i - x_j)^2}, \qquad T_{ii}^{(G)} = \frac{1}{3}(2N + 1 - x_i^2). \tag{F.19}$$

As concrete examples of the use of the LMM with the Hermite mesh we consider the ground state energies of quartic x^4 and sextic x^6 oscillators.

F.5.1.1 *Quartic oscillator* x^4

The goal of this section is to show how the accuracy of the ground state energy $E_0^{(4)}$ of the quartic oscillator depends on the number of used mesh points while carrying out the detailed LMM calculation. *For the presentation of the results it is convenient to introduce the following notation*

$$\underset{Y_Z}{X} : X = \textit{Digit},\ Y = \textit{Mesh Points},\ Z = \textit{Decimal Place of } X.$$

Digit X*, printed in bold, indicates the maximal correct digit in energy, which is reproduced with number of mesh points* Y.[3] Maximal accuracy of the energy of 204 correct decimal digits is reached with 2000 mesh points and then it is checked with 2020 mesh points. Non-optimized mesh is considered. Final result reads,

$$\begin{aligned} E_0^{(4)} = 1.&\underset{10_1}{\underline{\mathbf{0}}}\ 60\,362\,\underset{25_7}{\underline{\mathbf{0}}}90\,4\underset{50_{10}}{\underline{\mathbf{8}}}4\,182\,899\,6\underset{100_{20}}{\underline{\mathbf{4}}}7\,046\,016\,692\,663\,5\underset{200_{35}}{\underline{\mathbf{4}}}5\,515\,208\,728\,528\,977\,933\\ &216\,245\,241\,695\,943\,5\underset{500_{71}}{\underline{\mathbf{6}}}3\,044\,344\,421\,126\,896\,299\,134\,671\,703\,510\,546\,244\\ &358\,582\,525\,58\underset{1000_{120}}{\underline{\mathbf{0}}}879\,808\,210\,293\,147\,013\,176\,836\,373\,824\,935\,789\,226\,246\\ &004\,708\,175\,446\,960\,141\,637\,488\,417\,282\,25\underset{1900_{195}}{\underline{\mathbf{6}}}905\,935\,75\underset{2000_{204}}{\underline{\mathbf{7}}} \end{aligned} \tag{F.20}$$

Let us note a very important property of convergence of LMM procedure. The rate of convergence deteriorates slowly with the increase of the number of used mesh points. For example, with 100 mesh points we get 20 correct decimal digits, while by adding the 100 mesh points more this leads to extra 15 correct decimal digits amounting to 35 correct decimal digits; using 1000 mesh points allows us to get 120 correct decimal digits; coming from 1900 mesh points to 2000 mesh points provides 9 additional correct decimal points only. Finally, with 2000 mesh points we get 204 correct decimal digits in energy.

F.5.1.2 *Sextic oscillator* x^6

The goal of this section is to show how the accuracy of the ground state energy $E_0^{(6)}$ of the sextic oscillator depends on the number of mesh points. Maximal accuracy of the energy is calculated with 2000 mesh points and

[3] Accuracy of 300 digits in definition of mesh points is used. In MATHEMATICA-12 code, it corresponds to the command **WorkingPrecision**$\to$300. The results were verified in MATHEMATICA-13 code.

then it is checked with 2020 mesh points. Non-optimized mesh is considered. The same notations as in App.F.5.1.1 are used. Final result reads,

$$E_0^{(6)} = 1.\underset{10_1}{\underline{\mathbf{1}}}44\,802\underset{50_7}{\underline{\mathbf{4}}}53\,797\underset{150_{13}}{\underline{\mathbf{0}}}52\,763\underset{250_{19}}{\underline{\mathbf{7}}}65\,457\,5\underset{400_{26}}{\underline{\mathbf{3}}}4\,149\,549\,076\,537\,812\,528\,979\,8\underset{1000_{50}}{\underline{\mathbf{7}}}7\,729$$
$$205\,213\,231\,264\,923\,551\,780\,190\,2\underset{2000_{80}}{\underline{\mathbf{1}}} \tag{F.21}$$

Making a comparison of the energies (F.20) and (F.21) for quartic and sextic oscillators, respectively, one can note two important properties of convergence of LMM procedure:

(i) the rate of convergence for sextic oscillator is much slower than for quartic oscillator, with 2000 mesh points one can reach 80 and 204 accurate decimal digits in energy, respectively; it is natural to assume that this trend will continue with the increase in the degree of anharmonicity to octic, decatic etc oscillators,
(ii) the rate of convergence for sextic oscillator with the increase of the number of mesh points is much slower than for quartic oscillator.

Finally, we have to emphasize that with 2000 mesh points we reached 80 correct decimal digits in energy. Optimization of mesh lattice points (even with global dilatation factor h) allows us to decrease the CPU time of calculations.

F.6 Jacobi Mesh

If the problem is defined on the finite interval, it is natural to map it to the interval $[-1, 1]$ and use the Jacobi mesh, where the mesh points are defined by zeroes (roots) of the Jacobi polynomials. The Lagrange functions can be written explicitly in this case, cf. (F.17), (G.2). Since the problems with domain given by the finite interval are not explored in the text, these formulas are not presented. They can be found in [Baye (2015)].

F.7 Minimal Code for the LMM

We conclude this Appendix F by presenting the numerical realization of the LMM in MATHEMATICA-12. This Section focuses on the calculation of the energy spectra for potentials defined in the infinite domain $(-\infty, \infty)$.

We have chosen the quartic AHO potential

$$V(x) = x^2 + x^4$$

as the illustration of the numerical implementation of the LMM. The formulas presented in F.5.1 for the Hermite mesh are used. In concrete, the code is constructed to employ the mesh with 200 mesh points. In order to avoid the error accumulation the calculations with 100-digit arithmetic, **WorkingPrecision**$\rightarrow$ 100, is used.

```
n = 200;
sols = NSolve[HermiteH[n, x] == 0, x, WorkingPrecision → 100];
j = 0;
sols /. {r__Rule} :→ Set @@@ ({r} /. var : x → Subscript[var, ++j]);
V[x_] := x^2 + x^4
Do[V_{i,j} = V[x_i] KroneckerDelta[i, j], {i, 1, n}, {j, 1, n}];
Do[For[i = 1, i < j, i++, T_{i,j} = (-1)^(i-j) 2/(x_i - x_j)^2; T_{j,i} = T_{i,j}], {j, 1, n}];
Clear[i];
Clear[j];
Do[T_{i,i} = 2/6 × (2 n + 1 - x_i^2), {i, 1, n}];
H = Table[Table[T_{l,m} + V_{l,m}, {m, 1, n}], {l, 1, n}];
N[Eigenvalues[H, -1], 34]

{1.392351641530291855657507876609934}
```

The first two lines are used to define the Hermite mesh, while the rest of the code is the implementation of the formulas presented in (F.14) and (F.19).

In the case of the ground state, we ask MATHEMATICA for the lowest eigenvalue using the command **Eigevalues[H,-1]**. The output of the code, the ground state energy of the quartic AHO is printed on-screen, see the last line. An interested reader could check that the first 34 digits of this energy are stable with respect to an increase of the number of the mesh points and/or the increase of the number of digits in arithmetic by changing **WorkingPrecision**. Following the ideology of the Lagrange Mesh approach it guarantees the correctness of all 34 digits. Making comparison of this result with those presented in Table E.1, we can see that there is an excellent agreement in 34 significant digits. Here, we note that reproduction of the ground state energy obtained in LMM but by using the Non-Linearization Procedure requires the calculation of 8 corrections only.

The CPU time, required by running the code shown above, is less than one minute.

We point out that above minimal code can be easily modified by implementing the optimization of the mesh. It allows us to reach desired accuracy faster. In fact, the code allows a straightforward modification to tackle the energy spectra of any one-dimensional anharmonic oscillator, in general, an arbitrary confining, growing at large distances potential. It can be extended to consider the problems on half-line and finite intervals.

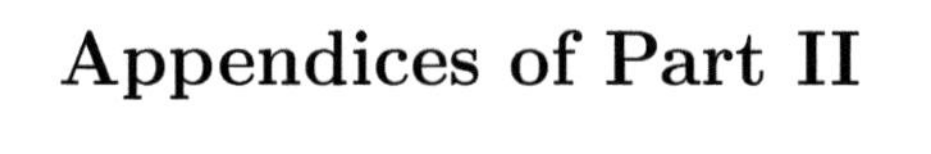

Appendices of Part II

Appendix G

d-dimensional Radial Oscillator: Lagrange Mesh Method

The radial Schrödinger equation is defined on the half-line, $r \in [0, \infty)$, and it is natural to use the Laguerre mesh, when the mesh points coincide with zeroes $\{r_i\}$ of the Laguerre polynomials. As we have already seen, in the radial case the kinetic energy operator of the radial Hamiltonian contains the centrifugal term, namely,

$$T = -\partial_r^2 + \frac{\alpha(\alpha-2)}{4r^2}. \tag{G.1}$$

Here, parameter α should be chosen appropriately, it depends on the dimension D and angular quantum number ℓ. Eventually, the Lagrange functions are

$$f_i(x) = (-1)^i \left(\frac{r_i}{h_N^{(\alpha)}}\right)^{1/2} \frac{L_N^{(\alpha)}(r)}{r - r_i}\, r^{\alpha/2}\, e^{-r/2}, \quad i = 1, 2, \ldots, N, \tag{G.2}$$

where $L_N^{(\alpha)}(r)$ is the associated Laguerre polynomial with index α in the standard notation [Abramowitz and Stegun (1964)] and

$$h_N^{(\alpha)} = \frac{\Gamma(N+\alpha+1)}{N!}.$$

The corresponding kinetic energy matrix elements $T_{ij}^{(G)}$ read

$$\begin{aligned} T_{i\neq j}^{(G)} &= (-1)^{i-j} \frac{1}{\sqrt{r_i r_j}} \left[\frac{N}{\alpha+1} + \frac{1}{2} - \left(\frac{1}{r_i} + \frac{1}{r_j}\right) + \frac{r_i + r_j}{(r_i - r_j)^2}\right], \\ T_{ii}^{(G)} &= -\frac{1}{12} + \frac{(2N+\alpha+1)(\alpha+4)}{6(\alpha+1)r_i} + \frac{(\alpha+2)(\alpha-5)}{6r_i^2}. \end{aligned} \tag{G.3}$$

G.1 Cubic Radial Oscillator r^3

Ground state energy for cubic radial oscillator r^3 found in LMM with up to 2000 mesh points. Maximal accuracy of the energy is calculated with 2000 mesh points and then it is checked with 2020 mesh points. Non-optimized mesh is considered. Similar but slightly different notations are used

$$\underset{Y_Z}{X} : X = \text{Digit}, Y = \text{Mesh Points},\ \ Z = \text{Significant Digit Place of } X.$$

Now Digit X, printed in bold, indicates the maximal correct *significant* digit in energy, which is reproduced with number of used mesh points Y. Final result reads,

$$\begin{aligned} E_0^{(3)} = &\underset{10_1}{\underline{\mathbf{3}}}.450\underset{50_5}{\underline{\mathbf{5}}}62\,689\,947\,4\underset{250_{15}}{\underline{\mathbf{4}}}6\,566\,551\,0\underset{500_{24}}{\underline{\mathbf{3}}}5\,222\,676\,476\,759 \\ &\underset{1000_{38}}{\underline{\mathbf{4}}}05\,866\,162\,611\,219\,538505\underset{2000_{59}}{\underline{\mathbf{0}}} \end{aligned} \tag{G.4}$$

Finally, we have to emphasize that with 2000 mesh points we reached 58 correct decimal digits in energy. This is much less than the correct decimal digits obtained for the ground state energy of the one-dimensional quartic (204 d.d., see (I.E.20) in Part I, App.E) or sextic (80 d.d., see (I.E.21) in Part I, App.E) oscillators.

Appendix H

First PT Corrections and Generating Functions $G_{3,4}$ for the Cubic Anharmonic Oscillator

Asymptotic expansions for the first three corrections $\mathcal{Y}_n(v), n = 1,2,3$ for the cubic AHO (3.1) at small and large v.

- At $v \to 0$

$$
\begin{aligned}
\mathcal{Y}_1(v) &= \epsilon_1 v + \frac{2\,\epsilon_1}{d+2} v^3 - \frac{1}{d+3} v^4 + \frac{\epsilon_1}{(d+2)(d+4)} v^5 + \cdots\,, \\
\mathcal{Y}_2(v) &= \epsilon_2\, v + \frac{\epsilon_1^2 + 2\,\epsilon_2}{d+2} v^3 + \frac{6\,\epsilon_1^2 + 4\,\epsilon_2}{(d+2)(d+4)} v^5 - \frac{2\,\epsilon_1}{(d+3)(d+5)} v^6 + \cdots\,, \\
\mathcal{Y}_3(v) &= \epsilon_3\, v + \frac{2(\epsilon_1\,\epsilon_2 + \epsilon_3)}{d+2} v^3 + \frac{2(\epsilon_1^3 + 6\,\epsilon_1\,\epsilon_2 + 2\,\epsilon_3)}{(d+2)(d+4)} v^5 \\
&\quad - \frac{2\,\epsilon_2}{(d+3)(d+5)} v^6 + \cdots\,.
\end{aligned}
\tag{H.1}
$$

where $\epsilon_n = \frac{\varepsilon_n}{d}$.

- At $v \to \infty$

$$
\begin{aligned}
\mathcal{Y}_1(v) &= \frac{1}{2} v^2 + \frac{1}{4}(d+1) - \frac{\varepsilon_1}{2} v^{-1} + \frac{1}{8}(d^2-1) v^{-2} + \cdots\,, \\
\mathcal{Y}_2(v) &= -\frac{1}{8} v^3 - \frac{1}{16}(3d+4) v + \frac{\varepsilon_1}{4} - \frac{1}{32}(6d^2 + 6d + 16\,\varepsilon_2 - 1) v^{-1} \\
&\quad + \frac{1}{8}(3d-2)\,\varepsilon_1 v^{-2} + \cdots\,,
\end{aligned}
$$

$$\mathcal{Y}_3(v) = \frac{1}{16}v^4 + \frac{1}{32}(5d+8)v^2 - \frac{3\,\varepsilon_1}{16}v + \frac{1}{64}(15d^2 + 26d + 16\,\varepsilon_2 + 10) \\ - \frac{1}{32}(15d\,\varepsilon_1 + 16\,\varepsilon_3)v^{-1} + \cdots . \tag{H.2}$$

Straightforward application of formulas (1.71) and (1.83) leads to the generating functions in the explicit form. In particular, G_3 and G_4 are given by

$$G_3 = -\frac{(2m)^{1/2}}{g^2}\left(\frac{\varepsilon_1}{2}\log\left[\frac{w-1}{w+1}\right]\right),$$

$$G_4 = -\frac{(2m)^{1/2}}{g^2}\left(\frac{5+(5+12d)w+(1-6d(d+1))(w+1)w^2\left(3w^2-2\right)}{48(w-1)(w+1)^2w^3}\right. \\ \left. \frac{(1-16\,\varepsilon_2-6d(d+1))}{32}\log\left[\frac{w-1}{w+1}\right]\right), \tag{H.3}$$

where $w = (1+gr)^{1/2}$.

Appendix I

First PT Corrections and Generating Functions $G_{4,6}$ for the Quartic Anharmonic Oscillator

We present here the corrections $\varepsilon_{4,6}$ and $\mathcal{Y}_{4,6}$ in explicit form for the quartic rAHO potential (10.1), which occur in the expansions (4.6) and (4.8), (4.9):

$$\varepsilon_4 = -\frac{1}{16}\, d(d+2)(2d+5),$$

$$\varepsilon_6 = \frac{1}{64}\, d(d+2)(8d^2+43d+60),$$

$$\mathcal{Y}_4(v) = -\frac{1}{8}v^5 - \frac{1}{16}(3d+8)v^3 + \frac{\varepsilon_4}{d}\, v,$$

$$\tilde{\mathcal{Y}}_4(v_2) = -\frac{1}{16}\left[2{v_2}^2 + (3d+8)v_2 + (d+2)(2d+5)\right],$$

$$\mathcal{Y}_6(v) = \frac{1}{16}v^7 + \frac{1}{32}(5d+16)v^5 + \frac{1}{16}(3d^2+17d+25)v^3 + \frac{\varepsilon_6}{d}\, v.$$

$$\tilde{\mathcal{Y}}_6(v_2) = \frac{1}{64}\left[4{v_2}^3 + 2(5d+16){v_2}^2 + 4(3d^2+17d+25)v_2 + (d+2)(8d^2+43d+60)\right].$$

In addition to the first two generating functions $G_0(r;g), G_2(r;g)$, see (4.27), for the phase, see expansion (1.17) and (1.82), we present two more generating functions:

$$G_4(r;g) = \frac{5}{24g^2w^3} + \frac{d\,(1+w+w^2)}{4\,g^2\,(w+1)w^2} + \frac{d^2}{8g^2w},$$

$$-G_6(r;g) = \frac{5}{16g^2w^6} + \frac{d\,(15+30w+20w^2+16w^3+20w^4 +30w^5+15w^6)}{48g^2\,(w+1)^2\,w^5}$$
$$+\frac{d^2\,(4+8w+8w^2+12w^3+18w^4+9w^5)}{32\,g^2\,(w+1)^2w^4} + \frac{d^3\,(1+3w^2)}{48g^2w^3},$$

where $w = \sqrt{1+g^2\,r^2}$. Two remarks in a row: (i) in the variable w all generating functions are rational functions, (ii) the polynomial structure in d of any generating function is evident.

Appendix J

First PT Corrections and Generating Functions $G_{8,12}$ for the Sextic Anharmonic Oscillator

We present explicitly the first corrections $\varepsilon_{8,12}$ and $\mathcal{Y}_{8,12}$ for the sextic anharmonic oscillator (5.1), see (5.8) and (5.10),

$$\varepsilon_8 = -\frac{1}{128}\,d\,(d+2)\,(d+4)\,(9d^2+72d+152),$$

$$\varepsilon_{12} = \frac{1}{1024}\,d\,(d+2)\,(d+4)\,(81d^4+1404d^3+9624d^2+31152d + 40384),$$

$$\mathcal{Y}_8(v) = -\frac{1}{8}v^9 - \frac{1}{16}(3d+16)v^7 - \frac{1}{16}(3d^2+27d+64)v^5 - \frac{1}{32}(4d^3+49d^2+204d+288)v^3 + \frac{\varepsilon_8}{d}\,v,$$

$$\tilde{\mathcal{Y}}_8(v_2) = -\frac{1}{8}{v_2}^4 - \frac{1}{16}(3d+16){v_2}^3 - \frac{1}{16}(3d^2+27d+64){v_2}^2 - \frac{1}{32}(4d^3+49d^2+204d+288)v_2 - \frac{1}{128}\,(d+2)\,(d+4)\,(9d^2+72d+152),$$

$$\mathcal{Y}_{12}(v) = \frac{1}{16}v^{13} + \frac{1}{32}(5d+32)v^{11} + \frac{5}{64}(3d^2+34d+104)v^9 + \frac{1}{32}(8d^3+125d^2+688d+1344)v^7 + \frac{1}{256}(55d^4+1038d^3$$

$$+7708d^2+26784d+36800)v^5+\frac{1}{256}(36d^5+783d^4+7040d^3$$

$$+32768d^2+78912d+78336)v^3+\frac{\varepsilon_{12}}{d}\,v,$$

$$\tilde{\mathcal{Y}}_{12}(v_2)=\frac{1}{16}\,v_2^6+\frac{1}{32}\,(5d+32)\,v_2^5$$

$$+\frac{5}{64}\,(3d^2+34d+104)\,v_2^4+\frac{1}{32}\,(8d^3+125d^2+688d+1344)\,v_2^3$$

$$+\frac{1}{256}\,(55d^4+1038d^3+7708d^2+26784d+36800)\,v_2^2$$

$$+\frac{1}{256}\,(36d^5+783d^4+7040d^3+32768d^2+78912d+78336)\,v_2$$

$$+\frac{1}{1024}\,(d+2)\,(d+4)\,(81d^4+1404d^3+9624d^2+31152d$$

$$+40384).$$

Now we present two generating functions $G_4(r), G_6(r)$ in the expansion (1.17) for sextic rAHO in the explicit form,

$$G_4(r)=\frac{r^2}{4w}\left(\frac{(5+w^2)}{3w^2}+\frac{d\,(1+w+w^2)}{w(w+1)}+\frac{d^2}{4}\right),$$

$$4g^2\,G_6(r)=\frac{5-3w^2}{w^6}+\frac{d\,(15+15w-4w^2+2w^3+6w^4+6w^5)}{6(w+1)w^5}$$

$$+\frac{d^2\,(2+2w+w^2+3w^3+3w^4)}{4(w+1)w^4}+\frac{d^3\,(1+3w^2)}{24w^3},$$

where $w=\sqrt{1+g^4\,r^4}$. Two remarks in a row: (i) in the variable w all generating functions are rational functions; (ii) The polynomial structure in d of all generating functions is evident.

Bibliography

Abramowitz, M. and Stegun, I. A. (1964). *Handbook of Mathematical Functions with Formulas, Graphs, and Mathematical Tables* (Dover).

Baker, G. A. and Graves-Morris, P. (2010). *Padé Approximants, 2nd edition (revised)* (Cambridge University Press, 2010).

Bateman, H. (1953). *Higher Transcendental Functions*, Vol. 1,2,3 (McGraw-Hill Book Company).

Baye, D. (2015). The Lagrange-Mesh Method, *Phys. Rep.* **565**, Supplement C, pp. 1–107.

Bender, C. and Orszag, S. (1978). *Advanced Mathematical Methods for Scientists and Engineers (Asymptotic Methods and Perturbation Theory)* (McGraw-Hill Book Company).

Bender, C. M. and Wu, T. T. (1969). Anharmonic oscillator, *Phys. Rev.* **184**, pp. 1231–1260.

Brezin, E., Le Guillou, J.-C., and Zinn-Justin, J. (1977a). Perturbation theory at large order. I. The ϕ^4 interaction, *Phys. Rev. D* **15**, pp. 1544–1557.

Brezin, E., Le Guillou, J.-C., and Zinn-Justin, J. (1977b). Perturbation theory at large order. II. Role of the vacuum instability, *Phys. Rev. D* **15**, pp. 1558–1564.

Del Valle, J. C. and Nader, D. J. (2018). Towards the theory of the Yukawa potential, *J. Math. Phys.* **59**, 10, p. 102103.

Del Valle, J. C. and Turbiner, A. V. (2019). Radial anharmonic oscillator: Perturbation theory, new semiclassical expansion, approximating eigenfunctions. I. Generalities, cubic anharmonicity case, *Int. J. Mod. Phys. A* **34**, 26, p. 1950143.

Del Valle, J. C. and Turbiner, A. V. (2020). Radial anharmonic oscillator: Perturbation theory, new semiclassical expansion, approximating eigenfunctions. ii. quartic and sextic anharmonicity cases, *Int. J. Mod. Phys. A* **35**, 01, p. 2050005.

Del Valle, J. C., Turbiner, A. V., and Escobar Ruiz, M. A. (2020). Two-body neutral coulomb system in a magnetic field at rest: from hydrogen atom to positronium, *Phys. Rev. A* **103**, p. 032820.

Dolgov, A. D. and Popov, V. S. (1978). Higher orders an structure of the perturbation series for the anharmonic oscillator, *Zh. Eksp. Teor. Fiz.* **75**, pp. 668–694.

Dolgov, A. D. and Popov, V. S. (1979). The anharmonic oscillator and its dependence on space dimensions, *Phys. Lett.* **86B**, pp. 185–188.

Dolgov, A. D. and Turbiner, A. V. (1980). A New Approach to the Stark Effect in Hydrogen, *Phys. Lett.* **77A**, pp. 15–22.

Doren, D. J. and Herschbach, D. R. (1986). Spatial dimension as an expansion parameter in quantum mechanics, *Phys. Rev. A* **34**, pp. 2654–2664.

Duffing, G. (1918). Erzwungene Schwingungen bei Veranderlicher Eigen-frequenz, *Buchbesprechung* **41/42**.

Dyson, F. J. (1952). Divergence of perturbation theory in quantum electrodynamics, *Phys. Rev.* **85**, pp. 631–632.

Efthimiou, C. and Frye, C. (2014). *Spherical Harmonics in p Dimensions* (World Scientific).

Eremenko, A. and Gabrielov, A. (2009). Analytic continuation of eigenvalues of a quartic oscillator, *Comm. Math. Phys.* **287**, pp. 431–457.

Eremenko, A., Gabrielov, A., and B., S. (2008). Zeros of eigenfunctions of some anharmonic oscillators, *Ann. Inst. Fourier, Grenoble* **58**, pp. 603–624.

Escobar-Ruiz, M. A., Shuryak, E., and Turbiner, A. V. (2016). Fluctuations in quantum mechanics and field theories from a new version of semiclassical theory. *Phys. Rev. D* **93**, p. 105039.

Escobar-Ruiz, M. A., Shuryak, E., and Turbiner, A. V. (2017). Fluctuations in quantum mechanics and field theories from a new version of semiclassical theory. II, *Phys. Rev. D* **96**, p. 045005.

Fernández, F. M. and Guardiola, R. (1997). The strong coupling expansion for anharmonic oscillators, *J. Phys. A* **30**, 20, p. 7187.

Feynman, R. (1972). *Statistical Mechanics: A Set of Lectures* (W. A. Benjamin Inc., Reading, MA).

Feynman, R. and Hibbs, A. (1965). *Quantum mechanics and path integrals, Emended edition by D.F. Styer, New York: Mc Graw-Hill (2005)* (Dover Publications, INC, Mineola, NY).

Graffi, S., Grecchi, V., and Simon, B. (1970). Borel Summability: Application to the Anharmonic Oscillator, *Phys. Lett. B* **B 32**, pp. 631–634.

Krylov, N. and Bogoliubov, N. (1950). *Introduction to Non-linear Mechanics* (Princeton University Press).

Landau, L. D. and Lifshitz, M. (1976). *Mechanics* (Butterworth-Heinemann).

Landau, L. D. and Lifshitz, M. (1977). *Quantum Mechanics, Non-Relativistic Theory* (Pergamon).

Lipatov, L. N. (1976). Calculation of the Gell-Mann-Low Function in a scalar field theory with strong nonlinearity, *Sov. Phys. — JETP* **44**, 3, p. 1055.

Meißner, H. and Steinborn, E. O. (1997). Quartic, sextic, and octic anharmonic oscillators: Precise energies of ground state and excited states by an iterative method based on the generalized Bloch equation, *Phys. Rev. A* **56**, pp. 1189–1200.

Müller, C. (1966). *Spherical Harmonics* (Springer-Verlag).

Olivares-Pilón, H. and Turbiner, A. V. (2018). H_2^+, HeH and H_2: Approximating potential curves, calculating rovibrational states, *Ann. Phys* **393**, pp. 335–357.

Shifman, M. A. (2015). Resurgence, operator product expansion, and remarks on renormalons in supersymmetric Yang-Mills theory, *JETP* **120**, pp. 386–398.

Shuryak, E. and Torres-Rincon, J. M. (2020). Baryon preclustering at the freeze-out of heavy-ion collisions and light-nuclei production, *Phys. Rev. C* **101**, p. 034914.

Shuryak, E. and Turbiner, A. V. (2018). Transseries for the ground state density and generalized Bloch equation: Double-well potential case, *Phys. Rev. D* **98**, p. 105007.

Shuryak, E. V. (1988). Toward the Quantitative Theory of the 'Instanton Liquid' 4. Tunneling in the Double Well Potential, *Nucl. Phys. B* **302**, pp. 621–644.

Taşeli, H. and Eid, R. (1996). Eigenvalues of the two-dimensional Schrödinger equation with nonseparable potentials, *Int. J. Quantum Chem.* **59**, 3, pp. 183–201.

Turbiner, A. and del Valle, J. (2021a). Anharmonic oscillator: a solution, *J. Phys. A* **54**, 295404, pp. 1–10.

Turbiner, A. and del Valle, J. (2021b). Comment on: *Uncommonly accurate energies for the general quartic oscillator*, Int. J. Quantum Chem., e26554 (2020), by P. Okun and K. Burke, *Int. Journ. Quant. Chem.* **122**, qua.26766, pp. 1–8.

Turbiner, A. and del Valle, J. (2022). From quartic anharmonic oscillator to double well potential, *Acta Polytecnica* **62**, 295404, pp. 208–210.

Turbiner, A. V. (1979). A new approach to finding levels of energy of bound states in quantum mechanics: Convergent perturbation theory, *Soviet Phys. — Pisma ZhETF* **30**, pp. 379–383.

Turbiner, A. V. (1980). On perturbation theory and variational methods in quantum mechanics, *Soviet Phys. — ZhETF* **79**, pp. 719–1745.

Turbiner, A. V. (1981). A new approach to the eigenvalue problem in quantum mechanics: convergent perturbation theory for rising potentials, *J. Phys. A* **14**, 7, p. 1641.

Turbiner, A. V. (1984). The eigenvalue spectrum in quantum mechanics and the nonlinearization procedure, *Phys. Usp.* **27**, 9, p. 668.

Turbiner, A. V. (1987). On finding the eigenfunctions in a potential quarkonium model (Perturbation theory and variational method), *Sov. Phys. — Yad. Fiz.* **46**, pp. 204–218.

Turbiner, A. V. (1988). Multi-Dimensional Anisotropic Anharmonic Oscillator (Quantitative Approach), *Soviet Scientific Reviews* **10**, pp. 79–131.

Turbiner, A. V. (2005). Anharmonic oscillator and double-well potential: Approximating eigenfunctions, *Lett. Math. Phys.* **74**, 2, pp. 169–180.

Turbiner, A. V. (2010). Double well potential: perturbation theory, tunneling, WKB (beyond instantons), *Int. J. Mod. Phys. A* **25**, 2–3, pp. 647–658.

Turbiner, A. V. and del Valle, J. (2021c). Power-like potentials: From the Bohr-Sommerfeld energies to exact ones, *Int. J. Mod. Phys. A* **36**, 2150221, pp. 1–14.

Turbiner, A. V. and Shuryak, E. (2021). On connection between perturbation theory and semiclassical expansion in quantum mechanics, *Arxiv: 2102.04623v2*, pp. 1–16, *Nucl. Phys. B* **988** (2023) (to be published).

Turbiner, A. V. and Ushveridze, A. G. (1988). Anharmonic oscillator: Constructing the strong coupling expansions, *J. Math. Phys.* **29**, 9, pp. 2053–2063.

Voros, A. (1994). Exact quantization condition for anharmonic oscillators (in one dimension), *J. Phys. A* **27**, pp. 4653–4661.

Weniger, E. J. (1996a). A convergent renormalized strong coupling perturbation expansion for the ground state energy of the quartic, sextic, and octic anharmonic oscillator, *Ann. Phys.* **246**, 1, pp. 133–165.

Weniger, E. J. (1996b). Construction of the strong coupling expansion for the ground state energy of the quartic, sextic, and octic anharmonic oscillator via a renormalized strong coupling expansion, *Phys. Rev. Lett.* **77**, pp. 2859–2862.

Witwit, M. R. M. (1992). Energy levels of two-dimensional anharmonic oscillators with sextic and octic perturbations, *J. Math. Phys.* **33**, 12, pp. 4196–4205.